Communications in Computer and Information Science 1605

More information about this series at https://link.springer.com/bookseries/7899

Alexander Dudin · Anatoly Nazarov ·
Alexander Moiseev (Eds.)

Information Technologies and Mathematical Modelling

Queueing Theory and Applications

20th International Conference, ITMM 2021
Named after A.F. Terpugov, Tomsk, Russia, December 1–5, 2021
Revised Selected Papers

 Springer

Editors
Alexander Dudin ⬤
Belarusian State University
Minsk, Belarus

Anatoly Nazarov ⬤
Tomsk State University
Tomsk, Russia

Alexander Moiseev ⬤
Tomsk State University
Tomsk, Russia

ISSN 1865-0929   ISSN 1865-0937 (electronic)
Communications in Computer and Information Science
ISBN 978-3-031-09330-2   ISBN 978-3-031-09331-9 (eBook)
https://doi.org/10.1007/978-3-031-09331-9

This Springer imprint is published by the registered company Springer Nature Switzerland AG
The registered company address is: Gewerbestrasse 11, 6330 Cham, Switzerland

Preface

The series of scientific conferences on Information Technologies and Mathematical Modelling (ITMM) was started in 2002. In 2012, the series acquired an international status, and selected revised papers have been published in Communications in Computer and Information Science since 2014. The conference series was named after Alexander Terpugov, one of the first organizers of the conference, an outstanding scientist of the Tomsk State University and a leader of the famous Siberian school on applied probability, queueing theory, and applications.

Traditionally, the conferences have about ten sections in various fields of mathematical modelling and information technologies. Throughout the years, the sections on probabilistic methods and models, queueing theory, and communication networks have been the most popular ones at the conference. These sections gather many scientists from different countries. Many foreign participants come to this Siberian conference every year because of our warm welcome and serious scientific discussions. In 2021, the 20th ITMM conference was held online due to the ongoing COVID-19 pandemic.

This volume presents selected papers from the 20th ITMM conference. The papers are devoted to new results in queueing theory and its applications. Its target audience includes specialists in probabilistic theory, random processes, and mathematical modeling, as well as engineers engaged in logical and technical design and operational management of data processing systems, communication, and computer networks.

December 2021

Alexander Dudin
Anatoly Nazarov
Alexander Moiseev

Organization

The conference was organized by the International Computer Science Continuous Professional Development Center of the National Research Tomsk State University, the Peoples' Friendship University of Russia (RUDN University), and the Trapeznikov Institute of Control Sciences of the Russian Academy of Sciences.

Program Committee Chairs

Alexander Dudin	Belarusian State University, Belarus
Anatoly Nazarov	Tomsk State University, Russia

Publication Chairs

Alexander Dudin	Belarusian State University, Belarus
Alexander Moiseev	Tomsk State University, Russia

Program Committee

Khalid Al-Begain	Kuwait College of Science and Technology, Kuwait
Ivan Atencia	University of Malaga, Spain
Pedro Cabral	Universidade Nova de Lisboa, Portugal
Pau Fonseca i Casas	Universitat Politècnica de Catalunya, Spain
Srinivas Chakravarthy	Kettering University, USA
Bong Dae Choi	National Institute for Mathematical Sciences, South Korea
Tadeusz Czachórski	Institute of Theoretical and Applied Informatics, Polish Academy of Sciences, Poland
Rui Dinis	Universidade Nova de Lisboa, Portugal
Dmitry Efrosinin	Johannes Kepler University Linz, Austria
Mais Farhadov	Institute of Control Sciences, Russian Academy of Sciences, Russia
Yulia Gaydamaka	Peoples' Friendship University of Russia (RUDN University), Russia
Erol Gelenbe	Institute of Theoretical and Applied Informatics, Polish Academy of Sciences, Poland
Alexander Gortsev	Tomsk State University, Russia
Bara Kim	Korea University, South Korea
Che Soong Kim	Sangji University, South Korea
Udo Krieger	Universität Bamberg, Germany
B. Krishna Kumar	Anna University, Chennai, India

Achyutha Krishnamoorthy	Cochin University of Science and Technology, India
Krishna Kumar	Cochin University of Science and Technology, India
Quan-Lin Li	Yan Shan University, China
Yury Malinkovsky	Francisk Skorina Gomel State University, Belarus
Agassi Melikov	National Aviation Academy of Azerbaijan, Azerbaijan
Svetlana Moiseeva	Tomsk State University, Russia
Dmitri Moltchanov	Tampere University, Finland
Paulo Montezuma-Carvalho	Universidade Nova de Lisboa, Portugal
Evsey Morozov	Institute of Applied Mathematical Research, Karelian Research Centre of the Russian Academy of Sciences, Russia
Valery Naumov	Service Innovation Research Institute, Finland
Rein Nobel	Vrije Universiteit Amsterdam, The Netherlands
Michele Pagano	University of Pisa, Italy
Tuan Phung-Duc	University of Tsukuba, Japan
Thomas B. Preußer	Technische Universität Dresden, Germany
Jacques Resing	Eindhoven University of Technology, The Netherlands
Vladimir Rykov	Gubkin Russian State University of Oil and Gas, Russia
Konstantin Samouylov	Peoples' Friendship University of Russia (RUDN University), Russia
Sergey Suschenko	Tomsk State University, Russia
Daniel Stamate	Goldsmiths, University of London, UK
János Sztrik	University of Debrecen, Hungary
Henk Tijms	Vrije Universiteit Amsterdam, The Netherlands
Oleg Tikhonenko	Cardinal Stefan Wyszynski University in Warsaw, Poland
Gurami Tsitsiashvili	Institute of Applied Mathematics, Far Eastern Branch of the Russian Academy of Sciences, Russia
Vladimir Vishnevsky	Institute of Control Sciences, Russian Academy of Sciences, Russia
Anton Voitishek	Institute of Computational Mathematics and Mathematical Geophysics, Siberian Branch of the Russian Academy of Sciences, Russia
Vladimir Zadorozhny	Omsk State Technical University, Russia
Alexander Zamyatin	Tomsk State University, Russia
Andrey Zorin	Lobachevsky State University of Nizhni Novgorod, Russia

Local Organizing Committee

Svetlana Moiseeva (Chair)	Tomsk State University, Russia
Svetlana Paul (Co-chair)	Tomsk State University, Russia
Valentina Broner	Tomsk State University, Russia
Elena Danilyuk	Tomsk State University, Russia
Ekaterina Fedorova	Tomsk State University, Russia
Yana Izmailova	Tomsk State University, Russia
Irina Kochetkova	Peoples' Friendship University of Russia (RUDN University), Russia
Ivan Lapatin	Tomsk State University, Russia
Ekaterina Lisovskaya	Tomsk State University, Russia
Olga Lizyura	Tomsk State University, Russia
Anna Morozova	Tomsk State University, Russia
Ekaterina Pankratova	Institute of Control Sciences, Russian Academy of Sciences, Russia
Svetlana Rozhkova	Tomsk Polytechnical University, Russia
Daria Semenova	Siberian Federal University, Russia
Maria Shklennik	Tomsk State University, Russia
Alexey Shkurkin	Tomsk State University, Russia
Konstantin Voytikov	Moscow Institute of Physics and Technology

Contents

Analysis of the Polling System with Two Markovian Arrival Flows, Finite Buffers, Gated Service and Phase-Type Distribution of Service and Switching Times

Alexander Dudin[1,2]([✉]) [iD] and Yuliya Sinyugina[3] [iD]

[1] Belarusian State University, 4 Nezavisimosti Avenue, 220030 Minsk, Belarus
dudin@bsu.by
[2] Applied Mathematics and Communications Technology Institute,
Peoples' Friendship University of Russia (RUDN University),
6 Miklukho-Maklaya St, 117198 Moscow, Russia
[3] Francisk Skorina Gomel State University,
104 Sovetskaya str., 246019 Gomel, Belarus
sinyugina@gsu.by

Abstract. The polling system with two Markovian Arrival Flows, finite buffers, gated service discipline and Phase-Type (PH) distribution of service and switching times is considered. Stationary distribution of the continuous-time multi-dimensional Markov chain defining the current state of the server, number of customers in the buffers, the number of customers that should obtain service during the residual time of service of customers from various buffers and underlying processes of service or switching time and of arrival process is computed. Expressions for Laplace-Stieltjes transforms of distribution of waiting times of customers in both buffers are obtained. Numerical results giving some insight into performance of the system are presented.

Keywords: Polling system · Markovian Arrival Process · Phase-Type Service Time Distribution

1 Introduction

Stochastic polling models are effectively used for performance evaluation, design and optimization of telecommunication systems and networks, transport systems and road management systems, traffic, production systems and inventory management systems. In the recent review of the state of art in [1] the authors gave the extensive survey of the basic notions and existing results in polling models. For more references see, e.g., [2–13]. In particular, in [1] the authors separately discuss the importance of analysis and the existing in the literature results for two-queue systems as a special case of polling systems. In our paper, polling system with two Markovian Arrival Processes ($MAPs$), buffers of finite

A. Dudin et al. (Eds.): ITMM 2021, CCIS 1605, pp. 1–15, 2022.
https://doi.org/10.1007/978-3-031-09331-9_1

capacity, gated service discipline and Phase-Type (PH) distribution of service and switching times is considered. Consideration of such quite general arrival, service and switching process is the main contribution of our paper. Especially, this concerns analysis of waiting times distribution.

In Sect. 2, we describe the model under study. In Sect. 3, the continuous-time multi-dimensional Markov chain describing behavior of the system is described. A finite system of equations for the steady-state distribution of the chain is derived. Short Sect. 4 contains formulas for computation of the average number of customers and loss probabilities in the buffers. In Sect. 5, analysis of the stationary distribution of waiting times in the buffers is presented. Section 6 contains some illustrative numerical results.

2   Mathematical Model

We consider a single server polling queueing system the structure of which is shown in Fig. 1.

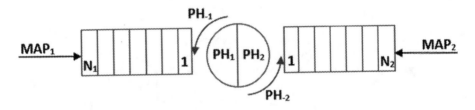

Fig. 1. Queueing system under study

The system has two queues with finite buffers of capacities N_1 and N_2, correspondingly. Each queue receives its own flow of customers, which is defined by the MAP (Markovian Arrival Process), see, e.g., [14–16]. The process of arrival to the kth queue is defined by the irreducible continuous-time Markov chain $\nu_t^{(k)}$, $t \geq 0$, having a finite state space $\{0, 1, ..., W_k\}$. The underlying process $\nu_t^{(k)}$ stays in the state ν during an exponentially distributed time interval with parameter $\lambda_\nu^{(k)}$, $\nu = \overline{0, W_k}$. After that, with probability $p_l^{(k)}(\nu, \nu')$ the underlying process transits to the state ν' with generation of l customers, $l = 0, 1$.

The behavior of the kth MAP is described by matrices $D_0^{(k)}$ and $D_1^{(k)}$ of size $\bar{W}_k = W_k + 1$, which are defined by formulas:

$$
(D_0^{(k)})_{\nu, \nu'} = \begin{cases} -\lambda_\nu^{(k)}, & \nu = \nu', \\ \lambda_\nu^{(k)} p_0^{(k)}(\nu, \nu'), & \nu \neq \nu', \end{cases}
$$

$$
(D_1^{(k)})_{\nu, \nu'} = \lambda_\nu^{(k)} p_1^{(k)}(\nu, \nu'), \quad \nu, \nu' = \overline{0, W_k}.
$$

The matrix $D^{(k)} = D_0^{(k)} + D_1^{(k)}$ is the infinitesimal generator of the Markov chain $\nu_t^{(k)}$. The average intensity λ_k of customers arrival to the kth system is defined by the formula $\lambda_k = \boldsymbol{\chi}^{(k)} D_1^{(k)} \mathbf{e}$, where $\boldsymbol{\chi}^{(k)}$ is the row vector of the stationary probabilities of the Markov chain $\nu_t^{(k)}$. The vector $\boldsymbol{\chi}^{(k)}$ is the unique solution to the system $\boldsymbol{\chi}^{(k)} D^{(k)} = \mathbf{0}$, $\boldsymbol{\chi}^{(k)} \mathbf{e} = 1$. Here and throughout this paper, $\mathbf{e}$ is a column vector of appropriate size consisting of ones, and $\mathbf{0}$ is a row vector of appropriate size consisting of zeroes.

The service time of an arbitrary customer from the kth buffer has a PH distribution, given by the irreducible representation $(\boldsymbol{\beta}^{(k)}, S^{(k)})$, $k = 1, 2$, and the underlying process $\eta_t^{(k)}$, $t \geq 0$, with the state space $\{1, ..., M_k, M_k+1\}$, where the state $M_k + 1$ is the absorbing one. The initial state of the process $\eta_t^{(k)}$ is chosen among the transient states in accordance with a stochastic row vector $\boldsymbol{\beta}^{(k)} = (\beta_1^{(k)}, \beta_2^{(k)}, ..., \beta_{M_k}^{(k)})$. The intensities of the transition of the process $\eta_t^{(k)}$ between transient states are defined by the matrix $S^{(k)}$. The intensities of the transition to the absorbing state $M_k + 1$ is defined by the entries of the column vector $\mathbf{S}_0^{(k)} = -S^{(k)} \mathbf{e}$. More information about the PH distribution can be found in [16, 17]. Switching of the server between the queues is not instantaneous. The switching time of the server to the service of customers located in the kth buffer has a PH distribution given by the irreducible representation $(\boldsymbol{\beta}^{(-k)}, S^{(-k)})$, $k = 1, 2$.

We assume the gated discipline of service. This means that the server provides service only to those customers that are presenting in the buffer immediately after completion of the server switching to this buffer. All customers that arrive after completion of the switching will receive service only after the next switching of the server to this buffer.

3 Process of System States

We describe the operation of the system by the process

$$\xi_t = \{r_t, j_t, i_t^{(1)}, i_t^{(2)}, m_t, \nu_t^{(1)}, \nu_t^{(2)}\}, \ t \geq 0,$$

where, at the time instant t,

- $i_t^{(k)}$ is the number of customers at the kth buffer, $k = 1, 2$;
- r_t characterizes the state of the server:

$$r_t = \begin{cases} k, & \text{if the server is processing the customer from the } k\text{th queue,} \\ -k, & \text{if the server is switching to the } k\text{th queue, } k = 1, 2; \end{cases}$$

- j_t is the number of customers from the current queue that still need to be serviced (including one in service). This component is absent in definition of ξ_t if the server is currently switching to another queue;
- m_t is the state of the underlying process of PH distributed ongoing service or switching time;
- $\nu_t^{(k)}$, $k = 1, 2$, is the state of the underlying process of the customers arrival in the kth MAP, $k = 1, 2$.

The process ξ_t, $t \geq 0$, is a regular irreducible continuous time Markov chain and has a finite state space. Thus, the following limits (stationary probabilities) exist:

$$\pi^{(r)}\left(j, i_1,\ i_2, m,\ \nu^{(1)},\ \nu^{(2)}\right) =$$

$$\lim_{t\to\infty} P\left\{r_t = r, j_t = j, i_t^{(1)} = i_1, i_t^{(2)} = i_2, m_t = m, \nu_t^{(1)} = \nu^{(1)}, \nu_t^{(2)} = \nu^{(2)}\right\}.$$

Let us form the row vectors of these probabilities enumerated in the direct lexicographical order of components r_t, j_t, $i_t^{(1)}$, $i_t^{(2)}$, m_t, $\nu_t^{(1)}$, $\nu_t^{(2)}$:

$$\boldsymbol{\pi}^{(r)}(j, i_1,\ i_2) = \left(\pi^{(r)}(j, i_1,\ i_2,\ 1,\ 0, 0),\ ...,\ \pi^{(r)}(j, i_1,\ i_2,\ M_r,\ W_1, W_2)\right),$$

$$\boldsymbol{\pi} = \left(\boldsymbol{\pi}^{(1)}(1, 0, 0),...,\ \boldsymbol{\pi}^{(1)}(N_1, N_1, N_2),\boldsymbol{\pi}^{(2)}(1, 0, 0),...,\boldsymbol{\pi}^{(2)}(N_2, N_1, N_2),\right.$$

$$\left.\boldsymbol{\pi}^{(-1)}(0, 0),...,\ \boldsymbol{\pi}^{(-1)}(N_1, N_2),\ \boldsymbol{\pi}^{(-2)}(0, 0),...,\ \boldsymbol{\pi}^{(-2)}(N_1, N_2)\right).$$

Let us denote

$$R_{i_1, i_2}^{(r)} = I_{M_r} \otimes D_0^{(1)} \otimes I_{\bar{W}_2}(1 - \delta_{i_1 N_1}) + I_{M_r} \otimes D^{(1)}\delta_{i_1 N_1} \otimes I_{W_2}$$

$$+ I_{M_r} \otimes I_{\bar{W}_1} \otimes D_0^{(2)}(1 - \delta_{i_2 N_2}) + I_{M_r} \otimes I_{\bar{W}_1} \otimes D^{(2)}\delta_{i_2 N_2} + S^{(r)} \otimes I_{\bar{W}_1 \bar{W}_2},\ i_k = \overline{0, N_k},$$

$$\hat{D}_1^{(1)} = D_1^{(1)} \otimes I_{\bar{W}_2},\ \hat{D}_1^{(2)} = I_{\bar{W}_1} \otimes D_1^{(2)},$$

where I is the identity matrix size of which is indicated by the suffix, $\otimes$ is the symbol of the Kronecker product of matrices, see [18] δ_{ij} is the Kronecker delta, $\bar{\delta}_{ij} = 1 - \delta_{ij}$.

The probability vector $\boldsymbol{\pi}$ satisfy the following system of linear algebraic equations, called equilibrium or Chapman-Kolmogorov equations:

$$\boldsymbol{\pi}^{(1)}(j, i_1,\ i_2) R_{i_1, i_2}^{(1)} + \boldsymbol{\pi}^{(1)}(j, i_1 - 1,\ i_2)\left(I_{M_1} \otimes \hat{D}_1^{(1)}\right)\bar{\delta}_{i_1 0}$$

$$+ \boldsymbol{\pi}^{(1)}(j, i_1,\ i_2 - 1)\left(I_{M_1} \otimes \hat{D}_1^{(2)}\right)\bar{\delta}_{i_2 0} + \boldsymbol{\pi}^{(1)}(j + 1, i_1,\ i_2)\bar{\delta}_{j N_1} S_0^{(1)}\boldsymbol{\beta}^{(1)} \otimes I_{\bar{W}_1 \bar{W}_2}$$

$$+ \boldsymbol{\pi}^{(-1)}(j,\ i_2) S_0^{(-1)}\boldsymbol{\beta}^{(1)}\delta_{i_1 0} \otimes I_{\bar{W}_1 \bar{W}_2} = \mathbf{0},\ j = \overline{1, N_1},$$

$$\boldsymbol{\pi}^{(2)}(j, i_1,\ i_2) R_{i_1, i_2}^{(2)} + \boldsymbol{\pi}^{(2)}(j, i_1 - 1,\ i_2)\left(I_{M_2} \otimes \hat{D}_1^{(1)}\right)\bar{\delta}_{i_1 0}$$

$$+ \boldsymbol{\pi}^{(2)}(j, i_1,\ i_2 - 1)\left(I_{M_2} \otimes \hat{D}_1^{(2)}\right)\bar{\delta}_{i_2 0} + \boldsymbol{\pi}^{(2)}(j + 1, i_1,\ i_2)\bar{\delta}_{j N_2} S_0^{(2)}\boldsymbol{\beta}^{(2)} \otimes I_{\bar{W}_1 \bar{W}_2}$$

$$+ \boldsymbol{\pi}^{(-2)}(i_1,\ j) S_0^{(-2)}\boldsymbol{\beta}^{(2)}\delta_{i_2 0} \otimes I_{\bar{W}_1 \bar{W}_2} = \mathbf{0},\ j = \overline{1, N_2},$$

$$\boldsymbol{\pi}^{(-1)}(i_1,\ i_2) R_{i_1, i_2}^{(-1)} + \boldsymbol{\pi}^{(-1)}(i_1 - 1,\ i_2)\left(I_{M_{-1}} \otimes \hat{D}_1^{(1)}\right)\bar{\delta}_{i_1 0}$$

$$+ \boldsymbol{\pi}^{(-1)}\left(i_1,\ i_2 - 1\right)\left(I_{M_{-1}} \otimes \hat{D}_1^{(2)}\right)\bar{\delta}_{i_2 0} + \boldsymbol{\pi}^{(2)}\left(1, i_1,\ i_2\right)\mathbf{S}_0^{(2)}\boldsymbol{\beta}^{(-1)} \otimes I_{\bar{W}_1 \bar{W}_2}$$

$$+ \boldsymbol{\pi}^{(-2)}\left(i_1,\ 0\right)\mathbf{S}_0^{(-2)}\boldsymbol{\beta}^{(-1)}\delta_{i_2 0} \otimes I_{\bar{W}_1 \bar{W}_2} = \mathbf{0},$$

$$\boldsymbol{\pi}^{(-2)}\left(i_1,\ i_2\right)R_{i_1, i_2}^{(-2)} + \boldsymbol{\pi}^{(-2)}\left(i_1 - 1,\ i_2\right)\left(I_{M_{-2}} \otimes \hat{D}_1^{(1)}\right)\bar{\delta}_{i_1 0}$$

$$+ \boldsymbol{\pi}^{(-2)}\left(i_1,\ i_2 - 1\right)\left(I_{M_{-2}} \otimes \hat{D}_1^{(2)}\right)\bar{\delta}_{i_2 0} + \boldsymbol{\pi}^{(1)}\left(1, i_1,\ i_2\right)\mathbf{S}_0^{(1)}\boldsymbol{\beta}^{(-2)} \otimes I_{\bar{W}_1 \bar{W}_2}$$

$$+ \boldsymbol{\pi}^{(-1)}\left(0,\ i_2\right)\mathbf{S}_0^{(-1)}\boldsymbol{\beta}^{(-2)}\delta_{i_1 0} \otimes I_{\bar{W}_1 \bar{W}_2} = \mathbf{0}.$$

The matrix of the Chapman-Kolmogorov system is degenerate according to the properties of the infinitesimal generator. In order to find the vector $\boldsymbol{\pi}$, add the normalization condition $\boldsymbol{\pi}\mathbf{e} = 1$ and remove one of the equations of the system. Thus, we obtain a system, the only solution of which is the vector of stationary probabilities of the states of the system. As a numerically stable algorithm for solving such a system, the algorithm from [19] is recommended.

4   Performance Measures

Having computed the vectors of the stationary probabilities $\boldsymbol{\pi}_i$, $i \geq 0$, defined by the partition $\boldsymbol{\pi} = (\boldsymbol{\pi}_0, \boldsymbol{\pi}_1, \boldsymbol{\pi}_2, \dots)$, it is possible to compute a variety of the performance measures of the system.

The average number of customers in the kth buffer, $k = 1, 2$, is computed by

$$L_k = \sum_{i=1}^{N_k} i\boldsymbol{\pi}_k(i)\mathbf{e},$$

where

$$\boldsymbol{\pi}_1(i)\mathbf{e} = \sum_{k=1}^{2}\sum_{i_2=0}^{N_2}\left(\sum_{j=1}^{N_k}\boldsymbol{\pi}^{(k)}(j, i, i_2)\mathbf{e} + \boldsymbol{\pi}^{(-k)}(i, i_2)\mathbf{e}\right),$$

$$\boldsymbol{\pi}_2(i)\mathbf{e} = \sum_{k=1}^{2}\sum_{i_1=0}^{N_1}\left(\sum_{j=1}^{N_k}\boldsymbol{\pi}^{(k)}(j, i_1, i)\mathbf{e} + \boldsymbol{\pi}^{(-k)}(i_1, i)\mathbf{e}\right).$$

The probability $P_k^{(loss)}$ that an arbitrary customer arriving to the kth buffer $k = 1, 2$, will be lost is computed by

$$P_1^{(loss)} = \frac{1}{\lambda_1}\sum_{k=1}^{2}\sum_{i_2=0}^{N_2}\left(\sum_{j=1}^{N_k}\boldsymbol{\pi}^{(k)}(j, N_1, i_2)(I_{M_k} \otimes \hat{D}_1^{(1)})\mathbf{e} + \boldsymbol{\pi}^{(-k)}(N_1, i_2)(I_{M_{-k}} \otimes \hat{D}_1^{(1)})\mathbf{e}\right),$$

$$P_2^{(loss)} = \frac{1}{\lambda_2}\sum_{k=1}^{2}\sum_{i_1=0}^{N_1}\left(\sum_{j=1}^{N_k}\boldsymbol{\pi}^{(k)}(j, i_1, N_2)(I_{M_k} \otimes \hat{D}_1^{(2)})\mathbf{e} + \boldsymbol{\pi}^{(-k)}(i_1, N_2)(I_{M_{-k}} \otimes \hat{D}_1^{(2)})\mathbf{e}\right).$$

5   Distribution of the Waiting Time

Let $V_k(x)$, $x \geq 0$, be distribution function of the waiting time of an arbitrary customer in the kth buffer and $v_k(s)$ be its Laplace-Stieltjes transform (LST):

$$v_k(s) = \int_0^\infty e^{-st} dV_k(t), \ Re \ s > 0.$$

We assume that the customers are served in the order of their arrival into the buffers ($FCFS$ service discipline).

We will derive expression for the LST $v_k(s)$ by means of the method of catastrophes. We interpret the variable s as the intensity of some virtual stationary Poisson flow of so-called catastrophes. It is easy to see that the LST $v_k(s)$ is equal to probability that no one catastrophe arrives during the waiting time. The possible scenarios of the waiting time of an arbitrary customer are as follows.

1) The customer arrives to the kth buffer and the buffer is full. In that case the customer is lost and $v_k(s) = 1$.
2) The customer arrives when the server is switching to the kth queue. In that case waiting time consists of the remaining switching time and the service time of customers which arrived before the tagged customer.
3) The customer arrives when the server is servicing customers from another queue. In that case waiting time consists of the remaining service time, the service time of customers from another queue that still need to be serviced, the switching time to the kth queue, the service time of customers which arrived to the kth queue before the tagged customer.
4) The customer arrives when the server is switching to another queue. In that case waiting time consists of the remaining switching time to another queue, the service time of customers which have been staying in another buffer and which arrived during the remaining switching time, the switching time to the kth queue and the service time of customers which arrived before the tagged customer.
5) The customer arrives when the server is servicing customers from the kth queue. In that case, waiting time consists of the remaining service time, the service time of customers from the kth buffer that still need to be serviced, the switching time to another queue, the service time of customers which have been staying in another buffer and which arrived during the switching time, the switching time to the kth queue and the service time of customers which arrived to this buffer before the tagged customer arrival.

Thus, to calculate the LST $v_k(s)$ of the waiting time of an arbitrary customer, we need to analyse all the listed above scenarios.

Let us introduce the following functions: $L^{(k)}(s) = \left(sI - S^{(k)}\right)^{-1} \mathbf{S}_0^{(k)}$ is the vector consisting of LSTs of the remaining service time of a customer from the kth queue, if $k = 1, 2$ (or of switching time to kth queue, if $k = -1, -2$) with a fixed current state of the corresponding underlying process; $\beta^{(k)}(s) = \boldsymbol{\beta}^{(k)} L^{(k)}(s)$ is the LST of the full service (or switching) time; $P_m(l, t)$ is the matrix of probabilities that l customers arrive to the mth queue during time t.

Lemma 1. *The LST of the column vector of remaining service times of a customer from the rth queue, $r = 1, 2$, (or remaining switching time to the rth buffer, $-r = 1, 2$) during which l customers from the mth flow will arrive to the system, is calculated as follows:*

$$F_l^{(r)}(m, s) = z_l^{(r)}(m, s)\left(\mathbf{S}_0^{(r)} \otimes I_{\bar{W}_m}\right),$$

the LST of the total service time during which l customers from the mth flow will arrive in the system, is calculated as follows:

$$P_l^{(r)}(m, s) = k_l^{(r)}(m, s)\left(\mathbf{S}_0^{(r)} \otimes I_{\bar{W}_m}\right),$$

where

$$z_0^{(r)}(m, s) = -(\Delta(s, r) \otimes I_{\bar{W}_m})\Psi(s, r, m),$$

$$z_l^{(r)}(m, s) = -\sum_{i=0}^{l-1} z_i^{(r)}(m, s)(\Delta(s, r) \otimes D_{l-i}^{(m)})\Psi(s, r, m),$$

$$k_0^{(r)}(m, s) = -(\boldsymbol{\beta}^{(r)}(\Delta(s, r) \otimes I_{\bar{W}_m})\Psi(s, r, m),$$

$$k_l^{(r)}(m, s) = -\sum_{i=0}^{l-1} k_i^{(r)}(m, s)(\Delta(s, r) \otimes D_{l-i}^{(m)})\Psi(s, r, m),$$

$$\Psi(s, r, m) = (I + \Delta(s, r) \otimes D_0^{(m)})^{-1}, \quad \Delta(s, r) = (-sI + S^{(r)})^{-1}.$$

Proof. By definition we have

$$F_l^{(r)}(m, s) = \int_0^\infty e^{-st} e^{S^{(r)} t} \mathbf{S}_0^{(r)} \otimes P_m(l, t) I_{\bar{W}_m} dt$$

$$= \int_0^\infty e^{-st} e^{S^{(r)} t} \otimes P_m(l, t) dt (\mathbf{S}_0^{(r)} \otimes I_{\bar{W}_m}) = z_l^{(r)}(m, s)(\mathbf{S}_0^{(r)} \otimes I_{\bar{W}_m}).$$

In turn,

$$z_l^{(r)}(m, s) = \int_0^\infty e^{-st} e^{S^{(r)} t} \otimes P_m(l, t) dt = \int_0^\infty e^{(S^{(r)} - sI)t} \otimes P_m(l, t) dt$$

$$= -(\Delta(s, r) \otimes I_{\bar{W}_m})\delta_{l,0} - \int_0^\infty e^{(S^{(r)} - sI)t} \Delta(s, r) \otimes \sum_{i=0}^l P_m(i, t) D_{l-i}^{(m)} dt$$

$$= -(\Delta(s, r) \otimes I_{\bar{W}_m})\delta_{l,0} - \sum_{i=0}^l z_i^{(r)}(m, s)(\Delta(s, r) \otimes D_{l-i}^{(m)}).$$

From where we get the formulas for $F_l^{(r)}(m, s)$ and $z_l^{(r)}(m, s)$ under proof. In a similar way, we obtain formulas for $P_l^{(r)}(m, s)$ and $k_l^{(r)}(m, s)$.

Lemma 2. *The LST of the total service time of n customers, $n \geq 1$, from the rth queue, $r = 1, 2$, during which l customers, $l \geq 0$, from the mth flow, $m = 1, 2$, will arrive to the system, is calculated as follows:*

$$P_l^{(*n,r)}(m, s) = h_{l,n}^{(r)}(m, s)\left(\Gamma_{0,r}^{(n)} \otimes I_{\bar{W}_m}\right),$$

where

$$h_{0,n}^{(r)}(m, s) = -\left(\gamma_r^{(n)}\left(-sI + \Gamma_r^{(n)}\right)^{-1} \otimes I_{\bar{W}_m}\right)\Phi(s, r, m, n),$$

$$h_{l,n}^{(r)}(m, s) = -\sum_{i=0}^{l-1} h_{i,n}^{(r)}(m, s)\left(\left(-sI + \Gamma_r^{(n)}\right)^{-1} \otimes D_{l-i}^{(m)}\right)\Phi(s, r, m, n),$$

$$\Phi(s, r, m, n) = (I + (-sI + \Gamma_r^{(n)})^{-1}) \otimes D_0^{(m)})^{-1}.$$

Here $\gamma_r^{(n)}$ and $\Gamma_r^{(n)}$ are parameters of the phase-type distribution of the sum of n independent random variables having a phase-type distribution with the irreducible representation $\left(\boldsymbol{\beta}^{(r)},\ S^{(r)}\right)$, and $\gamma_r^{(n)} = \left(\boldsymbol{\beta}^{(r)}, \mathbf{0}, ..., \mathbf{0}\right)$, where $\mathbf{0}$ is a null row vector of the same size as $\boldsymbol{\beta}^{(r)}$, and

$$\Gamma_r^{(n)} = \begin{pmatrix} S^{(r)} & \mathbf{S}_0^{(r)}\boldsymbol{\beta}^{(r)} & \mathbf{O} & \cdots & \mathbf{O} \\ \mathbf{O} & S^{(r)} & \mathbf{S}_0^{(r)}\boldsymbol{\beta}^{(r)} & \cdots & \mathbf{O} \\ \mathbf{O} & \mathbf{O} & S^{(r)} & \cdots & \mathbf{O} \\ \vdots & \vdots & \vdots & \ddots & \vdots \\ \mathbf{O} & \mathbf{O} & \mathbf{O} & \cdots & S^{(r)} \end{pmatrix}$$

where $\mathbf{O}$ is a null matrix of the same dimension as $S^{(r)}$, and

$$\Gamma_{0,r}^{(n)} = (\mathbf{0}^T, \ldots, \mathbf{0}^T, \mathbf{S}_0^{(r)})^T.$$

Lemma 3. *The LST of the conditional waiting time, provided that at the moment of arrival of tagged customer to the first buffer the server is switching to the first queue and there are i_1 customers in the first buffer, is calculated by the formula:*

$$v_1^{(-1)}(s, i_1) = L^{(-1)}(s)\left(\beta^{(1)}(s)\right)^{i_1}.$$

Proof. The probability that no one catastrophe arrives during the waiting time of the tagged customer is the product of the probability that no one catastrophe arrives during the remaining time of switching the server to the first queue $L^{(-1)}(s)$ by the probability that no one catastrophe arrives during the service time of i_1 customers $(\beta^{(1)}(s))^{i_1}$.

Lemma 4. *The LST of the conditional waiting time, provided that at the moment of arrival of tagged customer to the first buffer the server is servicing customers from the second queue, there are i_2 customers in the second buffer,* and j customers from second queue still need to be serviced, and there are i_1 customers in the first buffer, is calculated by the formula:

$$v_1^{(2)}(s, j, i_1, i_2) = L^{(2)}(s)\left(\beta^{(2)}(s)\right)^{j-1}\beta^{(-1)}v_1^{(-1)}(s, i_1).$$

Proof. The probability that no one catastrophe arrives during the waiting time of the tagged customer is the product of the following probabilities: the probability that no one catastrophe arrives during the remaining service time of the current customer $L^{(2)}(s)$; the probability that no one catastrophe arrives during the service time of $j-1$ customers $\left(\beta^{(2)}(s)\right)^{j-1}$; the probabilities of the states of the underlying process when the server starts switching to the first queue $\beta^{(-1)}$; the probability that no one catastrophe arrives during the remaining from the moment of switching start waiting time $v_1^{(-1)}(s, i_1)$.

Lemma 5. *The LST of the conditional waiting time, provided that at the moment of arrival of tagged customer to the first buffer, the server is switching to the second buffer, which contains i_2 customers, and the first buffer contains i_1 customers, is calculated as follows:*

$$v_1^{(-2)}(s, i_1, i_2) = \sum_{k=0}^{\infty} F_k^{(-2)}(2, s)\beta^{(2)}v_1^{(2)}(s, \min\{i_2 + k, N_2\}, i_1, 0).$$

Proof. The probability that no one catastrophe arrives during the waiting time is the product of probabilities: the probability that no one catastrophe arrives during the remaining switching time and k customers come to the second buffer $F_k^{(-2)}(2, s)$; the probabilities of the states of the underlying process for servicing the first customer from the second buffer $\beta^{(2)}$; the probability that no one catastrophe will arrive in the future $v_1^{(2)}(s, \min\{i_2 + k, N_2\}, i_1, 0)$.

Lemma 6. *The LST of the conditional waiting time, provided that at the moment of arrival of tagged customer in the first buffer, the server is servicing customer from the first queue, j customers are still need to be serviced, there are i_1 customers in the first buffer, and i_2 customers in the second buffer, is calculated as follows:*

$$v_1^{(1)}(s, j, \ i_1, i_2)$$

$$= \sum_{m=0}^{N_2-i_2-1}\sum_{k=0}^{N_2-i_2-1-m} F_m^{(1)}(2, s)P_k^{(*j-1,1)}(2, s)\beta^{(-2)}v_1^{(-2)}(s, i_1, i_2 + m + k)$$

$$+ \sum_{m=0}^{N_2-i_2-1}\sum_{k=N_2-i_2-m}^{\infty} F_m^{(1)}(2, s)P_k^{(*j-1,1)}(2, s)\beta^{(-2)}v_1^{(-2)}(s, i_1, N_2)$$

$$+ \sum_{m=N_2-i_2}^{\infty} F_m^{(1)}(2, s)(\beta^{(1)}(s))^{j-1}\beta^{(-2)}v_1^{(-2)}(s, i_1, N_2).$$

Proof. The probability that no one catastrophe arrives during the waiting time is the product of probabilities: the probability that no one catastrophe arrives during the remaining service time of customer and m customers arrive to the second buffer $F_m^{(1)}(2, s)$; the probability that no one catastrophe arrives during the service time of the remaining customers and k customers arrive to the second buffer $P_k^{(*j-1,1)}(2, s)$; the probabilities of the states of the underlying process of switching to the second queue $\beta^{(-2)}$; probability that no one catastrophe will arrive in the future $v_1^{(-2)}(s, i_1, i_2)$.

Theorem 1. *The LST of the waiting time of customer in the first buffer has the form*

$$v_1(s) = P_1^{(loss)} + \frac{1}{\lambda_1} \sum_{i_1=0}^{N_1-1} \sum_{i_2=0}^{N_2} \left(\pi^{(-1)}(i_1, i_2) \left(I_{M_{-1}} \otimes \hat{D}_1^{(1)} \right) \mathbf{e} \, v_1^{(-1)}(s, i_1) \right.$$

$$+ \pi^{(-2)}(i_1, i_2) \left(I_{M_{-2}} \otimes \hat{D}_1^{(1)} \right) \mathbf{e} \, v_1^{(-2)}(s, i_1, i_2)$$

$$+ \left. \sum_{k=1}^{2} \sum_{j=1}^{N_k} \pi^{(k)}(j, i_1, i_2) \left(I_{M_k} \otimes \hat{D}_1^{(1)} \right) \mathbf{e} \, v_1^{(k)}(s, j, i_1, i_2) \right).$$

The proof follows from the above lemmas and the total probability formula.

Theorem 2. *The LST of the waiting time of customer in the second buffer has the form*

$$v_2(s) = P_2^{(loss)} + \frac{1}{\lambda_2} \sum_{i_2=0}^{N_2-1} \sum_{i_1=0}^{N_1} \left(\pi^{(-2)}(i_1, i_2) \left(I_{M_{-2}} \otimes \hat{D}_1^{(2)} \right) \mathbf{e} \, v_2^{(-2)}(s, i_2) \right.$$

$$+ \pi^{(-1)}(i_1, i_2) \left(I_{M_{-1}} \otimes \hat{D}_1^{(2)} \right) \mathbf{e} \, v_2^{(-1)}(s, i_1, i_2)$$

$$+ \left. \sum_{k=1}^{2} \sum_{j=1}^{N_k} \pi^{(k)}(j, i_1, i_2) \left(I_{M_k} \otimes \hat{D}_1^{(2)} \right) \mathbf{e} \, v_2^{(k)}(s, j, i_1, i_2) \right),$$

where the corresponding functions are defined similarly to the above:

$$v_2^{(-2)}(s, i_2) = L^{(-2)}(s)(\beta^{(2)}(s))^{i_2},$$

$$v_2^{(1)}(s, j, i_1, i_2) = L^{(1)}(s)(\beta^{(1)}(s))^{j-1}\beta^{(-2)}v_2^{(-2)}(s, i_2),$$

$$v_2^{(-1)}(s, i_1, i_2) = \sum_{k=0}^{\infty} F_k^{(-1)}(1, s)\beta^{(1)}v_2^{(1)}(s, \min\{i_1 + k, N_1\}, 0, i_2),$$

$$v_2^{(2)}(s, j, i_1, i_2)$$

$$= \sum_{m=0}^{N_1-i_1-1} \sum_{k=0}^{\infty} F_m^{(2)}(1, s)P_k^{(*j-1,2)}(1, s)\beta^{(-1)}v_2^{(-1)}(s, \min\{i_1 + m + k, N_1\}, i_2)$$

$$+ \sum_{m=N_1-i_1}^{\infty} F_m^{(2)}(1, s)(\beta^{(2)}(s))^{j-1}\beta^{(-1)}v_2^{(-1)}(s, N_1, i_2).$$

Proof. The proof follows from the above lemmas and the total probability formula.

Corollary 1. *The average waiting time of an arbitrary customer in the kth buffer V_k, $k = 1, 2$, is calculated by the formula $V_k = -\dfrac{dv_k(s)}{ds}\Big|_{s=0}$.*

The average waiting time of an accepted customer in the kth buffer $V_k^{(accept)}$ is calculated by the formula $V_k^{(accept)} = V_k(1 - P_k^{(loss)})^{-1}$.

Proof. Note that the average waiting time for an arbitrary customer in the kth buffer, $k = 1, 2$, also takes into account lost customers, the waiting time of which is equal to zero:

$$V_k = V_k^{(loss)} P_k^{(loss)} + V_k^{(accept)} P_k^{(accept)},$$

where $V_k^{(loss)} = 0$ is the average waiting time for a lost customer in the kth buffer, $P_k^{(loss)}$ is the probability of loss of a customer when it arrives in the kth buffer. Note also that $P_k^{(loss)} + P_k^{(accept)} = 1$, then

$$V_k^{(accept)} = V_k(P_k^{(accept)})^{-1} = V_k(1 - P_k^{(loss)})^{-1}.$$

6   Numerical Examples

Now we consider numerical examples. Let us assume that the arrival flow of customers to the first queue MAP_1 is defined by the following matrices:

$$D_0^{(1)} = \begin{pmatrix} -10.08 & 0 \\ 0.003 & -0.327 \end{pmatrix}, \quad D_1^{(1)} = \begin{pmatrix} 9.975 & 0.105 \\ 0.036 & 0.288 \end{pmatrix}.$$

The average intensity of customers arrival is $\lambda_1 = 2.96625$. The coefficient of correlation of successive inter-arrival times in this arrival process is $cor = 0.4$, and the squared coefficient of variation of inter-arrival times is 12.39.

The arrival flow of customers to the second queue MAP_2 is defined by the following matrices:

$$D_0^{(1)} = \begin{pmatrix} -5.4104 & 0 \\ 0 & -0.17564 \end{pmatrix}, \quad D_1^{(1)} = \begin{pmatrix} 5.3744 & 0.036 \\ 0.09784 & 0.0778 \end{pmatrix}.$$

The average intensity of customers arrival is $\lambda_2 = 4$. The coefficient of correlation of successive inter-arrival times is $cor = 0.2$, and the squared coefficient of variation of inter-arrival times is 12.34.

We assume that the capacity of the first buffer is $N_1 = 4$ and the capacity of the second buffer is $N_2 = 5$.

The PHs distributions characterizing the service and switching processes are defined by the row vectors $\beta^{(k)} = (1,\ 0)$, $k = \pm 1, \pm 2$, and the sub-generators $S^{(k)} = \begin{pmatrix} -\alpha c_k & \alpha c_k \\ 0 & -\alpha c_k \end{pmatrix}$, where $c_1 = 1$, $c_2 = 1.2$, $c_{-1} = 0.3$, $c_{-2} = 0.2$, α is the parameter which we will vary.

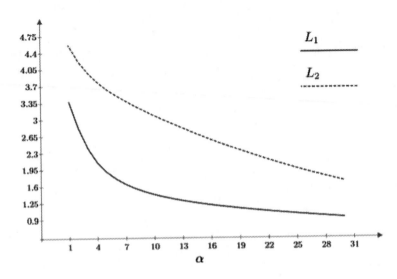

Fig. 2. The dependence of L_1 and L_2 on α.

Figure 2 shows that the queue length decreases with an increase in the parameter α which affects the speed of growth of the service and switching rates. Figure 3 shows that the probability of losing a customer also decreases with an increase in the parameter α.

To illustrate the importance of account of correlation in arrival process, now let us assume that the arrival flow of customers to the first queue MAP_1 is defined by the following matrices:

$$D_0^{(1)} = \begin{pmatrix} -5.25 & 2.25 \\ 3.75 & -6.6 \end{pmatrix}, \quad D_1^{(1)} = \begin{pmatrix} 3 & 0 \\ 0 & 2.85 \end{pmatrix}.$$

The average intensity of customers is practically the same, as in the MAP_1 used in the first example, $\lambda_1 = 2.94375$. But the coefficient of correlation is $cor = 0$. The squared coefficient of variation is 1.

The arrival flow of customers to the second queue MAP_2 and the PHs of service and switching processes are the same as above.

Figure 4 shows the dependence of the queue length L_1 on the parameter α with various correlations in the process MAP_1. Figure 5 shows the dependence of the probability of losing a customer $P_1^{(loss)}$ on the parameter α with various correlations in the process MAP_1. Figures 4 and 5 allow us to conclude that

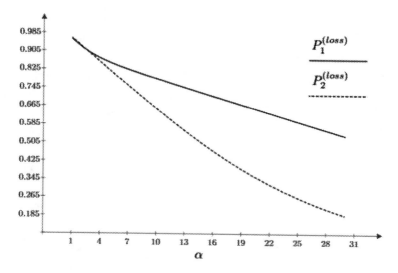

Fig. 3. The dependence of $P_1^{(loss)}$ and $P_2^{(loss)}$ on α.

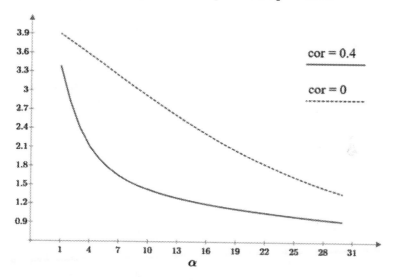

Fig. 4. The dependence of L_1 on α at different correlation coefficients.

ignoring the effect of correlation can lead to an essentially incorrect assessment of the effectiveness of a real system that may be described by the model under consideration.

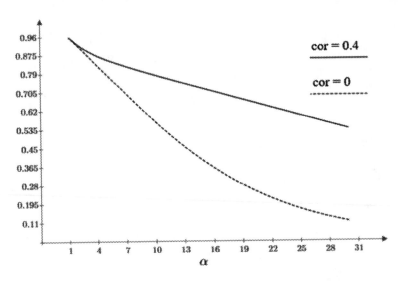

Fig. 5. The dependence of $P_1^{(loss)}$ on α at different correlation coefficients.

7   Conclusion

Polling system with two queues is analyzed. We considered the model under assumption that the input flows are described by the $MAPs$ and the service and switching times have phase-type distributions. This model can be applied to obtain the characteristics of a polling model with an arbitrary number of queues under the general assumptions about input flows and service and switching times distributions.

Acknowledgment. This paper has been supported by the RUDN University Strategic Academic Leadership Program.

References

1. Vishnevsky, V., Semenova, O.: Polling systems and their application to telecommunication networks. Mathematics **9**(117) (2021). https://doi.org/10.3390/math9020117
2. Takagi, H.: Queuing analysis of polling models: progress in 1990–1994. In: Dshalalow, J.H. (ed.) Frontiers in Queueing: Models, Methods and Problems, pp. 119–146. CRC Press, Boca Raton (1997)
3. Perel, E., Yechiali, U.: Two-queue polling systems with switching policy based on the queue that is not being served. Stoch. Model. **33**(3), 430–450 (2017)
4. Vishnevsky, V.M., Dudin, A.N., Semenova, O.V., Klimenok, V.I.: Performance analysis of the $BMAP/G/1$ queue with gated servicing and adaptive vacations. Perform. Eval. **68**, 446–462 (2011)
5. Vishnevsky, V., Dudin, A.N., Klimenok, V.I., Semenova, O.: Approximate method to study $M/G/1$-type polling system with adaptive polling mechanism. Qual. Technol. Quant. Manag. **9**(2), 211–228 (2012)

6. Levy, H.: Analysis of cyclic polling systems with binomial gated service. In: Hasegawa, T., Takagi, H., Takahashi, Y. (eds.) Performance of Distributed and Parallel Systems, pp. 127–139. North-Holland, Amsterdam (1989)
7. Altman, E., Konstantopoulos, P., Liu, Z.: Stability, monotonicity and invariant quantities in general polling systems. Queueing Syst. **11**, 35–57 (1992)
8. Altman, E.: Gated polling with stationary ergodic walking times, Markovian routing and random feedback. Ann. Oper. Res. **198**, 145–164 (2012)
9. Weiss, T., Hillenbrand, J., Krohn, A., Jondral, F.K.: Mutual interference in OFDM-based spectrum pooling systems. IEEE (2004)
10. Ohanissian, A.: Systems and methods for dynamic currency pooling interfaces. IEEE (2018)
11. Albrecht, F.: Resource pooling and sharing using distributed ledger systems. IEEE (2017)
12. Chakravarthy, S.R.: Analysis of a priority polling system with group services. Stoch. Model. **14**, 25–49 (1998)
13. Almasi, B., Sztrik, J.: A queueing model for a nonreliable multiterminal system with polling scheduling. J. Math. Sci. **92**, 3974–3981 (1998)
14. Lucantoni, D.M.: New results on the single server queue with a batch Markovian arrival process. Commun. Stat. Stoch. Models **7**(1), 1–46 (1991)
15. Chakravarthy, S.: The batch Markovian arrival process: a review and future work. Adv. Probab. Theory Stoch. Process. **1**, 21–39 (2001)
16. Dudin, A.N., Klimenok, V.I., Vishnevsky, V.M.: The Theory of Queuing Systems with Correlated Flows. Springer, Heidelberg (2020). ISBN 978-3-030-32072-0
17. Neuts, M.F.: Matrix-Geometric Solutions in Stochastic Models. The Johns Hopkins University Press, Baltimore (1981)
18. Graham, A.: Kronecker Products and Matrix Calculus with Applications. Ellis Horwood, Cichester (1981)
19. Dudin, S.A., Dudina, O.S.: Call center operation model as a $MAP/PH/N/R - N$ system with impatient customers. Probl. Inf. Transm. **47**(4), 364–377 (2011)

Simulation of Finite-Source Retrial Queueing Systems with Impatient Customers Using Different Failure Modes

János Sztrik[ID], Ádám Tóth$^{(\boxtimes)}$[ID], Ákos Pintér[ID], and Zoltán Bács[ID]

University of Debrecen, University Square 1, Debrecen 4032, Hungary
{sztrik.janos,toth.adam}@inf.unideb.hu,
apinter@science.unideb.hu, bacs.zoltan@econ.unideb.hu

Abstract. In this paper, a finite-source retrial queueing system is considered with impatient customers and catastrophic breakdowns. The characteristic of the system includes collision which occurs when a new job arrives in the system and the service facility is occupied with a job, they will collide. Both jobs will be forwarded to the virtual waiting room the so-called orbit. Here, the customers initiate other attempts to reach the server after a random time. But they give up retrying after staying in the orbit a while and leave the system which is the impatient attribute of the customers. In case of a negative event, a catastrophic breakdown takes place meaning that all the customers at the server and in the orbit depart from the system. The novelty of this paper is to investigate that feature in a collision environment with impatient customers using different distributions of the service time.

Keywords: Simulation · Catastrophic breakdown · Retrial queuing system · Collision · Impatience · Sensitivity analysis

1 Introduction

Designing info-communication systems are essential because of understanding how to optimize a system and also how to handle increasing network traffic. Many tools and mechanisms are available for modeling different systems, and among them, one of the most popular ones is retrial queuing systems. To illustrate real-life problems arising in main telecommunication systems, like telephone switching systems, call centers, computer networks, and computer systems, retrial queues can be effectively applied. In many publications, retrial-queuing systems with repeated calls are utilized to depict their models like in [2,5,6,9]. The specialty of retrial queuing systems relies on the orbit which is assumed to be a virtual waiting room with enough capacity to take in every customer. In this way, a job - whose service can not start - is not lost and may launch

The research was supported by the Thematic Excellence Programme (TKP2020-IKA-04) of the Ministry for Innovation and Technology in Hungary.

© Springer Nature Switzerland AG 2022
A. Dudin et al. (Eds.): ITMM 2021, CCIS 1605, pp. 16–27, 2022.
https://doi.org/10.1007/978-3-031-09331-9_2

numerous attempts to get its service requirement. The source is considered to be finite mainly because in many situations a finite number of entities participate in the operation of the system. Naturally, researchers have studied models with an infinite source but these are not suitably describing real-life applications in many cases. Results in connection with finite-source retrial queuing systems can be viewed in [1, 14, 18, 19]. Impatient behaviour is a natural characteristic of the customers provoking earlier departure without obtaining its service demand. This phenomenon is experienced in many fields of our life and here are some examples: healthcare applications, call centers, telecommunication networks. Not to mention all the papers where the behaviour of impatience is intensively examined, see for example [8, 11, 17]. Real-life systems tend to be subjected to random breakdowns which can be caused by a power outage, human negligence, or other sudden act. Thus, it is important to examine its effect on the operation of the system and the performance measures because it alters significantly the behaviour of a model. Many papers have studied models having service units assumed to be available all the time which is quite unrealistic. These types of systems have been investigated by many authors for example in [4, 10, 20]. In technologies, like in Ethernet or in communication sessions where the resources are constrained, the probability of collisions of the jobs occurs. Several individuals in the source may commence uncoordinated attempts leading to the interference of the signals resulting in the necessity for retransmissions. Consequently, it is important to include this phenomenon as part of the investigation creating effective policies preventing conflicts and corresponding message delays. Results that are in connection with collisions can be found in the following publication [12, 13, 15].

The objective of our investigation is to carry out a sensitivity analysis using different distributions of service times on the main performance measures while catastrophic breakdowns eventuate. In the case of these types of events, customers are forced to leave the system due to sudden acts which can be mechanical failures or power outages. Until repair, it is not allowed for any customer to enter the system and detailed studies on catastrophic breakdowns have been examined by several papers. Because we utilize different distributions for the service time of the customers the results are obtained by our simulation program that is based on Simpack [7]. The basic building blocks of the code are used in which we have the opportunity to calculate any desired measure using numerous values of input parameters. Graphical illustrations are provided depicting the effect of different parameters and distributions on the main performance metrics.

2   System Model

A finite-source retrial queueing system of type $M/G/1//N$ is considered with an unreliable service unit, impatient customers, the appearance of collisions, and blocking. This model has one service unit and a finite-source where every individual (altogether N) may generate a request towards the system according to exponential law with parameter λ/N meaning that the inter-arrival times are exponentially distributed with mean λ/N. As there are no queues the service of an arriving job starts immediately following gamma, hypo-exponential, hyperexponential, Pareto, and lognormal distribution with different parameters but with the same mean and variance value. In the case of a busy server, an arriving customer brings about a collision with the customer under service, and both are moved into orbit. Jobs residing in the orbit after an exponentially random time with parameter σ/N initiate other tries to be engaged with the server. Since random breakdowns emerge the failure time is also an exponential random variable with parameter γ_0 when the server is occupied and with γ_1 if idle. Two scenarios are distinguished:

- general breakdown: the service of a job is interrupted and it is forwarded back to the orbit, other jobs initiated by the individuals of the source can not enter the system until the service unit is functional.
- catastrophic breakdown: the service of a job is interrupted but instead of arriving at the orbit it leaves the system as the others from the orbit, no customers are allowed by the system until the server fully recovers.

The repair process starts instantly upon the failure of the service unit which follows an exponential distribution with parameter γ_2. Customers are characterized by impatience implicating that jobs can decide to leave the system after spending an exponentially distributed time with parameter τ in the orbit. These requests return to the source being unserved. In the paper of [16] similar models are analyzed by an asymptotic method where N tends to infinity this is why rates λ/N and σ/N are used. For example, it was proved that the number of customers in the system follows a normal distribution. All the random variables in the model creation are assumed to be totally independent of each other.

3   Simulation Results

3.1   First Scenario

To obtain the desired results, our self-developed simulation tool was used in which almost all the performance measures can be estimated. Its statistics package utilizes the batch means method where the useful run is divided into a certain number of batches. Batches are long enough in that way sample averages of the batches are approximately independent thus we have a valid estimation. The following article contains more information about that method [3]. The simulations are performed with a confidence level of 99.9%. The relative half-width of the confidence interval required to stop the simulation run is 0.00001. The size of a batch used to detect the initial transient duration is 1000.

Table 1 consists of every parameter that is applied for all the following figures. The parameters of service time of the customers can be found at Table 2, every chosen parameter is listed resulting in the same mean and variance in every used distribution. The reason for selecting these values is focusing on the interesting

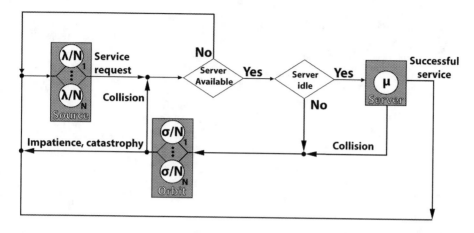

Fig. 1. System model

Table 1. Numerical values of model parameters

N	γ_0	γ_1	γ_2	σ/N	τ
100	0.05	0.05	1	0.05	0.001

situations and it must be noted that this model was tested with other values as well, and in most of the cases, the same phenomenon appeared. It is totally intentional that the squared coefficient of variation is more than one, later on in another scenario we will run the simulations when it is less than one (Fig. 1).

Table 2. Parameters of service time of primary customers

Distribution	Gamma	Hyper-exponential	Pareto	Lognormal
Parameters	$\alpha = 0.054$	$p = 0.473$	$\alpha = 2.027$	$m = -1.839$
	$\beta = 0.077$	$\lambda_1 = 1.353$	$k = 0.355$	$\sigma = 1.722$
		$\lambda_2 = 1.5$		
Mean	0.7			
Variance	9			
Squared coefficient of variation	18.367			

On Fig. 2 and 3 on the X-axes i represents the number of customers located in the system, and on the Y-axes $P(i)$ denotes the probability that exactly i customer are situated at the server and in the orbit altogether. In both Fig. 2 and 3 the distribution of the number of customers in the system is displayed when λ/N is 0.1 using various distributions of service time. Catastrophic breakdown feature is applied and interestingly the mean number of customers in the system differs from each other. In the case of the gamma distribution, customers tend to spend less time in the system compared to Pareto distribution. It is also noticeable that for both types of breakdowns the distribution of the number of customers tends to follow Gaussian distribution.

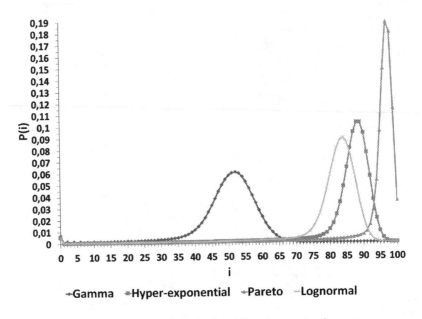

Fig. 2. Distribution of the number of customers in the system

Figure 3 depicts the comparison of different failure modes besides gamma and hyper-exponential distributions. Naturally more customers are in the system using the general breakdown method but the shape of the curves curiously are slightly disparate. In case of catastrophic breakdown, the peak is not that high and the mean number is fewer but other than that curves follow the same tendency.

The mean response time of an arbitrary customer is presented in the function of the arrival intensity of incoming customers in Fig. 4. Even though the mean and the variance are identical huge gaps develop among the applied distributions. With the increment of the arrival intensity, the mean response time of an arbitrary customer increases as well until λ/N equals 0.05 when the maximum is reached then it starts to decrease. The same tendency is observable for the other distributions, as well. The usage of gamma distribution results in a lower mean response time compared to the others, especially versus Pareto distribution.

Figure 5 demonstrates the development of the mean response time of a successfully served customer besides increasing arrival intensity. This measure shows

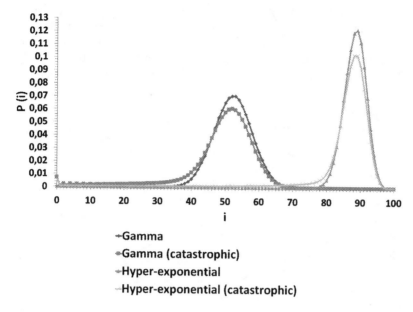

Fig. 3. Comparison of distribution of the number of customers in the system using different failure modes

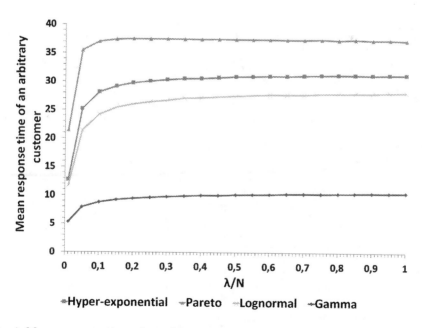

Fig. 4. Mean response time of an arbitrary customer vs. arrival intensity using various distributions.

the average response time of those customers who do not leave the system because of impatience or catastrophic event. As λ/N increases, the value of this performance measure raises as well which is true for every used distribution but the difference is quite high among them. At gamma distribution that value is much fewer than the others especially compared to Pareto distribution.

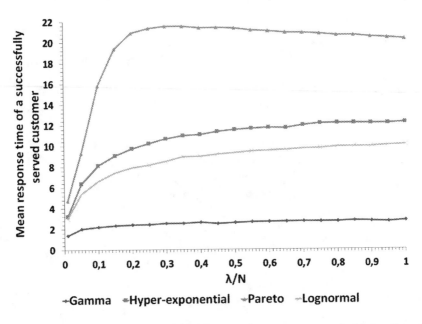

Fig. 5. Mean response time of a successfully served customer vs. arrival intensity using various distributions.

3.2 Second Scenario

In this section after analysing the obtained results of the previous scenario, we were curious to see what happens besides applying another parameter setting on the performance measures. In scenario 1 the squared coefficient of variation was greater than one and in this particular case, the parameters are selected in a way that the squared coefficient of variation is less than one. This also implies that the hyper-exponential distribution can not be used and instead of it we replace it with the hypo-exponential distribution. Table 3 contains the exact values of the parameters of the service time of primary customers in the case of this scenario, the other parameters remain unchanged which is shown in Table 1. Basically, our intention is to check that whether we get back the same tendencies of the previous section or it greatly changes the behaviour of the system and the performance measures with these modified parameters of service time of the incoming customers.

Table 3. Parameters of service time of incoming customers

Distribution	Gamma	Hypo-exponential	Pareto	Lognormal
Parameters	$\alpha = 1.69$	$\mu_1 = 2$	$\alpha = 2.64$	$m = -0.589$
	$\beta = 2.41$	$\mu_2 = 5$	$k = 0.435$	$\sigma = 0.682$
Mean	0.7			
Variance	0.29			
Squared coefficient of variation	0.592			

First, we will examine the figures in connection to the steady-state distribution. Analyzing the curves in more detail the obtained values are much closer to each other. As regards the shape of the curves they correspond to normal distribution. The mean number of customers is higher in the case of every distribution compared to the previous section. Not much difference is experienced though. In Fig. 6 regarding the mean values, they are very close to each other as well the shape of the curves, but in this case, the obtained graphs do not tend to correspond to Gaussian distribution.

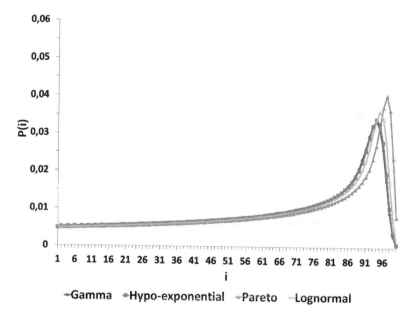

Fig. 6. Distribution of the number of customers in the system using various distributions, $\lambda = 0.1$.

Figure 7 emphasizes the difference between the applied failure modes. The results are depicted when gamma and hypo-exponential distribution are used but it is worth mentioning that the same tendencies occur utilizing the other two remaining ones. The difference is quite obvious even though the peak points are located in the same place but the value of possibility is much higher when catastrophe does not take place.

The next two figures are related to the mean response time of an arbitrary and a successfully served customer. First, in Fig. 8 it can be seen slight differences, in the case of Pareto distribution the values are a little bit higher, otherwise, the graphs almost overlap each other. Here, the same tendency develops as the mean response time increases with the increment of arrival intensity. Obviously, this maximum value feature is a specialty of finite-source retrial queuing systems under a suitable parameter setting.

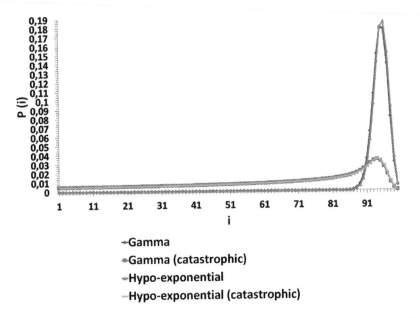

Fig. 7. Distribution of the number of customers in the system using various distributions, $\lambda/N = 0.1$.

Figure 9 demonstrates the comparison of the mean response time of a successfully served customer versus the arrival intensity. Not surprisingly after seeing the curves of the previous figure, the difference in the obtained values are very similar and it can be stated that the same maximum value feature appears in every case. The lowest values are obtained when the service time follows gamma distribution and the highest when Pareto distribution is applied.

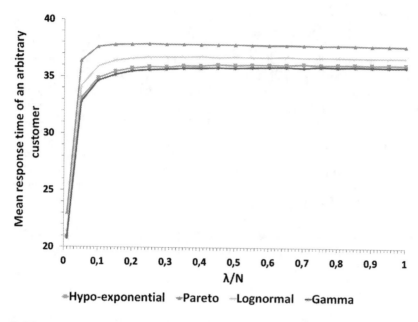

Fig. 8. Mean response time of an arbitrary customer vs. arrival intensity using various distributions.

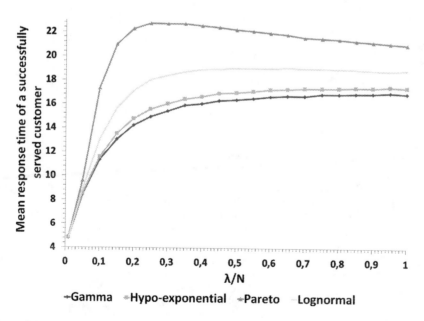

Fig. 9. Mean response time of a successfully served customer vs. arrival intensity using various distributions.

4   Conclusion

We simulated a retrial queueing system of type $M/G/1//N$ with impatient customers in the orbit and with an unreliable server using two different failure mechanisms when blocking is applied. Results are obtained by our program to carry out a sensitivity analysis on different performance measures like the distribution of the number of customers in the system. Under various parameter settings, the most interesting measures were chosen which were graphically illustrated. When the squared coefficient of variation is more than one significant deviation is experienced between the distributions in almost every aspect of the investigated measures. Consistently, it was also revealed that besides catastrophic breakdown less customer is in the system than in the case of a normal breakdown which is an expected phenomenon but the shape of the curves follows the same tendencies. In future works, the authors aim to carry on investigating the effect of catastrophic breakdown in other models and performing sensitivity analysis for other variables like the failure rate or the impatience of the customers.

References

1. Ali, A.A., Wei, S.: Modeling of coupled collision and congestion in finite source wireless access systems. In: 2015 IEEE Wireless Communications and Networking Conference (WCNC), pp. 1113–1118. IEEE (2015)
2. Artalejo, J., Corral, A.G.: Retrial Queueing Systems: A Computational Approach. Springer, Cham (2008). https://doi.org/10.1007/978-3-540-78725-9
3. Chen, E.J., Kelton, W.D.: A procedure for generating batch-means confidence intervals for simulation: checking independence and normality. Simulation **83**(10), 683–694 (2007)
4. Dragieva, V.I.: Number of retrials in a finite source retrial queue with unreliable server. Asia-Pac. J. Oper. Res. **31**(2), 23 (2014). https://doi.org/10.1142/S0217595914400053
5. Falin, G., Artalejo, J.: A finite source retrial queue. Eur. J. Oper. Res. **108**, 409–424 (1998)
6. Fiems, D., Phung-Duc, T.: Light-traffic analysis of random access systems without collisions. Ann. Oper. Res. **277**(2), 311–327 (2017). https://doi.org/10.1007/s10479-017-2636-7
7. Fishwick, P.A.: Simpack: getting started with simulation programming in C and C++. In: 1992 Winter Simulation Conference, pp. 154–162 (1992)
8. Gupta, N.: A view of queue analysis with customer behaviour and priorities. In: IJCA Proceedings on National Workshop-Cum-Conference on Recent Trends in Mathematics and Computing 2011, vol. RTMC, no. 4, May 2012
9. Kim, J., Kim, B.: A survey of retrial queueing systems. Ann. Oper. Res. **247**(1), 3–36 (2015). https://doi.org/10.1007/s10479-015-2038-7
10. Krishnamoorthy, A., Pramod, P.K., Chakravarthy, S.R.: Queues with interruptions: a survey. TOP **22**(1), 290–320 (2012). https://doi.org/10.1007/s11750-012-0256-6
11. Kumar, R., Jain, N., Som, B.: Optimization of an $M/M/1/N$ feedback queue with retention of reneged customers. Oper. Res. Decis. **24**, 45–58 (2014). https://doi.org/10.5277/ord140303

12. Kvach, A., Nazarov, A.: Sojourn time analysis of finite source Markov retrial queuing system with collision. In: Dudin, A., Nazarov, A., Yakupov, R. (eds.) ITMM 2015. CCIS, vol. 564, pp. 64–72. Springer, Cham (2015). https://doi.org/10.1007/978-3-319-25861-4_6

13. Kvach, A., Nazarov, A.: Numerical research of a closed retrial queueing system M/GI/1//N with collision of the customers. In: Proceedings of Tomsk State University. A Series of Physics and Mathematics. Tomsk. Materials of the III All-Russian Scientific Conference, vol. 297, pp. 65–70. TSU Publishing House (2015). (in Russian)

14. Lakaour, L., Aïssani, D., Adel-Aissanou, K., Barkaoui, K.: M/M/1 retrial queue with collisions and transmission errors. Methodol. Comput. Appl. Probab. **21**(4), 1395–1406 (2018). https://doi.org/10.1007/s11009-018-9680-x

15. Nazarov, A., Kvach, A., Yampolsky, V.: Asymptotic analysis of closed Markov retrial queuing system with collision. In: Dudin, A., Nazarov, A., Yakupov, R., Gortsev, A. (eds.) ITMM 2014. CCIS, vol. 487, pp. 334–341. Springer, Cham (2014). https://doi.org/10.1007/978-3-319-13671-4_38

16. Nazarov, A., Sztrik, J., Kvach, A., Bérczes, T.: Asymptotic analysis of finite-source M/M/1 retrial queueing system with collisions and server subject to breakdowns and repairs. Ann. Oper. Res. **277**(2), 213–229 (2018). https://doi.org/10.1007/s10479-018-2894-z

17. Panda, G., Goswami, V., Datta Banik, A., Guha, D.: Equilibrium balking strategies in renewal input queue with Bernoulli-schedule controlled vacation and vacation interruption. J. Ind. Manag. Optim. **12**, 851–878 (2015). https://doi.org/10.3934/jimo.2016.12.851

18. Peng, Y., Liu, Z., Wu, J.: An M/G/1 retrial G-queue with preemptive resume priority and collisions subject to the server breakdowns and delayed repairs. J. Appl. Math. Comput. **44**(1–2), 187–213 (2014). https://doi.org/10.1007/s12190-013-0688-7

19. Takeda, T., Yoshihiro, T.: A distributed scheduling through queue-length exchange in CSMA-based wireless mesh networks. J. Inf. Process. **25**, 174–181 (2017)

20. Tóth, A., Sztrik, J., Pintér, A., Bács, Z.: Reliability analysis of finite-source retrial queuing system with collisions and impatient customers in the orbit using simulation. In: 2021 International Conference on Information and Digital Technologies (IDT), pp. 230–234 (2021). https://doi.org/10.1109/IDT52577.2021.9497567

Markovian Models of Queuing Systems with Positive and Negative Replenishment Policies

Agassi Melikov[1]([✉])([iD]) and Mammad Shahmaliyev[2]

[1] Institute of Control System of National Academy of Science, st. B. Vahabzadeh 9,
1141 Baku, Azerbaijan
`agassi.melikov@gmail.com`
[2] National Aviation Academy, Mardakan Avenue, 30, Baku, Azerbaijan

Abstract. The Markov models of queuing-inventory systems with infinite buffer were analyzed under different replenishment policies. Besides traditional positive replenishment, the negative replenishment is considered after which inventory level instantly decreases. Some customers are assumed to leave the system without acquiring an item after the service completion. The ergodicity conditions of the introduced systems, as well as, formulas for stationary distributions and performance measures were developed. Total cost minimization problems were solved for the different replenishment policies.

Keywords: Queuing-inventory systems · Markovian models · Positive and negative replenishment · Matrix-geomteric method

1 Introduction

Systems where the serving process consists of releasing (selling) resource units to incoming customers are called Queuing-Inventory Systems (QIS) [1]. The reason for that naming is that such systems have properties both of Queuing and Inventory systems. First papers on this subject are known to be [2,3]. QIS subject has been widely studying by different authors during last three decades. The current state of QIS theory and its applications were extensively discussed in review paper [4].

In the most papers on QIS the replenishment is assumed to be positive, that is upon its completion the inventory goes up by the given positive amount that is defined by the accepted policy. But in practice, due to different reasons (technical errors, human errors, etc.) the inventory level may immediately decrease. We call such QIS with negative replenishment (like in case with negative customers). To our best knowledge, this kind of models were not studied in the available literature.

It should be noted that these models look similar to QIS models with perishable inventory. But the main difference is that in latter models items perish

© Springer Nature Switzerland AG 2022
A. Dudin et al. (Eds.): ITMM 2021, CCIS 1605, pp. 28–39, 2022.
https://doi.org/10.1007/978-3-031-09331-9_3

after some time and inventory goes down, while in our models inventory level decreases immediately due to negative replenishment. So in our paper we introduce separate class of QIS models with positive and negative replenishment.

2   Model Description

We consider Markov models of QIS system with one server under one of the three replenishment policies: (s, S), (s, Q), $(S - 1, S)$. Besides the traditional positive replenishment, we assume negative replenishment after which inventory level instantly decreases due to unexpected events. The negative replenishment events are described by Poisson point process with parameter κ. We assume that negative replenishment affect the items reserved for service as well. In each policy lead time is exponentially distributed with average ν^{-1}.

The customer income in all models are described by Poisson process with intensity λ. We assume that all customers require the identical item amount.

The customers are accepted for service if upon arrival the server is idle and inventory level is positive, otherwise customer joins the unlimited queue. Customers are assumed to join queue even if the inventory level is 0, i.e. according to Bernoulli scheme customer joins queue with probability ϕ_1 or leaves the system with complementary probability ϕ_2, where $\phi_1 + \phi_2 = 1$.

Customers in queue are considered impatient, when inventory level drops down to zero, customers leave the system independently after randomly distributed time that has exponential distribution with parameter τ^{-1}.

After the service completion customer according to Bernoulli scheme either acquires the item with probability σ_1 or leaves the system empty handed with probability σ_2, where $\sigma_1 + \sigma_2 = 1$. Average service times for both cases have exponential distribution with averages μ_1 and μ_2 accordingly.

3   Calculation of Stationary Distributions Under the Different Replenishment Policies

First let's consider the system under the (s, S) replenishment policy. The system is described with Two Dimensional Markov Chain, (2-D MC) with state vectors (m, n), where n represents the number of customers in the queue, $n = 0, 1, 2, ...,$ while m represents the inventory level, $m = 0, 1, ..., S$. The state space is defined as follows:

$$E = \bigcup_{n=0}^{\infty} L(n)$$

where $L(n) = \{(n, 0), (n, 1), ..., (n, S)\}$ called the nth level, $n = 0, 1, 2, ...$

Let's rearrange state space E in lexicographical order as follows

$$(0, 0), (0, 1), ..., (0, S), (1, 0), (1, 1), ..., (1, S), ...$$

In that case we obtain Level Independent Quasi-Birth-Death Process (LIQBD) with the following generator:

$$G = \begin{pmatrix} B & A_0 & . & . & . \\ A_2 & A_1 & A_0 & . & . \\ . & A_2 & A_1 & A_0 & . \\ . & . & . & . & . \end{pmatrix} \tag{1}$$

All block matrices in (1) are square matrices of dimension $S+1$ and their elements $B = ||b_{ij}||$ and $A_k = ||a_{ij}^{(k)}||, i, j = 0, 1, ..., S$ are calculated as follows:

$$b_{ij} = \begin{cases} \nu, & \text{if } i \leq s, j = S \\ \kappa & \text{if } i > s, j = i - 1 \\ -(\nu + \lambda\phi_1), & \text{if } i = j = 0 \\ -(\nu + \kappa + \lambda), & \text{if } 0 < i \leq s, j = i \\ -(\kappa + \lambda), & \text{if } s < i \leq S, j = i \\ 0, & \text{in other cases} \end{cases} \tag{2}$$

$$a_{ij}^{(0)} = \begin{cases} \lambda\phi_1, & \text{if } i = j = 0 \\ \lambda, & \text{if } i \neq 0, i = j \\ 0, & \text{in other cases} \end{cases} \tag{3}$$

$$a_{ij}^{(1)} = \begin{cases} \nu, & \text{if } 0 \leq i \leq s, j = S \\ \kappa, & \text{if } i > 0, j = i - 1 \\ -(\tau + \nu + \lambda\phi_1), & \text{if } i = j = 0 \\ -(\nu + \kappa + \lambda + \mu_1\sigma_1 + \mu_2\sigma_2), & \text{if } 0 < i, j = i \\ 0, & \text{in other cases} \end{cases} \tag{4}$$

$$a_{ij}^{(2)} = \begin{cases} \tau & \text{if } i = j = 0 \\ \mu_1\sigma_1, & \text{if } i \neq 0, i = j \\ \mu_2\sigma_2, & \text{if } i > 0, j = i - 1 \\ 0, & \text{in other cases} \end{cases} \tag{5}$$

Theorem 1. *Under the (s, S) replenishment policy system is ergodic if and only if the following inequality holds true:*

$$\lambda(1 - (1 - \phi_1)\pi(0)) < \tau\pi(0) + (\mu_1\sigma_1 + \mu_2\sigma_2)(1 - \pi(0)) \tag{6}$$

where

$$\pi(0) = \left(1 + (1 + a^{-1})((1 + a)^{s+1} - 1) + (S - s - 1)(1 + a)\right)^{-1},$$

$$a = \frac{\nu}{\mu_2\sigma_2 + \kappa}.$$

Proof. Let's designate stationary distribution corresponding to the generator $A = A_0 + A_1 + A_2$ by $\pi = (\pi(0), \pi(1), ..., \pi(S))$. These variables satisfies the following system of equations:

$$\pi A = 0, \pi e = 1 \tag{7}$$

where 0 is null row vector of dimension $S+1$ and e is column vector of dimension $S+1$ that contains only 1's. $\pi(m), m = 0, 1, ..., S$ is the probability of the state with inventory level equal to $m, m = 0, 1, ..., S$.

We conclude from (3)–(5) that elements of generator $A = ||a_{ij}|| i, j = 0, 1, ..., S$ are calculated as follows:

$$a_{ij} = \begin{cases} -\nu, & \text{if } i = j = 0 \\ \nu, & \text{if } 0 \le i \le s, j = S \\ \mu_2\sigma_2 + \kappa & \text{if } i > 0, j = i - 1 \\ -(\mu_2\sigma_2 + \kappa + \nu), & \text{if } 0 < i \le s, j = i \\ -(\mu_2\sigma_2 + \kappa), & \text{if } i > s, j = i \\ 0, & \text{in other cases} \end{cases} \tag{8}$$

We conclude from (8) that system of linear equations (7) gets the following form:

$$(\nu + (\kappa + \mu_2\sigma_2)(1 - \delta_{m,0}))\pi(m) = (\kappa + \mu_2\sigma_2)\pi(m + 1), 0 \le m \le s; \tag{9}$$

$$(\kappa + \mu_2\sigma_2)\pi(m) = (\kappa + \mu_2\sigma_2)\pi(m+1)(1 - \delta_{m,S}) + \nu \sum_{i=0}^{S} \pi(i)\delta_{m,S}, s+1 \le m \le S. \tag{10}$$

Here and in later formulas $\delta_{x,y}$ designates Kronecker symbols.

$$a_{ij} = \begin{cases} (1 + a)^m \pi(0), & \text{if } 1 \le m \le s + 1 \\ (1 + a)^{s+1} \pi(0), & \text{if } s + 1 < m \le S \end{cases} \tag{11}$$

where $\pi(0)$ is derived from the normalizing condition, $\pi(0) + \pi(1) + ... + \pi(S) = 1$. According [5] (Chapter 3, p. 81–83) LIQBD we are studying is ergodic iff:

$$\pi A_0 e < \pi A_2 e \tag{12}$$

Then from (3), (5) and (11) after applying some mathematical transformations we get (6) from (12).

Note 1. Ergodicity condition (6) has probabilistic meaning. Total summed intensity of incoming requests should be smaller than total summed intensity of outgoing requests. Condition (6) could be replaced with rough but easily checked condition: $\lambda < min(\tau, \mu_1\sigma_1 + \mu_2\sigma_2)$.

Let's replace stationary distribution corresponding to generator G with $p = (p_0, p_1, ...)$ where $p_n = (p(n, 0), p(n, 1), ..., p(n, S)), n = 0, 1, ...$ Assuming that ergodicity condition (6) holds true, stationary distributions may be calculated as follows:

$$p_n = p_0 R^n, n = 0, 1, \dots. \tag{13}$$

where R is minimal nonnegative solution of the following quadratic equation:

$$R^2 A_2 + R A_1 + A_0 = 0 \tag{14}$$

Probability of border states p_0 is calculated from the following system of equations:

$$p_0(B + RA_2) = 0 \tag{15}$$

$$p_0(I - R)^{-1} e = 1 \tag{16}$$

where I is unit matrix of size $S + 1$.

Now let's consider model with (s, Q) policy. State space of this model is also given by E, but corresponding generator matrix $\widetilde{G}$ is determined as follows:

$$\widetilde{G} = \begin{pmatrix} \widetilde{B} & A_0 & \cdot & \cdot & \cdot \\ A_2 & \widetilde{A_1} & A_0 & \cdot & \cdot \\ \cdot & A_2 & \widetilde{A_1} & A_0 & \cdot \\ \cdot & \cdot & \cdot & \cdot & \cdot \end{pmatrix}$$

where elements of matrices $\widetilde{B}$ and $\widetilde{A_1}$ are calculated as follows:

$$\widetilde{b}_{ij} = \begin{cases} \nu, & \text{if } j = i + S - s \\ \kappa, & \text{if } i > 0, j = i - 1 \\ -(\nu + \lambda\phi_1), & \text{if } i = j = 0 \\ -(\nu + \kappa + \lambda), & \text{if } 0 < i \le s, j = i \\ -(\kappa + \lambda), & \text{if } s < i \le S, j = i \\ 0, & \text{in other cases} \end{cases} \tag{17}$$

$$\widetilde{a}_{ij}^{(1)} = \begin{cases} \nu, & \text{if } 0 \le i \le s, j = i + S - s \\ \kappa, & \text{if } i > 0, j = i - 1 \\ -(\tau + \nu + \lambda\phi_1), & \text{if } i = j = 0 \\ -(\nu + \kappa + \lambda + \mu_1\sigma_1 + \mu_2\sigma_2), & \text{if } 0 < i, j = i \\ 0, & \text{in other cases} \end{cases} \tag{18}$$

Theorem 2. *Under the (s, Q) replenishment policy system is ergodic if and only if the inequality (6) holds true, where*

$$\pi(0) = (1 + a)^{-(s+1)} \left(\frac{(1 + a)^{s+1} - 1}{a(1 + a)} + S - s - a^{-1}(1 - (1 + a)^{-s}) \right)^{-1}.$$

Proof. The elements of generator $\tilde{A} = A_0 + \tilde{A}_1 + A_2$ are calculated as follows:

$$\tilde{a}_{ij} = \begin{cases} -\nu, & \text{if } i = j = 0 \\ \nu, & \text{if } 0 \leq i \leq s, j = i + S - s \\ \mu_2\sigma_2 + \kappa, & \text{if } i > 0, j = i - 1 \\ -(\mu_2\sigma_2 + \kappa + \nu), & \text{if } 0 < i \leq s, j = i \\ -(\mu_2\sigma_2 + \kappa), & \text{if } i > s, j = i \\ 0, & \text{in other cases} \end{cases} \tag{19}$$

We conclude from (19) that system of linear equations (7) corresponding to generator $\tilde{A}$ has the following form:

$$(\nu + (\kappa + \mu_2\sigma_2)(1 - \delta_{m,0}))\pi(m) = (\kappa + \mu_2\sigma_2)(\pi(m + 1), 0 \leq m \leq s; \tag{20}$$

$$(\kappa + \mu_2\sigma_2)\pi(m) = (\kappa + \mu_2\sigma_2)(\pi(m + 1)(1 - \delta_{m,0})$$
$$+ \nu\pi(m - S + s)\delta_{m,S}, s + 1 \leq m \leq S; \tag{21}$$

Then from (24) and (21) we get:

$$\pi_m = \begin{cases} (1 + a)^{m-(s+1)}\pi(s + 1) & \text{if } 0 \leq m \leq s \\ \pi(s + 1), & \text{if } s + 1 \leq m \leq S - s \\ (1 - (1 + a)^{m-(S-1)})\pi(s + 1), & \text{if } S - s + 1 \leq m \leq S \end{cases} \tag{22}$$

where $\pi(s + 1)$ is calculated from normalizing condition:

$$\pi(s + 1) = \left(\frac{(1 + a)^{s+1} - 1}{a(1 + a)} + S - s - a^{-1}(1 - (1 + a)^{-s}) \right)^{-1}.$$

Taking into consideration (3), (5) and (22) and after applying some transformations from (12) we conclude that Theorem 2 is true.

Finally, let's consider model with $(S-1, S)$ policy. Elements for corresponding generator matrix $\tilde{\tilde{G}}$ is calculated as follows:

$$\tilde{\tilde{G}} = \begin{pmatrix} \tilde{\tilde{B}} & A_0 & \cdot & \cdot & \cdot \\ A_2 & \tilde{\tilde{A}}_1 & A_0 & \cdot & \cdot \\ & \cdot & A_2 & \tilde{\tilde{A}}_1 & A_0 & \cdot \\ & & \cdot & \cdot & \cdot & \cdot \end{pmatrix}$$

where elements of matrices $\tilde{\tilde{B}}$ and $\tilde{\tilde{A}}_1$ are calculated as follows:

$$\tilde{\tilde{b}}_{ij} = \begin{cases} (S - i)\nu, & \text{if } 0 \leq i \leq S - 1, j = i + 1 \\ \kappa, & \text{if } i > 0, j = i - 1 \\ -(S\nu + \lambda\phi_1), & \text{if } i = j = 0 \\ -((S - i)\nu + \kappa + \lambda), & \text{if } 0 < i \leq s, j = i \\ 0, & \text{in other cases} \end{cases}$$

$$
\widetilde{a}_{ij}^{(1)} =
\begin{cases}
(S-i)\nu, & \text{if } 0 \leq i \leq S-1 j = i+1 \\
\kappa, & \text{if } i > 0, j = i-1 \\
-(\tau + S\nu + \lambda\phi_1), & \text{if } i = j = 0 \\
-((S-i)\nu + \kappa + \lambda + \mu_1\sigma_1 + \mu_2\sigma_2), & \text{if } 0 < i, j = i \\
0, & \text{in other cases}
\end{cases}
$$

Theorem 3. *Under the* $(S, S-1)$ *replenishment policy system is ergodic if and only if the inequality (6) holds true, where*

$$
\pi(0) = \left(\sum_{m=0}^{S} \frac{S! a^m}{(S-m)!} \right)^{-1}.
$$

Proof. The elements of generator $\widetilde{A} = A_0 + \widetilde{A}_1 + A_2$ are calculated as follows:

$$
\widetilde{a}_{ij} =
\begin{cases}
-(S\nu + \lambda\phi_1), & \text{if } i = j = 0 \\
-(\mu_2\sigma_2 + \kappa + (S-i)\nu), & \text{if } 0 < i \leq S, j = i \\
(S-i)\nu, & \text{if } 0 \leq i \leq S-1, j = i+1 \\
\mu_2\sigma_2 + \kappa, & \text{if } 0 < i \leq S, j = i-1 \\
0, & \text{in other cases}
\end{cases}
\tag{23}
$$

We conclude from (23) that system of linear equations (7) corresponding to generator $\widetilde{A}$ is the same as balance equations for one-dimensional birth-death process, where death intensity is equal to $\mu_2\sigma_2 + \kappa$ and birth intensity of state m is equal to $(S-m)\nu, m = 0, 1, ...S$. Therefore, we get the following:

$$
\pi(m) = \frac{S!}{(S-m)!} a^m \pi(0), m = 0, 1, ..., S
\tag{24}
$$

where $\pi(0)$ is calculated from normalizing condition.

Then taking into consideration (3), (5) and (23) after applying some transformation to (12) we conclude that Theorem 3 is true.

4   Calculation of Performance Measures

In each replenishment policy the performance measures are calculated through corresponding state probabilities. So average inventory level S_{av} is calculated as follows:

$$
S_{av} = \sum_{m=1}^{S} m \sum_{n=0}^{\infty} p(n, m)
\tag{25}
$$

Average reorder quantity V_{av} under (s, S) policy:

$$
V_{av} = \sum_{m=S-s}^{S} m \sum_{n=0}^{\infty} p(n, S-m)
\tag{26}
$$

Note 2. Average reorder quantities under (s, Q) and $(S, S-1)$ policies are constants and equal to $Q = S - s$ and 1 correspondingly. Average queue length L_{av} under all policies is calculated as follows:

$$L_{av} = \sum_{n=1}^{\infty} n \sum_{m=0}^{S} p(n, m) \tag{27}$$

Average reorder rate RR under (s, Q) and (s, S) policies is determined as follows:

$$RR = \kappa p(0, s+1) + (\mu_2 \sigma_2 + \kappa) \sum_{n=1}^{\infty} p(n, s+1) \tag{28}$$

RR under $(S, S-1)$ policy is calculated as follows:

$$RR = \kappa \sum_{m=1}^{S} p(0, m) + (\mu_2 \sigma_2 + \kappa) \sum_{m=1}^{S} \sum_{n=1}^{\infty} p(n, m) \tag{29}$$

Total loss probability PL is calculated as follows:
Under (s, S) and (s, Q) policies:

$$PL = \phi_2 \sum_{n=0}^{\infty} p(n, 0) + \frac{\tau}{\tau + \phi_2 \lambda + \nu} \sum_{n=1}^{\infty} p(n, 0) \tag{30}$$

Under $(S-1, S)$ policy:

$$PL = \phi_2 \sum_{n=0}^{\infty} p(n, 0) + \frac{\tau}{\tau + \phi_2 \lambda + S\nu} \sum_{n=1}^{\infty} p(n, 0) \tag{31}$$

First operand in formulas (30) and (31) refers to the loss due to the empty inventory, while the second operand refers to the loss due to customer impatience.

5   Numerical Results

In this section results of numerical experiments will be presented and discussed. The behavior of performance measures vs s under (s, S) and (s, Q) policies are depicted in Fig. 1 and Fig. 2.

We used the following parameters for numerical experiments:

$$\lambda = 30, \phi_1 = 0.5, \phi_2 = 0.5, \sigma_1 = 0.4, \sigma_2 = 0.6, \mu_1 = 45, \mu_2 = 35,$$

$$\nu = 8, \kappa = 6, \tau = 20, S = 20$$

S_{av} under (s, S) policy is increasing with the increase of s and is a little bit higher than (s, Q). This behavior is expected as with higher s the inventory is replenished more frequently up to S which results in higher average inventory level. But under (s, Q) the replenishment amount is fixed $(S - s)$ and becomes

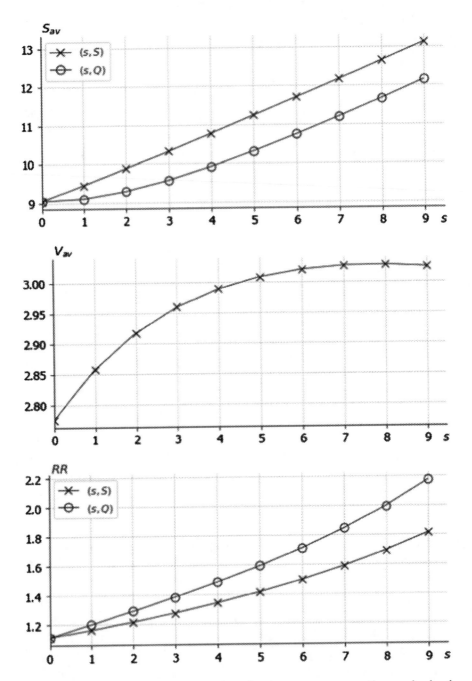

Fig. 1. Dependence of inventory related performance measures on the reorder level s under (s, S), (s, Q) policies

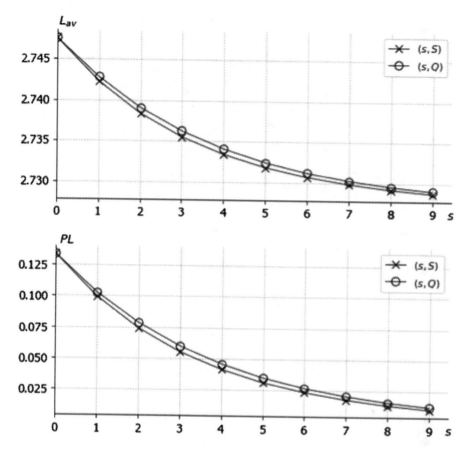

Fig. 2. Dependence of customer related performance measures on the reorder level s under (s, S), (s, Q) policies

lower with higher s which in turn results in lower average inventory level. Average order size V_{av} is also proportional to s which is reflected in graph. We excluded (s, Q) series from V_{av} as it is fixed for given s. RR is also lower under (s, S) policy due to higher average inventory level.

The average number of customers L_{av} in queue is almost the same for both policies and increase with s. Customer loss probabilities decrease for higher values of s due to higher S_{av} under both policies.

Behavior of the performance measures against maximum inventory size S under $(S - 1, S)$ policy is depicted in Fig. 3. The inventory related performance measures S_{av} and RR intuitively increases, while L_{av} and PL decreases because with larger inventory system could serve more customers.

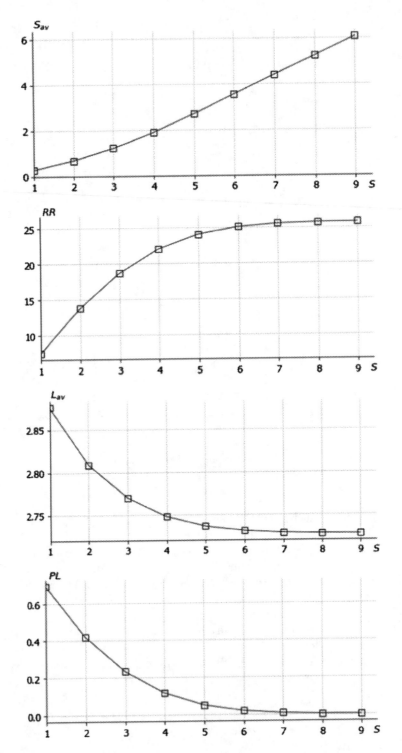

Fig. 3. Performance measures vs inventory size S under $(S-1, S)$ policy

6   Conclusion

The models of queuing-inventory systems with impatient customers and infinite buffer were studied under (s, S), (s, Q) and $(S, S - 1)$ replenishment policies. The negative replenishment were considered that decreases the inventory level. Customer enters the system even when the inventory level is zero. We assume that customers after being served according to Bernoulli scheme either leaves the system empty handed or with an item from inventory. We used 2D Markov chains with tridiagonal generator matrices for mathematical modeling of the system. Ergodicity conditions were found and the algorithm for calculation of system performance measure was developed. Numerical experiments were performed and behavior of performance measures was analyzed under different policies.

References

1. Schwarz, M., Sauer, C., Daduna, H., Kulik, R., Szekli, R.: M/M/1 queueing systems with inventory. Queueing Syst. Their Appl. **54**(1), 55–78 (2006)
2. Sigman, K., Simchi-Levi, D.: Light traffic heuristic for an M/G/1 queue with limited inventory. Ann. Oper. Res. **40**, 371–380 (1992)
3. Melikov, A.Z., Molchanov, A.A.: Stock optimization in transport/storage systems. Cybernetics **27**(3), 484–487 (1992)
4. Krishnamoorthy, A., Shajin, D., Narayanan, W.: Inventory with positive service time: a survey. In: Anisimov, V., Limnios, N. (eds.) Advanced Trends in Queueing Theory. Series of Books Mathematics and Statistics Sciences, vol. 2, pp. 201–238. ISTE and Wiley, London (2021)
5. Neuts, M.F.: Matrix-Geometric Solutions in Stochastic Models: An Algorithmic Approach. John Hopkins University Press, Baltimore (1981)

Analysis of Open Queueing Networks with Batch Services

Elena Stankevich[1](✉)(iD), Igor Tananko[1](iD), and Michele Pagano[2](iD)

[1] Saratov State University, 83 Astrakhanskaya Street, Saratov 410012, Russia
zsysan@gmail.com
[2] University of Pisa, Via G. Caruso 16, 56122 Pisa, Italy
michele.pagano@unipi.it

Abstract. In this work we analyze an open queueing network with batch services. In more detail, the arrival process is Poissonian and each node consists of a single server and an infinite waiting queue. Arrivals are served in fixed-size batches: if the number of customers in a node is less than the predefined batch size, the server remains idle, otherwise he will select the required number of customers, which then will be served as a unique batch with exponentially distributed service time. In this paper we show that, under suitable conditions on the routing matrix, such queueing network is equivalent, in terms of stationary distribution, to a Jackson network with single-server nodes and state-dependent service rates. Finally, the goodness of the proposed approach is confirmed by comparing analytical and simulation results.

Keywords: Open queueing networks · Analysis · Batch service

1 Introduction

Queueing systems and networks with batch services attract the interest of many researchers, since they permit to model and analyze various multi-user systems [1,2], large scale semiconductor manufacturing systems [3], cloud computing systems [4] and wireless sensor networks [5].

The analysis of any queueing network is aimed at obtaining expressions for its stationary characteristics, the most important of which is the stationary probability distribution of the states of the system. Since the equilibrium equations for queuing networks with batch services have a high dimensionality, the calculation of the stationary distribution as a numerical solution of these equations is computationally difficult. Therefore, special attention has been devoted to the search for product-form solutions.

It is worth noticing that the fundamental works on queueing networks with batch services are relatively recent, as they were published in 1990 [6,7]. In more

This work was supported by the Ministry of science and education of the Russian Federation in the framework of the basic part of the scientific research state task, project FSRR-2020-0006.

A. Dudin et al. (Eds.): ITMM 2021, CCIS 1605, pp. 40–51, 2022.
https://doi.org/10.1007/978-3-031-09331-9_4

detail, in [6] a continuous-time Markov chain is introduced to model queueing networks with simultaneous changes due to batch services, or discrete-time structure and clustering processes such as those arising in polymer chemistry. It is shown that if multiple instantaneous state transitions of the process are allowed and the Markov chain is reversible, then its stationary distribution has a product-form. In [7] a discrete-time closed queueing network with batch services is considered and the state of the network is defined by a vector with dimension equal to the number of customers. Each element of the status vector is associated with a specific customer and indicates the node occupied by that customer. So, customers transitions are reduced to changes of the corresponding labels, and it is assumed that the change of a label does not depend on the status of the labels of the other customers. It is shown that on an irreducible set of states and for arbitrary given functions of service and routing, there is a product-form for the stationary probability distribution of the queueing network states. Chao [8] and Economou [9] considered networks, for which the quasi-reversibility conditions are met and the groups of customers at the end of the service in one node always pass to another node together.

To analyze queuing networks with batch services and an arbitrary distribution of the service time that do not admit a product-form of the stationary distribution, in [3,10] it was proposed to use the decomposition method. Finally, in [11,12] the stationary distribution was calculated as the normalized solution of the system of equilibrium equations.

In this paper, we consider open queueing networks with service of fixed-size batches of customers and independent routing. It is assumed that the batch size is significantly smaller than the number of nodes to which the customers can be routed at the end of the service. Thus, the network nodes work independently and this consideration permits to simplify the analysis of the queueing network, which is reduced to the investigation of the individual queues in isolation. In more detail it is proposed to calculate the stationary state probability distribution of the open network in a product-form, similar to the case of birth-death processes after recalculating the transition rates. To the best of our knowledge, this approach is new. Until now, indeed, the probability generating function [4,13–15], the Laplace-Stieltjes transform [16], and the direct calculation of the stationary distribution as a solution of the Kolmogorov equations [17] have been mainly used to calculate the stationary characteristics of the queueing network.

The rest of the paper is organized as follows. Section 2 introduces the model of the queueing network, while in Sect. 3 an equivalent (in terms of stationary distribution) Jackson network with single-server nodes is proposed. In more detail, for such equivalent system state-dependent service rates as well as expressions for the stationary probability distribution are derived. Then, Sect. 4 compares the values of the analytical expression with the simulation results, and analyses the dependence of the characteristics of open queueing networks on different system parameters (batch size, arrival rate, service rate).

2   Statement of the Problem

Consider a continuous-time open queueing network N consisting of L nodes S_i, $i \in I$, $I = \{1, \ldots, L\}$. Customers arrive to the queueing network N from an outside source (denoted in the following as S_0) according to a Poisson stream of rate λ_0. Customer transitions between nodes and the source are defined by the routing matrix $\Theta = (\theta_{ij})$, $i, j = 0, \ldots, L$, where θ_{ij} is the transition probability from node S_i to node S_j. The state of the network is defined by a vector $s = (s_1, \ldots, s_L)$, where s_i is the number of customers at node S_i. Denote by $X = \{s : s_i \geq 0\}$ the state space of the queueing network N.

Each node S_i, $i = 1, \ldots, L$, operates as an infinite capacity single-server queue. Arriving customers are placed in the waiting queue if the server is busy. Customers are served in batches, and let b_i be the customer batch size for node S_i. The server remains idle until the required number b_i of customers arrives at the node and then the service of the batch starts immediately; otherwise, b_i customers are selected in any order for service, while the others remain in the queue. The service times of batches at node S_i are exponentially distributed with parameter μ_i, $i = 1, \ldots, L$. After a batch finishes its service at node S_i, each customer will go, independently of the others, to node S_j with probability θ_{ij}, $i, j = 0, 1, \ldots, L$.

Our aim is to find the stationary distribution $\pi(s) = (\pi_1(s_1), \ldots, \pi_L(s_L))$, $s \in X$, for the queueing network N, where $\pi_i(s_i)$ represents the stationary distribution for node S_i, $s_i = 0, 1, \ldots$, $i = 1, \ldots, L$, starting from the analysis of a single node.

3   Analysis of the Model

In this paper we analyze large scale networks with individual routing of the customers, assuming that the number of possible destinations is significantly larger than the batch size. Hence the probability of the simultaneous arrival of two or more customers in a node can be neglected. Therefore, we will assume that each node in N is fed by a Poisson stream of customers.

First we will study the isolated node S_i, $i = 1, \ldots, L$. It is known that the equilibrium equations for this node have the form

$$\begin{cases} \lambda_i \pi_i(n) = \mu_i \pi_i(b_i), \ n = 0, \\ \lambda_i \pi_i(n) = \lambda_i \pi_i(n-1) + \mu_i \pi_i(b_i + n), \ 1 \leq n \leq b_i - 1, \\ (\lambda_i + \mu_i)\pi_i(n) = \lambda_i \pi_i(n-1) + \mu_i \pi_i(b_i + n), \ n \geq b_i. \end{cases} \quad (1)$$

where λ_i denotes the arrival rate to node S_i, $i = 1, \ldots, L$.

We define a birth-death process ξ_i, which will be equivalent in steady-state probabilities to the Markov process describing the node S_i. Let the process ξ_i be defined on a set of states $\{0, 1, \ldots\}$, let $\lambda_i = \lambda_i(n)$ be the transition rate of the process ξ_i from state n to state $n+1$, which does not depend on the state n, $n \in \{0, 1, \ldots\}$, and let $\tilde{\mu}_i(n)$ be the transition rate of the process ξ_i from state

n to state $n - 1$, where $n \in \{1, 2, \ldots\}$. The states $\{0, 1, \ldots\}$ and the parameter λ_i of the process ξ_i correspond to the states $\{0, 1, \ldots\}$ and the parameter λ_i of node S_i. Let us find the rates $\widetilde{\mu}_i(n)$, $n = 1, 2, \ldots$. To this aim, note that the steady-state probabilities of the birth–death process ξ_i are given by [18]

$$\pi_i(k) = \pi_i(0) \prod_{n=1}^{k} \frac{\lambda_i}{\widetilde{\mu}_i(n)}, \quad k = 1, 2, \ldots, \tag{2}$$

where

$$\pi_i(0) = \left(1 + \sum_{k=1}^{\infty} \prod_{n=1}^{k} \frac{\lambda_i}{\widetilde{\mu}_i(n)}\right)^{-1}, \quad i = 1, \ldots, L.$$

By substituting (2) in (1), we get the expressions that define $\widetilde{\mu}_i(n)$, $n = 1, 2, \ldots$,

$$\begin{cases} \widetilde{\mu}_i(n) = \lambda_i - \mu_i \dfrac{\lambda_i^{b_i}}{\widetilde{\mu}_i(n+1) \cdot \ldots \cdot \widetilde{\mu}_i(b_i + n)}, & 1 \leq n \leq b_i - 1, \\[2mm] \widetilde{\mu}_i(n) = \lambda_i + \mu_i - \mu_i \dfrac{\lambda_i^{b_i}}{\widetilde{\mu}_i(n+1) \cdot \ldots \cdot \widetilde{\mu}_i(b_i + n)}, & n \geq b_i. \end{cases} \tag{3}$$

Let $M_i = \lim_{n \to \infty} \widetilde{\mu}_i(n)$; if the limit exists, then:

$$\mu_i \lambda_i^{b_i} = (\lambda_i + \mu_i - M_i) M_i^{b_i}$$

or

$$M_i^{b_i + 1} - (\lambda_i + \mu_i) M_i^{b_i} + \lambda_i^{b_i} \mu_i = 0. \tag{4}$$

The existence of the equivalent birth-death process ξ_i requires that the previous equation has a positive solution, fulfilling the stability condition for each node S_i.

The answer is provided by the following theorem (without loss of generality we denote the generic M_i, for $i \in I$ by x).

Theorem 1. *The equation*

$$x^{b+1} - (\lambda + \mu)x^b + \lambda^b \mu = 0 \tag{5}$$

has two positive roots, the largest of which belongs to the interval

$$\left(\frac{b(\lambda + \mu)}{b + 1}, \frac{(\lambda + \mu)^{b+1} - \lambda^b \mu}{(\lambda + \mu)^b}\right).$$

Proof. Consider the function

$$f(x) = x^{b+1} - (\lambda + \mu)x^b + \lambda^b \mu$$

for $\lambda < b\mu$ and $b \geq 1$.

It is easy to verify that $f(x)$ is continuous for any $x \in R$ and $x_1 = \lambda$ is a root of $f(x)$. To determine the existence of other roots let us consider the first derivative of $f(x)$:

$$f'(x) = (b+1)x^{b-1}\left(x - \frac{b}{b+1}(\lambda+\mu)\right). \tag{6}$$

The equation $f'(x) = 0$ has only one positive root

$$x^* = \frac{b(\lambda+\mu)}{b+1},$$

with $x^* > x_1$. Indeed,

$$x^* - x_1 = \frac{b}{b+1}(\lambda+\mu) - \lambda = \frac{b\mu - \lambda}{b+1} > 0,$$

since $\lambda < b\mu$ and $b \geq 1$. Since $f'(x) > 0$ for

$$x \in \left(\frac{b(\lambda+\mu)}{b+1}, \lambda+\mu\right),$$

then the function $f(x)$ is increasing in such interval. Moreover,

$$f\left(\frac{b(\lambda+\mu)}{b+1}\right) < 0$$

and $f(\lambda+\mu) > 0$, hence in the interval $\left(\frac{b(\lambda+\mu)}{b+1}, \lambda+\mu\right)$ there is a value of x such that $f(x) = 0$.

To further refine the estimation of the root, let us note that in the above-mentioned interval the function $f(x)$ is convex, since

$$f''(x) = bx^{b-2}((b+1)x - (b-1)(\lambda+\mu)) > 0$$

for

$$x > \frac{b(\lambda+\mu)}{b+1} > \frac{(b-1)(\lambda+\mu)}{b+1}.$$

The tangent line to $f(x)$ at the point $x = \lambda+\mu$ is

$$y(x) = \lambda^b\mu + (\lambda+\mu)^b(x - (\lambda+\mu))$$

and its intersection with the horizontal axis is

$$x_0 = (\lambda+\mu) - \frac{\lambda^b\mu}{(\lambda+\mu)^b}.$$

Since the function $f(x)$ is convex, x_0 is an upper bound for the roots of $f(x)$, and this implies that the largest root of Eq. (5) belongs to the interval

$$\left(\frac{b(\lambda+\mu)}{b+1}, \frac{(\lambda+\mu)^{b+1} - \lambda^b\mu}{(\lambda+\mu)^b}\right).$$

Taking into account the previous theorem and the stability condition of the equivalent birth-death process, Eq. (4) has a unique root, located in the interval $(\lambda_i, \lambda_i + \mu_i)$, which can be determined numerically (explicit closed-for solutions can be easily derived only for $b = 1$ and $b = 2$). From the system of Eqs. (3) it follows

$$\widetilde{\mu}_i(b_i) = \widetilde{\mu}_i(b_i + 1) = \widetilde{\mu}_i(b_i + 2) = \cdots = M_i,$$

and then the service rates $\widetilde{\mu}_i(b_i - 1), \widetilde{\mu}_i(b_i - 2), \ldots, \widetilde{\mu}_i(1)$ can be easily calculated. Thus, the rates $\widetilde{\mu}_i(n)$ are determined for each state n of process ξ_i.

The results obtained for the process ξ_i can be applied to any node, and so we can create an open queueing network $\widetilde{N}$ with nodes $\widetilde{S}_i$ and service rates $\widetilde{\mu}_i(n)$, where n is the number of customers in the node $\widetilde{S}_i$, $n = 1, 2, \ldots$, $i = 1, \ldots, L$. The other parameters of $\widetilde{N}$ coincide with the corresponding parameters of the original queueing network N.

$\widetilde{N}$ is equivalent in stationary distribution to the queueing network N with batch services and is a Jackson network.

The arrival rates in nodes S_i are determined by the following equations

$$\lambda_i = \frac{\omega_i}{\omega_0} \lambda_0, \ i = 1, \ldots, L,$$

where the vector of visitation rates $\omega = (\omega_1, \ldots, \omega_L)$ is the solution of the equation $\omega\Theta = \omega$ with the normalization condition $\sum_{i=0}^{L} \omega_i = 1$.

The queueing network N and its equivalent network $\widetilde{N}$ are stable if the utilization coefficient in the node S_i, $i = 1, \ldots, L$,

$$\rho_i = \frac{\lambda_i}{b_i \mu_i} < 1,$$

and, under such conditions, we can compute the stationary distribution for $\widetilde{N}$. We obtain

$$\pi(s) = \prod_{i=1}^{L} \pi_i(s_i), \ s \in X,$$

where

$$\pi_i(s_i) = \pi_i(0) \prod_{n=1}^{s_i} \frac{\lambda_i}{\widetilde{\mu}_i(n)}.$$

Then, the average number of customers in the node S_i, $i = 1, \ldots, L$, is given by

$$\bar{s}_i = \sum_{n=1}^{\infty} n\pi_i(n),$$

the average sojourn time in the node S_i, $i = 1, \ldots, L$, is

$$\bar{u}_i = \frac{\bar{s}_i}{\lambda_i},$$

and the average response time of the queueing network is

$$\bar{\tau} = \frac{1}{\lambda_0} \sum_{i=1}^{L} \lambda_i \bar{u}_i.$$

4   Numerical Examples

Numerical examples are reported in this section to verify the goodness of the product-form approximation for complex networks and investigate the dependence of their characteristics on different system parameters (batch size, arrival rate, service rate). Although different topologies have been investigated, for sake of brevity just one network topology is considered, focusing on overall system performance parameters as well as on characteristics of single queues.

Consider the queueing network N with the following parameters (unless otherwise stated): $L = 14$, $b = (3, 2, 2, 3, 2, 2, 3, 3, 2, 2, 3, 2, 2, 3)$, $\mu = (0.8, 0.6, 0.9, 0.6, 0.8, 0.8, 0.9, 0.6, 0.7, 0.8, 0.9, 1.0, 0.7, 0.7)$, and

$$\Theta = \begin{pmatrix}
0.0 & 0.3 & 0.4 & 0.3 & 0.0 & 0.0 & 0.0 & 0.0 & 0.0 & 0.0 & 0.0 & 0.0 & 0.0 & 0.0 & 0.0 \\
0.0 & 0.0 & 0.0 & 0.0 & 0.1 & 0.1 & 0.1 & 0.1 & 0.1 & 0.1 & 0.1 & 0.1 & 0.1 & 0.1 & 0.0 \\
0.0 & 0.0 & 0.0 & 0.0 & 0.0 & 0.1 & 0.1 & 0.2 & 0.2 & 0.1 & 0.1 & 0.1 & 0.1 & 0.0 & 0.0 \\
0.0 & 0.0 & 0.0 & 0.0 & 0.0 & 0.0 & 0.1 & 0.1 & 0.1 & 0.2 & 0.1 & 0.2 & 0.1 & 0.1 & 0.0 \\
0.0 & 0.0 & 0.0 & 0.0 & 0.0 & 0.0 & 0.0 & 0.1 & 0.1 & 0.1 & 0.1 & 0.1 & 0.2 & 0.2 & 0.1 \\
0.2 & 0.0 & 0.0 & 0.0 & 0.0 & 0.0 & 0.0 & 0.0 & 0.1 & 0.1 & 0.1 & 0.1 & 0.1 & 0.1 & 0.2 \\
0.2 & 0.1 & 0.0 & 0.0 & 0.0 & 0.0 & 0.0 & 0.0 & 0.0 & 0.1 & 0.2 & 0.1 & 0.1 & 0.1 & 0.1 \\
0.2 & 0.1 & 0.1 & 0.1 & 0.1 & 0.0 & 0.0 & 0.0 & 0.0 & 0.0 & 0.0 & 0.1 & 0.1 & 0.1 & 0.1 \\
0.2 & 0.0 & 0.0 & 0.0 & 0.1 & 0.1 & 0.1 & 0.0 & 0.0 & 0.1 & 0.1 & 0.1 & 0.1 & 0.0 & 0.1 \\
0.1 & 0.1 & 0.1 & 0.1 & 0.0 & 0.0 & 0.0 & 0.0 & 0.0 & 0.0 & 0.1 & 0.1 & 0.2 & 0.1 & 0.1 \\
0.3 & 0.1 & 0.1 & 0.1 & 0.1 & 0.1 & 0.1 & 0.0 & 0.0 & 0.0 & 0.0 & 0.0 & 0.0 & 0.1 & 0.0 \\
0.3 & 0.0 & 0.0 & 0.0 & 0.0 & 0.1 & 0.1 & 0.1 & 0.1 & 0.1 & 0.0 & 0.0 & 0.0 & 0.1 & 0.1 \\
0.3 & 0.1 & 0.1 & 0.1 & 0.1 & 0.0 & 0.0 & 0.0 & 0.0 & 0.1 & 0.1 & 0.1 & 0.0 & 0.0 & 0.0 \\
0.3 & 0.0 & 0.0 & 0.0 & 0.0 & 0.1 & 0.1 & 0.1 & 0.1 & 0.0 & 0.0 & 0.1 & 0.1 & 0.0 & 0.1 \\
0.4 & 0.0 & 0.0 & 0.0 & 0.0 & 0.0 & 0.0 & 0.0 & 0.1 & 0.1 & 0.1 & 0.1 & 0.1 & 0.1 & 0.0
\end{pmatrix}.$$

The considered network satisfies the assumptions introduced above. Indeed, the network consists of a relatively large number of nodes, the size of the batches that are served together is significantly less than the number of possible output nodes and the routing probabilities are of the same order of magnitude (there is no privileged path through the network). Hence, the Poissonian assumption can be reasonably assumed for any node of the network.

The first two sets of tests investigated the accuracy of the developed method by comparing the analytical values with the results of discrete-event simulation. In more detail, in the first experiment we analysed the (overall) average response time as a function of the input rate λ_0.

Table 1 shows that the largest difference in the values of $\bar{\tau}$ is observed for $\lambda_0 = 0.1$ and does not exceed 10.2%, while for the other values of λ_0, the

Table 1. Average response time of the queueing network.

λ_0	0.1	0.5	1.0	1.5	2.0	2.5	2.7
Approximation	107.32	27.57	18.29	16.20	16.76	20.62	24.06
Simulation	118.25	28.78	18.61	16.35	16.86	20.82	25.24

deviation is no more than 5%. Note that the intensity of the flow $\lambda_0 = 2.7$ is almost the maximum for the network under consideration, since for such value the stability condition for node S_9 is still met.

In the second example we focused on a specific node (the queue S_7), considering the average number of customers (Tables 2) as well as the average sojourn time in the node (Tables 3) for different values of the service rate μ_7 with fixed arrival rate $\lambda_0 = 1.5$.

Table 2. Average number of customers in the node S_7.

μ_7	0.2	0.4	0.6	0.8	0.9	1.0	1.2
Approximation	8.72	2.53	1.89	1.63	1.56	1.5	1.41
Simulation	8.66	2.53	1.91	1.66	1.58	1.52	1.43

Table 3. Average sojourn time in the node S_7.

μ_7	0.2	0.4	0.6	0.8	0.9	1.0	1.2
Approximation	18.26	5.31	3.95	3.42	3.26	3.13	2.95
Simulation	18.13	5.30	3.99	3.47	3.31	3.18	3.00

The characteristics of the node S_7, derived by discrete-event simulation, were calculated in stationary conditions with a confidence interval of 0.001 and a confidence level higher than 0.95.

In the third experiment we investigated the dependence of the stationary characteristics of the nodes S_2, S_6 and S_{11} on the intensity of the incoming flow λ_0 (see Fig. 1 and 2). The characteristics of the other nodes are not shown in the graphs for sake of clarity, since their behavior does not differ qualitatively form the reported ones.

Figure 1 shows that the average number of customers in all systems monotonically increases with λ_0. Instead, the average (node) sojourn time reaches a minimum for some value of λ_0 as highlighted by Fig. 2. This can be explained as follows. When λ_0 is close to zero, the device is idle for a long time, and the customers forming an "incomplete" batch have to wait in the buffer until

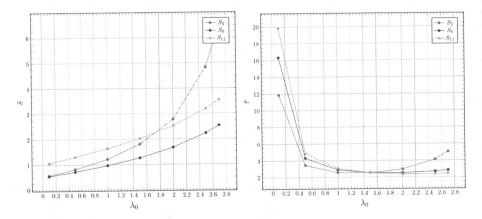

Fig. 1. Average number of customers in the nodes S_2, S_6 and S_{11}.

Fig. 2. Average sojourn time in the nodes S_2, S_6 and S_{11}.

the last element of the batch enters the system. Instead, when the arrival rate into the considered system approaches its service rate, the average waiting time increases significantly. Thus, there is an optimal value of the arrival rate, at which the average sojourn time in the node is minimal.

The fourth experiment is devoted to the study of stationary characteristics of the nodes S_7 and S_9 for different sizes b of the batch in these systems. The input rate in this experiment is again $\lambda_0 = 1.5$.

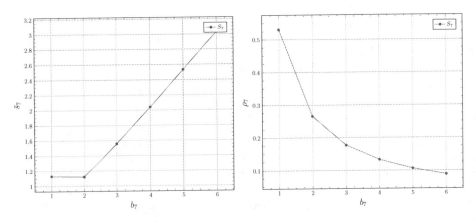

Fig. 3. Average number of customers in the node S_7.

Fig. 4. Utilization coefficient of the node S_7.

The minimum value of the average number of customers in both nodes is achieved when the batch size is two (Fig. 3 and 5), while the utilization coefficient is a monotone decreasing function of b (Fig. 4 and 6), but its numerical value

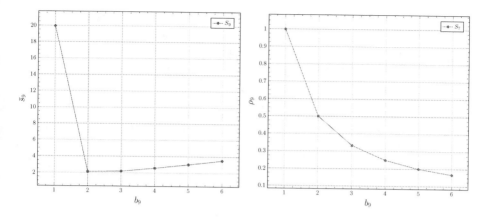

Fig. 5. Average number of customers in the node S_9.

Fig. 6. Utilization coefficient of the node S_9.

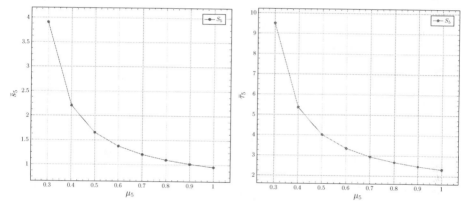

Fig. 7. Average number of customers in the node S_5.

Fig. 8. Average sojourn time in the node S_5.

depends on the arrival rate at the considered node (in our example ρ_9 is almost twice ρ_7). It is worth noticing (see Fig. 6) that for $b_9 = 1$, the utilization ρ_9 of the node S_9 is close to 1 and this is confirmed by the high value of the number of customers in the system ($\bar{s}_9 \approx 20$ as shown in Fig. 5). When $b_9 = 2$, then $\bar{s}_9 \approx 2$, while the increment of b_9 leads to a slight increase in $\bar{s}_9$. Thus, the increase of the batch size can significantly improve the basic average characteristics of service systems. Actually, as shown by numerical experiments, the minimum value of both the average sojourn time and average number of customers in the system can be assumed at different values b, depending on the network topology and the routing matrix.

Finally, we calculated the stationary characteristics of the nodes S_5 and S_9 for different values of the service rate in these systems (assuming, as before, $\lambda_0 = 1.5$).

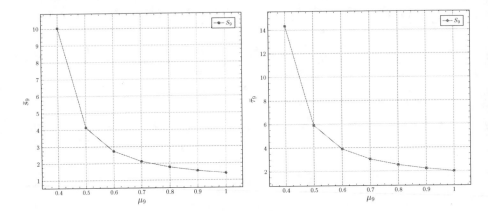

Fig. 9. Average number of customers in the node S_9.

Fig. 10. Average sojourn time in the node S_9.

The graphs shown in Fig. 7, 8, 9 and 10 decrease monotonically with the growth of μ and asymptotically tend to their limit values.

5   Conclusions

In this paper large-size open queueing networks with batch services are considered. Under the assumption that the number of output nodes is significantly more than the batch size, it is shown that the stationary distribution of the queueing network can be expressed in product-form. Then, the parameters of the equivalent queueing network are derived and the goodness of the approximation is verified by means of discrete-event simulation.

References

1. Baetens, J., Steyaert, B., Claeys, D., Bruneel, H.: System occupancy in a multiclass batch-service queueing system with limited variable service capacity. Ann. Oper. Res. **293**(1), 3–26 (2019). https://doi.org/10.1007/s10479-019-03470-1
2. Bellalta, B., Oliver, M.: A space-time batch-service queueing model for multi-user MIMO communication systems. In: Proceedings of the 12th ACM International Conference on Modeling, Analysis and Simulation of Wireless and Mobile Systems, pp. 357–364. Association for Computing Machinery, New York (2009). https://doi.org/10.1145/1641804.1641866
3. Hanschke, T., Zisgen, H.: Queueing networks with batch service. Eur. J. Ind. Eng. **5**(3), 313–326 (2011). https://doi.org/10.1504/EJIE.2011.041619
4. Santhi, K., Saravanan, R.: Performance analysis of cloud computing bulk service using queueing models. Int. J. Appl. Eng. Res. **12**(17), 6487–6492 (2017)
5. Mitici, M., Goseling, J., van Ommeren, J.-K., de Graaf, M., Boucherie, R.J.: On a tandem queue with batch service and its applications in wireless sensor networks. Queueing Syst. **87**(1), 81–93 (2017). https://doi.org/10.1007/s11134-017-9534-1

6. Boucherie, R.J., Dijk, N.M.: Spatial birth-death processes with multiple changes and applications to batch service networks and clustering processes. Adv. Appl. Probab. **22**(2), 433–455 (1990). https://doi.org/10.2307/1427544

7. Henderson, W., Pearce, C.E.M., Taylor, P.G., Dijk, N.M.: Closed queueing networks with batch services. Queueing Syst. **6**, 59–70 (1990). https://doi.org/10.1007/BF02411465

8. Chao, X., Pinedo, M., Shaw, D.: Networks of queues with batch services and customer coalescence. J. Appl. Probab. **33**(3), 858–869 (1996). https://doi.org/10.2307/3215364

9. Economou, A.: An alternative model for queueing systems with single arrivals, batch services and customer coalescence. Queueing Syst. **40**, 407–432 (2002). https://doi.org/10.1023/A:1015089518876

10. Klünder, W.: Decomposition of open queueing networks with batch service. In: Fink, A., Fügenschuh, A., Geiger, M.J. (eds.) Operations Research Proceedings 2016. ORP, pp. 575–581. Springer, Cham (2018). https://doi.org/10.1007/978-3-319-55702-1_76

11. Stankevich, E.P., Tananko, I.E., Dolgov, V.I.: Analysis of closed queueing networks with batch service. Izv. Saratov Univ. (N. S.) Ser. Math. Mech. Inform. **20**(4), 527–533 (2020). https://doi.org/10.18500/1816-9791-2020-20-4-527-533

12. Stankevich, E., Tananko, I., Osipov, O.: Analysis of closed unreliable queueing networks with batch service. In: Dudin, A., Nazarov, A., Moiseev, A. (eds.) ITMM 2020. CCIS, vol. 1391, pp. 352–362. Springer, Cham (2021). https://doi.org/10.1007/978-3-030-72247-0_26

13. Gupta, G.K., Banerjee, A.: Steady state analysis of system size-based balking in $M/M^b/1$ queue. Int. J. Math. Oper. Res. **14**(3), pp. 319–337(2019). https://doi.org/10.1504/IJMOR.2019.099383

14. Harrison, P.G.: Product-form queueing networks with batches. In: Bakhshi, R., Ballarini, P., Barbot, B., Castel-Taleb, H., Remke, A. (eds.) EPEW 2018. LNCS, vol. 11178, pp. 250–264. Springer, Cham (2018). https://doi.org/10.1007/978-3-030-02227-3_17

15. Chaudhry, M.L., Templeton, J.G.C.: A First Course in Bulk Queues. Wiley, New York (1983)

16. Nain, P., Panigrahy, N.K., Basu, P., Towsley, D.: One-dimensional service networks and batch service queues. Queueing Syst. **98**(1), 181–207 (2021). https://doi.org/10.1007/s11134-021-09703-0

17. Krishnamoorthy, A., Ushakumari, P.V.: A queueing system with single arrival bulk service and single departure. Math. Comput. Model. **31**(2–3), 99–108 (2000). https://doi.org/10.1016/S0895-7177(99)00226-5

18. Medhi, J.: Stochastic Models in Queueing Theory, 2nd edn. Elsevier Science (2003)

N-Policy for a Production Inventory System with Positive Service Time

N. J. Thresiamma[1] and K. P. Jose[2(✉)]

[1] Government Women's Polytechnic College, Thrissur 680007, Kerala, India
[2] Department of Mathematics, St. Peter's College, Kolenchery 682311, Kerala, India
kpjspc@gmail.com

Abstract. This paper presents an optimal control policy that minimizes the long-run cost in an (s, S) production inventory system with positive service time. The Matrix Geometric method is used to analyze the system. A necessary and sufficient condition for system stability is obtained. Some significant system performance measures are defined, and the effect of system parameters on performance measures is illustrated numerically. The Optimal (s, S) pair is determined for the specific set of parameter values, and the effects of the parameters on the cost function are graphically illustrated.

Keywords: N-policy · Production inventory · Service time · Matrix Geometric Method

1 Introduction

In most of the inventory models, it is crucial for the server to decide when to start its service, as an intermittent setup can greatly increase operating costs. A company's inventory control policies determine how the company manages the movement of inventory under its control. Proper inventory control policies and procedures reduce the cost associated with the inventory. In some production and manufacturing systems, the high switching costs associated with inactive servers often make it uneconomical to provide service immediately after the first customer arrives. In such cases, it is better to begin the service only when a few customers arrive, say N, so that excessive setups can be avoided. In this work, we introduce N policy to a production inventory with positive service time. According to the policy, the status of the server is turned ON only when there are N or more customers encountered in the system and the inventory level is positive and is turned OFF when the system is empty.

Inventory with positive service time is first investigated by Berman et al. [2] where demands and service formed two distinct deterministic processes. A detailed review of inventory models involving positive service time is given by Krishnamoorthy et al. [8]. Krishnamoorthy et al. [5] dealt with production inventory with positive service time. The authors discussed the stochastic decomposition of the system by considering the assumption that the customer does not

© Springer Nature Switzerland AG 2022
A. Dudin et al. (Eds.): ITMM 2021, CCIS 1605, pp. 52–66, 2022.
https://doi.org/10.1007/978-3-031-09331-9_5

join the system when the inventory level is zero. Krishnamoorthy and Jose [6] analyzed and compared three production inventory systems with positive service time and retrial of customers. The inter arrival time of customers, service time, production time and inter retrial times are assumed to follow exponential distributions. The authors arrive at the conclusion that overall and successful retrial rate of customers increases with the increase of arrival rate and decreases with the increase of production rate or service rate. Jose and Rejitha [11] analysed a stochastic inventory system with two modes of service rate and retrials. They derived several important performance measures of the system in the steady state and a suitable cost function is constructed and analyzed numerically for the expected minimum cost. Jose and Salini [4] studied a $MAP/PH/1$ production inventory model with varying service rates. They assumed that, when the inventory level decreases to s, service is given at a reduced rate and an arriving customer who identifies the server busy or inventory level zero, proceeds to an orbit of infinite capacity and retries from there. They computed some of the system performance measures and constructed a suitable cost function. Jose and Beena [3] studied a production inventory system with two heterogeneous servers involving multiple vacations. By assuming poisson arrival rate and exponential server vacation rate, they obtained the stability condition and performance measures of the system.

Over the past decade, an increasing attention can be seen in queuing scenario to control the queue by applying the concept of N-policy. The concept of N-policy is most commonly used for controlling service. This has been widely accepted due to their applicability for modeling purposes of any production and manufacturing system as well as computer and telecommunication system. N-policy was first introduced in 1963 by Yadin and Naor [12] in queueing literature to minimize the total operational cost in a cycle. Artalejo [1] compared N, T, D policies on $M/G/1$ queueing system. The author showed that the D-policy is superior to the N-policy when the cost function is based on the mean work-load, whilst the average queue length is used to show the superiority of the N-policy over the T-policy. The author also showed that the T-policy is the worst policy under both cost structures and the relation between the optimum N and D policies depends on the employed cost function. Krishnamoorthy et al. [7] was the first to introduce N-policy in (s, S) inventory system with positive service time. They assumed that the lead time is zero and showed that the cost function is separately convex in the variables S and N. They also proved that the cost is minimum at $s = 0$.

The technical aspects of this paper are presented in four parts. The first part offers a description of the model. In the second part, it moves on to the steady state analysis and computation of system performance measures. Numerical and graphical illustrations are given in the third section. Finally, it is concluded by computing the optimal (s, S) pair and the optimal value of N.

2 Description of the Model

Consider an (s, S) production inventory system with a single server and positive service time. Customers arrive according to the Poisson distribution with rate λ and service rate and production rate follow Exponential distributions with rate μ and β respectively. Each production is of 1 unit and the production process is ON when the inventory level reaches s and is switched OFF when the inventory level reaches S. Whenever the server is idle, it is switched off and is activated only when N customers accumulate and when there is a positive on-hand inventory. The following assumptions and notations are used in this model.

Assumptions

- The arrival of customers follows Poisson distribution with parameter λ.
- The service pattern and production process follow exponential distributions with parameters μ and β respectively.
- The server is switched OFF when the system is empty and it is turned ON at the instant when there are N customers in the waiting line; and there is a positive on-hand inventory.

Notations

$N(t)$: Number of customers in the system at time t.

$I(t)$: Inventory level at time t.

$$C(t) : \begin{cases} 0, & \text{if server is idle at time t;} \\ 1, & \text{if server is busy at time t.} \end{cases}$$

$$J(t) : \begin{cases} 0, & \text{if the production is OFF mode;} \\ 1, & \text{if the production is ON mode.} \end{cases}$$

$e : (1, 1, 1, ..., 1)^T$, column vector of appropriate dimension.

Then $Z(t) = \{(N(t), C(t), J(t), I(t)), t \geq 0\}$ is a Quasi Birth Death Process on the state space $S = \bigcup_{i=0}^{\infty} L(i)$ and is independent for $i \geq N + 1$, where,

$$L(0) = \{(0, 0, 0, j); s + 1 \leq j \leq S\} \bigcup \{0, 0, 1, j); 0 \leq j \leq S - 1\},$$

For $1 \leq i \leq N - 1$,

$$L(i) = \{(i, 0, 0, j); s + 1 \leq j \leq S\} \bigcup \{i, 0, 1, j); 0 \leq j \leq S - 1\}$$
$$\bigcup \{(i, 1, 0, j); s + 1 \leq j \leq S\} \bigcup \{(i, 1, 1, j); 1 \leq j \leq S - 1\},$$

For $i \geq N$,

$$L(i) = (i, 0, 1, 0) \bigcup \{(i, 1, 0, j); s + 1 \leq j \leq S\} \bigcup \{(i, 1, 1, j); 1 \leq j \leq S - 1\}.$$

Arranging the states in the lexicographic order, infinitesimal generator of the process $\{Z(t)|t \geq 0\}$ is a block tridiagonal matrix given by,

$$
G = \begin{array}{c} 0 \\ 1 \\ 2 \\ \\ N_{-1} \\ \underline{N} \\ \underline{N+1} \end{array}
\begin{bmatrix}
B_{00} & B_{01} & & & & & & \\
B_{10} & A_1^* & A_0^* & & & & & \\
 & A_2^* & A_1^* & A_0^* & & & & \\
 & & \ddots & \ddots & \ddots & & & \\
 & & & A_2^* & A_1^* & A_0^{**} & & \\
 & & & & A_2^{**} & A_1 & A_0 & \\
 & & & & & A_2 & A_1 & A_0 \\
 & & & & & & A_2 & A_1 & A_0 \\
 & & & & & & & \ddots & \ddots & \ddots
\end{bmatrix}
$$

where the block matrices are obtained as follows:

$$
[B_{00}](pq) = \begin{cases}
\lambda, & \text{if } p = q; \quad p = 1, 2, ..., S - s, \\
-(\lambda + \beta), & \text{if } p = q; \quad p = S - s + 1, ..., 2S - s, \\
\beta, & \text{if } p = 2S - s \quad \& \quad q = S - s, \\
& q = p + 1; \quad p = S - s + 1, ..., 2S - s - 1, \\
0, & \text{otherwise.}
\end{cases}
$$

$$
[B_{01}](pq) = \begin{cases}
\lambda, & \text{if } p = q; \quad p = 1, 2, ..., 2S - s, \\
0, & \text{otherwise.}
\end{cases}
$$

$$
[B_{10}](pq) = \begin{cases}
\mu, & \text{if } p = 2S - s + 1 \quad \& \quad q = S + 1, \\
& p = 2S - s + 1 + q; \quad q = 1, 2, ..., S - s - 1, \\
& p = 2S - s + q; \quad q = S - s + 1, ... 2S - s - 1, \\
0, & \text{otherwise.}
\end{cases}
$$

$$
[A_1^*](pq) = \begin{cases}
-\lambda, & \text{if } p = q; \quad q = 1, 2, ..., S - s, \\
-(\lambda + \beta), & \text{if } p = q; \quad q = S - s + 1, ...2S - s, \\
-(\lambda + \mu), & \text{if } p = q; \quad q = 2S - s + 1, , ..., 3S - 2s, \\
-(\beta + \lambda + \mu), & \text{if } p = q; \quad q = 3S - 2s + 1, , ..., 4S - 2s - 1, \\
\beta & \text{if } q = p + 1; \quad p = S - s + 1, ...2S - s - 1, \\
& p = 2S - s \quad \& \quad q = S - s, \\
& q = p + 1; \quad p = 3S - 2s + 1, , ..., 4S - 2s - 2, \\
& p = 4S - 2s - 1 \quad \& \quad q = 3S - 2s, \\
0. & \text{otherwise.}
\end{cases}
$$

$$
[A_0]^*(pq) = \begin{cases}
\lambda, & \text{if } p = q; \quad q = 1, 2, ..., 4S - 2s - 1, \\
0, & \text{otherwise.}
\end{cases}
$$

$$[A_2^*](pq) = \begin{cases} \mu, & \text{if } p = 2S - s + 1 \text{ and } q = 3S - s, \\ & q = p - 1; \quad p = 2S - s + 2, ..., 3S - 2s, \\ & p = 3S - 2s + 1; \quad q = S - s + 1, \\ & p = q - 1; \quad p = 3S - 2s + 2...4S - 2s - 1, \\ 0, & \text{otherwise.} \end{cases}$$

$$[A_0]^{**}(pq) = \begin{cases} \lambda, & \text{if } q = p + 1; \quad p = 1, 2, ..., S - s, \\ & p = S - s + 1 \quad \& \quad q = 1, \\ & q = p; \quad p = S - s + 2, ..., 2S - s, \\ & q = p - (2S - s - 1); \quad p = 2S - s + 1, ..., 4S - 2s - 1, \\ 0, & \text{otherwise.} \end{cases}$$

$$[A_2^{**}](pq) = \begin{cases} \mu, & \text{if } p = 2 \quad \& \quad q = 3S - s, \\ & q = 2S - s - 2 + p; \quad p = 3, 4, ..., S - s + 1, \\ & p = S - s + 2, \quad \& \quad q = S - s + 1, \\ & q = 2S - s - 2 + p, \quad p = S - s + 3, ..., 2S - s, \\ 0, & \text{otherwise.} \end{cases}$$

$$[A_0](pq) = \begin{cases} \lambda, & \text{if } p = q, \quad p = 1, 2, ..., 2S - s, \\ 0 & \text{otherwise.} \end{cases}$$

$$[A_1](pq) = \begin{cases} -(\lambda + \beta), & \text{if } p = q = 1, \\ -(\lambda + \mu), & \text{if } p = q; \quad p = 2, 3, ...S - s + 1, \\ -(\beta + \lambda + \mu), & \text{if } p = q; \quad p = S - s + 2, , ..., 2S - s, \\ \beta, & \text{if } p = 1 \quad \& \quad q = S - s + 2, \\ & q = p + 1, p = S - s + 2, ..., 2S - s - 1, \\ & p = 2S - s \quad \& \quad q = S - s + 1, \\ 0, & \text{otherwise.} \end{cases}$$

$$[A_2](pq) = \begin{cases} \mu & \text{if } p = 2 \quad \& \quad q = S + 1, \\ & q = p - 1; \quad p = 3, 4, ..., S - s + 1, \\ & p = S - s + 2 \quad \& \quad q = 1, \\ & p = q - 1; \quad p = S - s + 3, ..., 2S - s, \\ 0, & \text{otherwise.} \end{cases}$$

3 Steady State Analysis

Let $\boldsymbol{A}$ be the generator matrix $A_0 + A_1 + A_2$. The entries of A is given below:

$$[A](pq) = \begin{cases} -\beta, & \text{if } p = q = 1, \\ -\mu, & \text{if } p = q; \quad p = 2, 3, ...S - s + 1, \\ -(\beta + \mu), & \text{if} p = q; \quad p = S - s + 2, , ..., 2S - s, \\ \beta, & \text{if } p = 1 \quad \& \quad q = S - s + 2, \\ & q = p + 1; \quad p = S - s + 2, ..., 2S - s - 1, \\ & p = 2S - s \quad \& \quad q = S - s + 1, \\ \mu, & \text{if } p = 2 \quad \& \quad q = S + 1, \\ & q = p - 1; \quad p = 3, 4, ..., S - s + 1, \\ & p = S - s + 2 \quad \& \quad q = 1, \\ & p = q - 1; \quad p = S - s + 3, ..., 2S - s, \\ 0, & \text{otherwise.} \end{cases}$$

Theorem 1. *The steady state probability vector $\pi_A = (\pi_1, \pi_2, ..., \pi_{2S-s})$ corresponding to the generator matrix $A = A_0 + A_1 + A_2$ is given by $\pi_j = \psi_j \pi_{S-s+1}$, where*

$$\psi_j = \begin{cases} \frac{(\frac{\mu}{\beta})^{s+1}(1-(\frac{\mu}{\beta})^{S-s})}{1-\frac{\mu}{\beta}} & j = 1, \\ 1 & j = 2, ..., S - s + 1, \\ \frac{(\frac{\mu}{\beta})^{S+2-j}(1-(\frac{\mu}{\beta})^{S-s})}{1-\frac{\mu}{\beta}} & j = S - s + 2, ..., S + 1, \\ \frac{(\frac{\mu}{\beta})(1-(\frac{\mu}{\beta})^{2S-s+1-j})}{1-\frac{\mu}{\beta}} & j = S + 2, ..., 2S - s - 1, \\ \frac{\mu}{\beta} & j = 2S - s. \end{cases}$$

and $\pi_{S-s+1} = \frac{(1-\frac{\mu}{\beta})^2}{(S-s)(1-\frac{\mu}{\beta})+((\frac{\mu}{\beta})^{S+2}-(\frac{\mu}{\beta})^{s+2})}.$

Proof: We have $\pi_A A = 0$ and $\pi_A e = 1$.
From the equation $\pi_A A = 0$, we obtain the following system of equations.

$$\beta\pi_1 + \mu\pi_{S-s+2} = 0,$$

$$-\mu\pi_k + \mu\pi_{k+1} = 0, \quad k = 2, ..., S - s.$$

$$-\mu\pi_{S-s+1} + \beta\pi_{2S-s} = 0,$$

$$-(\mu + \beta)\pi_{S-s+2} + \beta\pi_1 + \mu\pi_{S-s+3} = 0,$$

$$-(\mu + \beta)\pi_k + \beta\pi_{k-1} + \mu\pi_{k+1} = 0, \quad k = S - s + 3, ..., S. \tag{1}$$

$$-(\mu + \beta)\pi_{S+1} + \mu\pi_2 + \beta\pi_S + \mu\pi_{S+2} = 0,$$

$$-(\mu + \beta)\pi_k + \beta\pi_{k-1} + \mu\pi_{k+1} = 0, \quad k = S + 2, ..., 2S - s - 1.$$

$$-(\mu + \beta)\pi_{2S-s} + \beta\pi_{2S-s-1} = 0.$$

Solving the system of Eqs. (1) and using the normalising condition $\pi_A e = 1$, we obtain the required result.

Theorem 2 *(Stability condition). The process $\{Z(t)|t \geq 0\}$ is stable if and only if $\lambda < (1 - \pi_1)\mu$, where*

$$\pi_1 = \frac{(\frac{\mu}{\beta})^{s+1}(1 - (\frac{\mu}{\beta})^{S-s})(1 - \frac{\mu}{\beta})}{(S - s)(1 - \frac{\mu}{\beta}) + ((\frac{\mu}{\beta})^{S+2} - (\frac{\mu}{\beta})^{s+2})}.$$

Proof: Since the process $\{Z(t)|t \geq 0\}$ is a level independent QBD, it will be stable if and only if $\pi_A A_0 e < \pi_A A_2 e$ (see Neuts [10]). Here $\pi_A A_0 e = \lambda$ and $\pi_A A_1 e = (1 - \pi_1)\mu$. Using Theorem 1 we get the required result.

3.1   The Steady State Probability Vector of G

Let the steady state probability vector **x** of G can be partitioned according to the levels as

$$\mathbf{x} = (x(0), x(1), ..., x(N - 1), x(N), ...),$$

where $x(i), 1 \leq i \leq N - 1$ contain $4S - 2s - 1$ elements and all other sub vectors contains $2S - s$ elements. The QBD process $Z(t)$ is state independent for $i \geq N + 1$. Therefore the steady state solution is of the form (see Latouche and Ramaswami [9].)

$$x_{N+1+j} = x_{N+1} R^j : j \geq 1.$$

where R is the minimal nonnegative solution of the matrix quadratic equation $R^2 A_2 + R A_1 + A_0 = 0$. R can be calculated from the iterative procedure (refer Neuts [10])

$$R_{n+1} = -(R_n^2 A_2 + A_0)A_1^{-1}.$$

Also **x** satisfies the equations $\mathbf{x}G = 0$ and $\mathbf{x} \, e = 1$.

Thus we obtain the following system of equations

$$x(0)B_{00} + x(1)B_{10} = 0,$$

$$x(0)B_{01} + x(1)A_1^* + x(2)A_2^* = 0,$$

$$x(i - 1)A_0^* + x(i)A_1^* + x(i + 1)A_2^* = 0, \quad 2 \leq i \leq N - 2.$$

$$x(N - 2)A_0^* + x(N - 1)A_1^* + x(N)A_2^{**} = 0, \tag{2}$$

$$x(N - 1)A_0^{**} + x(N)A_1 + x(N + 1)A_2 = 0,$$

$$x(N)A_0 + x(N + 1)(A_1 + R A_2) = 0,$$

$$x(0)e + \sum_{i=1}^{N-1} x(i)e + x(N)e + x(N + 1)(I - R)^{-1}e = 1. \tag{3}$$

Solving Eqs. (2) and (3) we get **x**.

3.2 System Performance Measures

The steady state probability vector for the system allows us to calculate the system's measures of effectiveness. We partition the components of the steady state probability vector $\mathbf{x}$ as

$$x(0) = (x(0,0,0,k), x(0,0,1,j)); k = s+1...S; j = 0,1....S-1.$$

For $1 \leq i \leq N-1$,

$$x(i) = (x(i,0,0,k), x(i,0,1,j), x(i,1,0,k), x(i,1,1,n)),$$

$$k = s+1...S; \quad j = 0,1....S-1; \quad n = 1,2,...,S-1.$$

For $i \geq N$,

$$x(i) = (x(i,0,1,0), x(i,1,0,j), x(i,1,1,k))j = s+1,...,S; k = 1,...,S-1.$$

With the above notation, we obtain the following system performance measures.

1. Expected Number of customers in the system,

$$EC = \sum_{i=1}^{\infty} ix(i)e$$

$$= \sum_{i=1}^{N-1} ix(i)e + Nx(N)e + \sum_{i=N+1}^{\infty} ix(i)e$$

$$= \sum_{i=1}^{N-1} ix(i)e + Nx(N)e + x(N+1)(N(I-R)^{-1} + (I-R)^{-2})e.$$

2. Expected Inventory Level.

$$EI = \sum_{j=s+1}^{S} jx(0,0,0,j) + \sum_{j=0}^{S-1} jx(0,0,1,j)$$

$$+ \sum_{i=1}^{N-1} (\sum_{j=s+1}^{S} jx(i,0,0,j) + \sum_{j=0}^{S-1} jx(i,0,1,j) + \sum_{j=s+1}^{S} jx(i,1,0,j)$$

$$+ \sum_{j=1}^{S-1} jx(i,1,1,j)) + \sum_{i=N}^{\infty} (\sum_{j=s+1}^{S} jx(i,1,0,j) + \sum_{j=1}^{S-1} jx(i,1,1,j)).$$

3. Probability that the server is idle,

$$PI = x(0)e + \sum_{i=1}^{N-1}(\sum_{j=s+1}^{S}(x(i,0,0,j) + \sum_{j=0}^{S-1}x(i,0,1,j))$$

$$+ \sum_{i=N}^{\infty}x(i,0,1,0).$$

4. Expected Number of customers in the system when the server is busy.

$$EC_{busy} = \sum_{i=1}^{N-1}i(\sum_{j=s+1}^{S}(x(i,1,0,j) + \sum_{j=1}^{S-1}x(i,1,1,j))$$

$$+ \sum_{i=N}^{\infty}i(\sum_{j=s+1}^{S}(x(i,1,0,j) + \sum_{j=1}^{S-1}x(i,1,1,j))$$

5. Expected Number of customers in the system when the server is idle,

$$EC_{idle} = \sum_{i=1}^{N-1}i(\sum_{j=s+1}^{S}(x(i,0,0,j) + \sum_{j=0}^{S-1}x(i,0,1,j))$$

$$+ \sum_{i=N}^{\infty}ix(i,0,1,0).$$

6. Expected inventory in the system when the server is busy,

$$EI_{busy} = \sum_{i=1}^{N-1}\sum_{j=s+1}^{S}jx(i,1,0,j) + \sum_{j=1}^{S-1}jx(i,1,1,j))$$

$$+ \sum_{i=N}^{\infty}(\sum_{j=s+1}^{S}jx(i,1,0,j) + \sum_{j=1}^{S-1}jx(i,1,1,j)).$$

7. Expected inventory in the system when the server is idle,

$$EI_{idle} = (\sum_{j=s+1}^{S}jx(0,0,0,j) + \sum_{j=0}^{S-1}jx(0,0,1,j))$$

$$+ \sum_{i=1}^{N-1}(\sum_{j=s+1}^{S}jx(i,0,0,j) + \sum_{j=0}^{S-1}jx(i,0,1,j))).$$

8. Expected number of items produced,

$$EP = \beta(\{\sum_{j=0}^{S-1}x(0,0,1,j) + \sum_{i=1}^{N-1}(\sum_{j=0}^{S-1}x(i,0,1,j) + \sum_{j=1}^{S-1}x(i,1,1,j))$$

$$+ \sum_{i=N}^{\infty}(x(i,0,1,0) + \sum_{j=1}^{S-1}x(i,1,1,j))\}).$$

9. Expected Switching Rate for production,

$$ES_1 = \mu \sum_{i=1}^{\infty} x(i, 1, 0, s+1).$$

10. Expected Switching Rate for service,

$$ES_2 = \lambda \left(\sum_{j=s+1}^{S} x(N-1, 0, 0, j) + \sum_{j=0}^{S-1} x(N-1, 0, 1, j) \right).$$

11. Expected Number of departures after completing the service,

$$ED = \mu \left(\sum_{i=1}^{\infty} \left(\sum_{j=1}^{S-1} x(i, 1, 1, j) + \sum_{j=s+1}^{S} x(i, 1, 0, j) \right) \right).$$

3.3 Cost Analysis

Now, we develop the following cost function by means of some of important performance measures given in Subsect. 3.2. The expected total cost per unit time,

$$ETC = c_0 ES_1 + c_1 EP + c_2 ES_2 + c_3 EI + c_4 EC + c_5 ED,$$

where,

c_0: fixed cost for production,
c_1: production cost/item/unit time,
c_2: reward cost of customer when the server is idle/customer/unit time,
c_3: holding cost of inventory/unit/unit time,
c_4: holding cost of customer/unit time,
c_5: cost of service/item/unit time.

4 Numerical and Graphical Illustrations

This section provides the details of numerical experiments that have been carried out for studying the effects of variation of different parameters on various performance measures. Figure 1 shows the plots of variation of expected total cost with respect to the different parameters S, s, N, λ, β and μ. Table 1 shows the effect of λ, β and μ on performance measures.

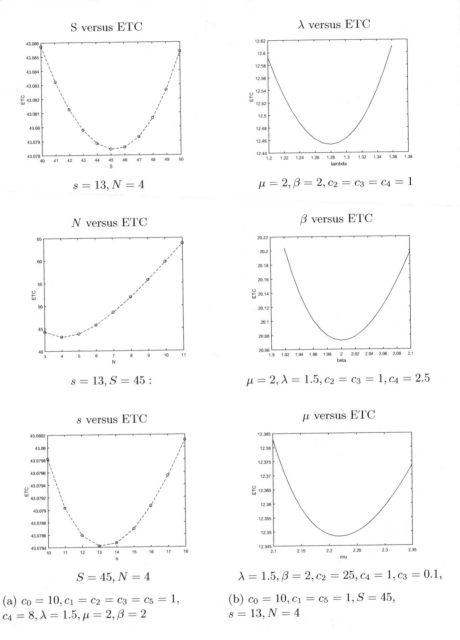

Fig. 1. Variation of ETC with respect to various parameters

From Table 1(a), it is clear that when the arrival rate λ increases, the expected number of customers, expected production rate and expected departure rate increases, while the mean on-hand inventory level decreases. As the mean arrival rate increases, more items are taken by customers from the inventory and hence

Table 1. Effect of parameters on performance measures

Table 1(a). Effect of λ on performance measures

λ	EC	EI	PI	EC_{busy}	EC_{idle}	EI_{busy}	EI_{idle}	EP	ES1	ES2	ED
1.2	2.5638	6.9215	0.5333	1.6439	0.9199	3.7093	3.2122	1.9979	1.00E-04	0.1753	0.9329
1.21	2.5944	6.852	0.5307	1.6742	0.9202	3.7048	3.1472	1.9979	1.00E-04	0.1764	0.9382
1.22	2.6259	6.7858	0.528	1.7055	0.9204	3.702	3.0839	1.998	1.00E-04	0.1774	0.9436
1.23	2.6583	6.723	0.5253	1.7378	0.9205	3.7008	3.0222	1.9981	1.00E-04	0.1783	0.949
1.24	2.6917	6.6634	0.5225	1.7712	0.9205	3.7013	2.9621	1.9981	1.00E-04	0.1792	0.9546
1.25	2.7261	6.6071	0.5197	1.8057	0.9205	3.7035	2.9036	1.9982	1.00E-04	0.1801	0.9602
1.26	2.7617	6.5541	0.5169	1.8414	0.9203	3.7075	2.8466	1.9982	1.00E-04	0.1809	0.9659
1.27	2.7984	6.5043	0.514	1.8783	0.9201	3.7132	2.791	1.9983	1.00E-04	0.1817	0.9717
1.28	2.8364	6.4577	0.511	1.9165	0.9198	3.7209	2.7368	1.9983	1.00E-04	0.1824	0.9776
1.29	2.8756	6.4144	0.508	1.9562	0.9195	3.7304	2.684	1.9983	1.00E-04	0.183	0.9836
1.3	2.9162	6.3743	0.505	1.9972	0.919	3.7418	2.6325	1.9983	1.00E-04	0.1836	0.9898

$$S = 45, s = 13, N = 4, \mu = 2, \beta = 2$$

Table 1(b). Effect of μ on Performance measures

μ	EC	EI	PI	EC_{busy}	EC_{idle}	EI_{busy}	EI_{idle}	EP	ES1	ES2	ED
2.2	3.5612	4.902	0.4892	2.5235	1.0377	3.1027	1.7993	1.9995	3.00E-05	0.207	1.1234
2.21	3.5397	4.8566	0.4918	2.4958	1.0438	3.0574	1.7992	1.9996	3.00E-05	0.208	1.1235
2.22	3.5187	4.8129	0.4943	2.4689	1.0499	3.0136	1.7992	1.9996	3.00E-05	0.209	1.1236
2.23	3.4983	4.7707	0.4968	2.4425	1.0558	2.9714	1.7993	1.9996	3.00E-05	0.21	1.1237
2.24	3.4785	4.7301	0.4992	2.4169	1.0616	2.9306	1.7995	1.9996	3.00E-05	0.211	1.1238
2.25	3.4592	4.6909	0.5016	2.3918	1.0674	2.8912	1.7997	1.9997	2.00E-05	0.212	1.1239
2.26	3.4404	4.653	0.504	2.3674	1.073	2.853	1.8	1.9997	2.00E-05	0.213	1.1239
2.27	3.422	4.6165	0.5063	2.3435	1.0786	2.8161	1.8004	1.9997	2.00E-05	0.2139	1.1240
2.28	3.4042	4.5813	0.5086	2.3201	1.0841	2.7804	1.8009	1.9997	2.00E-05	0.2149	1.1241
2.29	3.3867	4.5473	0.5109	2.2973	1.0894	2.7459	1.8014	1.9997	2.00E-05	0.2158	1.1242
2.3	3.3697	4.5144	0.5131	2.2749	1.0947	2.7124	1.802	1.9998	2.00E-05	0.2167	1.1243

$$S = 45, s = 13, N = 4, \lambda = 1.5, \beta = 2$$

Table 1(c). Effect of β on performance measures

β	EC	EI	PI	EC_{busy}	EC_{idle}	EI_{busy}	EI_{idle}	EP	ES1	ES2	ED
1.95	4.332	6.0141	0.426	3.4129	0.9191	4.3063	1.7078	1.9476	1.00E-04	0.1795	1.1379
1.96	4.2986	6.0703	0.4265	3.3858	0.9128	4.3426	1.7278	1.9574	1.00E-04	0.18	1.1402
1.97	4.2672	6.1263	0.427	3.3603	0.9069	4.3788	1.7475	1.9672	1.10E-04	0.1806	1.1408
1.98	4.2375	6.1819	0.4274	3.3361	0.9014	4.4149	1.7669	1.9771	1.10E-04	0.1811	1.1409
1.99	4.2094	6.2372	0.4279	3.3133	0.8961	4.451	1.7862	1.9869	1.20E-04	0.1815	1.1416
2	4.1829	6.2923	0.4283	3.2917	0.8912	4.487	1.8052	1.9966	1.30E-04	0.182	1.1432
2.01	4.1577	6.347	0.4287	3.2712	0.8865	4.523	1.8241	2.0064	1.30E-04	0.1824	1.1424
2.02	4.1339	6.4015	0.4291	3.2518	0.8821	4.5588	1.8427	2.0162	1.40E-04	0.1828	1.1425
2.03	4.1113	6.4557	0.4294	3.2334	0.8779	4.5945	1.8612	2.026	1.50E-04	0.1832	1.1426
2.04	4.0898	6.5096	0.4298	3.2159	0.874	4.6302	1.8794	2.0357	1.50E-04	0.1836	1.1430
2.05	4.0695	6.5632	0.4301	3.1992	0.8702	4.6657	1.8975	2.0455	1.60E-04	0.1839	1.1432

$$S = 45, s = 13, N = 4, \lambda = 1.5, \mu = 2.$$

the on-hand inventory level EI decreases. Furthermore, as this inventory level decreases, the fraction of the time that the production process is switched ON will increase and hence the mean production rate EP also increases. It is observed from Table 1(b) that expected inventory and expected number of customers decrease with the increase of μ, while the expected production rate and expected switching rate for service increases. This can be explained as follows. When the service rate increases, more customers are served and hence the inventory and number of customers in queue reduces. A decrease in the number of customers may result in the system being empty. So there may be more chances for the server to take a vacation which leads to an increase in the expected switching rate for service. In Table 1(c), one can see that the expected production rate, expected inventory level and expected departure rate increase with the increasing of the production rate β, while the expected number of customers decreases. This agrees with our intuition that as the production rate increases, more items are produced and replenished to the inventory which results in an increase of the mean production rate EP and the mean on-hand inventory level EI and a decrease in the expected number of waiting customers.

Optimal (s, S) Pair

The variation of the expected total cost with respect to (s, S) is shown in Table 2 and Fig. 2. The optimum value of (s, S) pair and N are obtained by considering suitable parameter values. For the set of parameters $S = 45, s = 13, N = 4, \lambda = 1.5, \mu = 2, \beta = 2, c_0 = 10, c_1 = c_2 = c_3 = c_5 = 1, c_4 = 8$, the optimal$(s, S)$ pair is found to be $(13, 45)$ and optimum value of N is 4. The minimum cost is obtained as 43.0784

Table 2. Variation of ETC with respect to (s, S)

S \ s	10	11	12	13	14	15
42	43.0833	43.0822	43.0815	**43.0812**	43.0812	43.0815
43	43.0816	43.0806	43.0800	**43.0797**	43.0798	43.0800
44	43.0804	43.0795	43.0790	**43.0788**	43.0788	43.0791
45	43.0798	43.0790	43.0785	**43.0784**	43.0784	43.0786
46	43.0803	43.0790	43.0787	**43.0785**	43.0786	43.0788
47	43.0816	43.0797	43.0794	**43.0792**	43.0793	43.0795

$N = 4, \lambda = 1.5, \mu = 2, \beta = 2, c_0 = 10, c_1 = c_2 = c_3 = c_5 = 1, c_4 = 8$

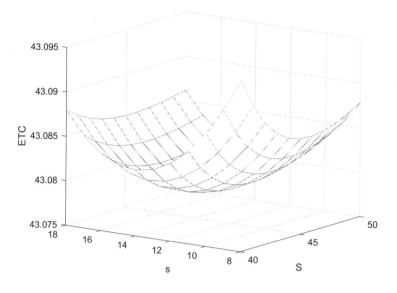

$$N = 4, \lambda = 1.5, \mu = 2, \beta = 2, c_0 = 10, c_1 = c_2 = c_3 = c_5 = 1, c_4 = 8$$

Fig. 2. Variation of ETC with respect to (s, S)

5 Conclusion

In this paper, we studied a production inventory system with $N-$ policy and positive service time. The production process added items one by one exponentially to the inventory and is governed by an (s, S) policy. Matrix Geometric Method is used to find the stationary probability vector, which makes it easy to obtain some key performance measures. The results are numerically and graphically illustrated to show the effect of various parameters on system performance measures. A suitable cost function is constructed and the optimal (s, S) pair is obtained. The optimal value of N is also obtained. This work can be extended in the future for multi-server production models.

References

1. Artalejo, J.R.: A unified cost function for M/G/1 queueing systems with removable server. Trabajos de Investigacion Operativa **7**, 95–104 (1992)
2. Berman, O., Kaplan, E.H., Shimshak, D.G.: Deterministic approximations for inventory management at service facilities. IIE Trans. **25**, 98–104 (1993)
3. Jose, K.P., Beena, P.: Investigation of production inventory model with two servers having multiple vacation. J. Math. Comput. Sci. **10**, 1214–1227 (2020)
4. Jose, K.P., Nair, S.S.: A MAP/PH/1 production inventory model with varying service rates. Int. J. Pure Appl. Math. **117**, 373–381 (2017)
5. Krishnamoorthy, A., Narayanan, V.C.: Stochastic decomposition in production inventory with service time. Eur. J. Oper. Res. **228**(2), 358–366 (2013)

6. Krishnamoorthy, A., Jose, K.P.: Three production inventory systems with service, loss and retrial of customers. Int. J. Inform. Manag. Sci. **19**, 367–389 (2015)
7. Krishnamoorthy, A., Narayanan, V.C., Deepak, T.G., Vineetha, P.: Control policies for inventory with service time. Stochast. Anal. Appl. **24**(4), 889–899 (2006)
8. Lakshmy, B., Krishnamoorthy, A., Manikandan, R.: A survey on inventory models with positive service time. OPSEARCH **48**, 153–169 (2011)
9. Latouche, G., Ramaswami, V.: Introduction to Matrix Analytic Methods in Stochastic Modelling. SIAM, Philadelphia
10. Neuts, M.F.: Matrix-Geometric Solutions in Stochastic Models - An Algorithmic Approach. John Hopkins University Press
11. Rejitha, K.R., Jose, K.P.: A stochastic inventory system with two modes of service and retrial of customers. Opsearch **55**(1), 134–149 (2017). https://doi.org/10.1007/s12597-017-0322-9
12. Yadin, M., Naor, P.: Queueing systems with a removable service station. OR **14**, 393–405 (1963)

Generalised Self-induced Service Interruption and Consequent Reduced Priority in Discrete-Time Queues

M. P. Anilkumar[1] and K. P. Jose[2][(✉)]

[1] Department of Mathematics, T. M. Government College,
Tirur 676 502, Kerala, India
[2] Department of Mathematics, St. Peter's College,
Kolenchery 682 311, Kerala, India
kpjspc@gmail.com

Abstract. This paper considers the analysis of discrete-time priority queues formed due to the customer's induced interruption during service. The customers who interrupted during service are moved to a lower priority queue. Both preemptive and non-preemptive disciplines for the service are considered. The Matrix-Analytic Method extended to the infinite phase is used to analyze the model. The stability condition for the system is derived. The marginal distributions of both higher and lower priority queue lengths in each discipline are studied. Numerical experiments are incorporated to illustrate the calculation of the rate matrix and queue lengths.

Keywords: Discrete-time queue · Peemptive · Non-preemptive · Markovian arrival process · Discrete phase-type distribution · Matrix-Analytic Method

1 Introduction

The concept of priority in queues was introduced by White and Christie [16] in 1958. The main classification of priority queues are i) Preemptive and ii) Non-preemptive. In preemptive priority, the service of the lower priority is interrupted on the arrival of high priority customer during the service whereas in non-preemptive, the arriving high priority customer during the service of the lower priority customer gets service only after the completion of the undergoing service. Jaiswal [10] discussed the service on non-priority unit when preemption occurred. The service is started at the point where it was interrupted. Further Jaiswal [9] described the development of priority queues until 1968. Recent works of Gated Batch Priority Queues and reservation in priority queues are found in Takagi [1]. A survey of priority queues is analysed by Brodal [5]. The author tried to list some of the directions research on priority queues that has gone over the last 50 years. Matrix-geometric method for discrete-time priority queue is

© Springer Nature Switzerland AG 2022
A. Dudin et al. (Eds.): ITMM 2021, CCIS 1605, pp. 67–79, 2022.
https://doi.org/10.1007/978-3-031-09331-9_6

discussed by Alfa [2] in which arrivals two classes are modeled by Markovian Arrival Process in which correlation of inter-arrival time within each class and between two classes of jobs are considered and service time of each class follows a phase-type distribution with different parameters. The author extended the structure of the rate matrix R obtained by Miller [14] to the discrete case.

The creation of high priority or low priority customers may occur during the service due to an emergency or interruption. The self-generated priority queue with MAP arrivals and phase-type service time distribution is analysed by Krishnamoorthi et al. [11] in which priority customer accommodating capacity is one and remaining generated priority customers are assumed to be lost. Interruption in a queue occurs due to many reasons such as server breakdowns, servers taking emergency breaks, and customers having incomplete information or getting distracted. Krishnamoorthy et al. [13] look at both continuous and discrete-time queueing models with interruptions in service. Jacob et al. [8] investigated an infinite capacity queueing system with a single server to which customers arrive according to a Poisson process and the service time follows an exponential distribution. The customer interruption occurs according to a Poisson process and the interruption duration follows an exponential distribution. The self-interrupted customers will enter a buffer of finite size and any interrupted customer, finding the buffer full, is considered to be lost. Dudin et al. [6] generalized the model with MAP arrivals and phase-type service in which two multi-server service systems are considered. Primary customers arrive at a multi-server queueing system-1 having an infinite buffer. An interruption removes one of the primary customers from the service and with some probability, the interrupted primary customer moves for service to system-2 and after completing this service, this customer becomes a priority customer. The ergodicity and various performance measures are analysed. The concept of self interruption infinite buffer for lower priority in continuous time was analysed by Krishnamoorthy and Manjunath [12]. Anillumar and Jose [4] generalized this model to discrete-time cases with MAP arrivals and phase-type service processes. A discrete-time priority queueing inventory model with customer-induced interruption was also analyzed by Anilkumar and Jose [3]. In this paper, we discus the generalized self-induced service interruption.

2   Modeling

We consider infinite capacity single server discrete-time queue in which arrival of customers is modeled by Markovian arrival process having n phases with representation (D_0, D_1). Then the arrival rate $\lambda = \psi D_1 e$, where ψ is the stationary probability vector of $(D_0 + D_1)$ and e is the column vector of 1's having dimension $n \times 1$. There are two types of priority queues $\mathcal{P}_1$ and $\mathcal{P}_2$. The customer who arrives in the system first enters into the high priority queue. During his service, he may or may not interrupt the service. The time taken for primary service is considered to be the time till absorption of a discrete-time Markov chain that has two absorbing states which are represented by the transition probability matrix

$$T = \begin{bmatrix} S(1) & S^0(1) & S^{02}(1) \\ 0 & 1 & 0 \\ 0 & 0 & 1 \end{bmatrix}$$

with initial probability $\beta(1)$ and transient matrix $S(1)$ having dimension m_1. If a customer interrupts the service, he is transferred to a low priority queue $\mathcal{P}_2$. The same server serves customers in these two queues one at a time according to their priority. Once an interrupted customer in P_2 receives service, he will have no further interruptions during service (except when $\mathcal{P}_1$ customers arrive in the preemptive case). After completing the service customers in both queues leave the system. A customer in $\mathcal{P}_2$ is taken for service only when no customer is present in $\mathcal{P}_1$ and no customer in $\mathcal{P}_2$ ahead of him. The arrival of customers in $\mathcal{P}_1$ during the service of P_2, may or may not affect the service. In other words, service P_2 is either according to preemptive or non-preemptive discipline. We study this separately. The processing time of customers in $\mathcal{P}_2$ is discrete phase-type distributed random variables with parameters $(\beta(2), S(2))$ with dimension m_2.

Notations

  (i) $N_1(n)$: Number of $\mathcal{P}_1$ customers in the system at an epoch n.
 (ii) $N_2(n)$: Number of $\mathcal{P}_2$ customers waiting for service an epoch n
(iii) $I(n)$: The arriving phase of a customer
 (iv) $J(n)$: The service phase
  (v) $\bar{a} = 1 - a$ where $0 \le a \le 1$
 (vi) e: Column vector of 1's of appropriate order
(vii) $e(k)$: Column vector of 1's of order k
(viii) $S^0(2) = e - S(2)e$
 (ix) For an $m \times n$ matrix Z given by $Z = [Z_1, Z_2, \ldots Z_n]$ where Z_j is the jth column of Z, $Vec(Z)$ is the $mn \times 1$ column vector defined by,

$$Vec(Z) = \begin{bmatrix} Z_1 \\ Z_2 \\ \vdots \\ Z_n \end{bmatrix}$$

3   Preemptive Priority

We assume that the priority of service in $\mathcal{P}_2$ is in the preemptive discipline. That is the service of nonpriority customers affects the arrival of customers in $\mathcal{P}_1$. Now, $\{(N_1(n), N_2(n), I(n), J(n)), n = 1, 2, 3, \ldots\}$ is a Level Independent Quasi-Birth Death process (LIQBD) on the state space $\Delta_1 \cup \Delta_2 \cup \Delta_3$, where

$$\Delta_1 = \{(0, j); 1 \le j \le n\},$$
$$\Delta_2 = \{(0, i_2, j, k_2); i_2 \ge 0, 1 \le j \le n, 1 \le k_2 \le m_2\},$$
$$\Delta_3 = \{(i_1, i_2, j, k_1); i_1 \ge 1, i_2 \ge 0, 1 \le j \le n, 1 \le k_1 \le m_1\}.$$

The transition probability matrix P of this process is given by

$$P = \begin{bmatrix} B_{00} & B_{01} & 0 \\ B_{10} & A_1 & A_0 \\ & A_2 & A_1 & A_0 \\ & & \ddots & \ddots & \ddots \\ & & & \ddots & \ddots & \ddots \end{bmatrix},$$

where the elements of P are square matrices of order S and are given by

$$B_{00} = \begin{bmatrix} B_{00}^{00} \\ B_{00}^{10} & B_{00}^{1} \\ & B_{00}^{2} & B_{00}^{1} \\ & & \ddots & \ddots \end{bmatrix}, \quad B_{01} = \begin{bmatrix} B_{01}^{00} \\ B_{01}^{2} & B_{01}^{1} \\ & B_{01}^{2} & B_{01}^{1} \\ & & \ddots & \ddots \end{bmatrix},$$

$$B_{10} = \begin{bmatrix} B_{10}^{00} & B_{10}^{0} \\ & B_{10}^{1} & B_{10}^{0} \\ & & B_{10}^{1} & B_{10}^{0} \\ & & & \ddots & \ddots \end{bmatrix}, \quad A_2 = \begin{bmatrix} A_2^{1} & A_2^{0} \\ & A_2^{1} & A_2^{0} \\ & & A_2^{1} & A_2^{0} \\ & & & \ddots & \ddots \end{bmatrix},$$

$$A_1 = \begin{bmatrix} A_1^{1} & A_1^{0} \\ & A_1^{1} & A_1^{0} \\ & & A_1^{1} & A_1^{0} \\ & & & \ddots & \ddots \end{bmatrix}, \quad A_0 = \begin{bmatrix} A_0^{1} \\ & A_0^{1} \\ & & A_0^{1} \\ & & & \ddots \end{bmatrix}.$$

$B_{00}^{00} = D_0, B_{00}^{10} = D_0 \otimes S^0(2), B_{00}^{1} = D_0 \otimes S(2), B_{00}^{2} = D_0 \otimes S^0(2)\beta(2),$

$B_{01}^{00} = D_1 \otimes \beta(1), B_{01}^{2} = D_1 \otimes S^0(2)\beta(1), B_{01}^{1} = D_1 \otimes S(2)e\beta(1),$

$B_{10}^{00} = D_0 \otimes S^0(1), B_{10}^{0} = D_0 \otimes S^{02}(1)\beta(2), B_{10}^{1} = D_0 \otimes S^0(1)\beta(2).$

$A_1^{1} = (D_0 \otimes S(1) + D_1 \otimes S^0(1)\beta(1)), A_1^{0} = D_1 \otimes S^{02}(1)\beta(1), A_0^{1} = D_1 \otimes S(1),$

$A_2^{1} = D_0 \otimes S^0(1)\beta(1), A_2^{0} = D_0 \otimes S^{02}(2)\beta(1).$

3.1   Stability

Theorem 1. *The system is stable if and only if*

$$\lambda\beta(1)(I - S(1))^{-1}e + \lambda\beta(1)(I - S(1))^{-2}S^{02}(1)\beta(2)(I - S(2))^{-1}e < 1. \quad (1)$$

Proof. The proof can follow using an intuitive argument. The server is always available to high priority job. Hence the portion of time the priority queue is empty is

$$\lambda\beta(1)(I - S(1))^{-1}e.$$

Now the arrival rate to low priority queue is

$$\lambda\beta(1)(I - S(1))^{-2}S^{02}(1).$$

Hence the portion of time the low priority queue is nonempty is

$$\lambda\beta(1)(I - S(1))^{-2}S^{02}(1)\beta(2)\,(I - S(2))^{-1}\,e.$$

Therefore, the portion of time both queues is nonempty is

$$\lambda\beta(1)(I - S(1))^{-1}e + \lambda\beta(1)(I - S(1))^{-2}S^{02}(1)\beta(2)\,(I - S(2))^{-1}\,e.$$

The system is stable if and only if this portion of time is less than 1.

3.2   Steady-State Analysis

Since the matrix P has the structure of quasi-birth and death process and the individual phases are infinite, we can use the generalization of matrix geometric method of Neuts [15] to the case of infinite submatrix by Miller [14]. For this, first find the minimal nonnegative solution R of the matrix quadratic equation,

$$R^2 A_2 + R A_1 + A_0 = R,$$

in which spectral radius is less than 1. Since A_0, A_1 and A_2 are of upper triangular structure, the rate matrix R also has the upper triangular structure which is given by,

$$R = \begin{bmatrix} R_0 & R_1 & R_2 & R_3 & \dots \\ & R_0 & R_1 & R_2 & \dots \\ & & R_0 & R_1 & \dots \\ & & & R_0 & \dots \\ & & & & \ddots \end{bmatrix}.$$

Then,

$$R^2 = \begin{bmatrix} R_0^{(2)} & R_1^{(2)} & R_2^{(2)} & R_3^{(2)} & \dots \\ & R_0^{(2)} & R_1^{(2)} & R_2^{(2)} & \dots \\ & & R_0^{(2)} & R_1^{(2)} & \dots \\ & & & R_0^{(2)} & \ddots \\ & & & & \ddots \end{bmatrix},$$

where $R_j^{(2)} = \sum_{v=0}^{j} R_j R_{j-v}$ for $j \geq 0$.
Substituting in (1), we get

$$R_0^{(2)} A_2^1 + R_0 A_1 + A_0^1 = R_0,$$

$$R^{(2)}_{j-1} A^0_2 + R^{(2)}_j A^1_2 + R_{j-1} A^0_1 + R_j A^1_1 = R_j, \text{for } j \geq 1.$$

R_0 is the minimal non negative solution of $R_0 = (A^1_0 + R^2_0)(I - A^1_1)^{-1}$, which can be calculated using the iterative method.

Now from (2), we have,

$$\sum_{v=0}^{j-1} R_v R_{j-v-1} A^0_2 + \sum_{v=0}^{j} R_v R_{j-v} A^1_2 + R_{j-1} A^0_1 + R_j A^1_1 = R_j, \text{for } j \geq 1.$$

Which can be re written as,

$$G_j + R_0 R_j G_2 = R_j G_0, \tag{2}$$

where $G_j = R_{j-1} A^0_1 + \sum_{v=0}^{j-1} R_v R_{j-v-1} A^0_2 + \sum_{v=1}^{j-1} R_v R_{j-v} A^1_2$, for $j \geq 1$, $G_2 = A^1_2$ and

$$G_0 = I - A^1_1 - R_0 A^1_2.$$

Using the property of kronecker product (see [7]), (2) is equivalent to

$$Vec(G_j) + (G^T_2 \otimes R_0)Vec(R_j) = (G^T_0 \otimes I)Vec(R_j), \text{ for } j \geq 1.$$

$$Vec(R_j) = ((G^T_0 \otimes I) - (G^T_2 \otimes R_0))^{-1} Vec(G_j), \text{ for } j \geq 1. \tag{3}$$

In order to compute R, first calculate R_0, then successively find $R_1, R_2, \ldots$ recursively using (3).

Since, under stability condition $(R_n)_{ij} \to 0$, one can truncate R. That is, we need only to consider low priority queue to a certain level. This generates a QBD having a finite set of phases, which can be easily analysed.

3.3   Steady-State Probability Vector

Let $\mathbf{x} = (x_0, x_1, x_2, \ldots)$ be the steady state probability vector of P. That is $\mathbf{x}P = \mathbf{x}$ and $\mathbf{x}e = 1$, where e is the infinite column vector of 1's

Then $x_{i+1} = x_i R$, for $i \geq 1$.

To find the boundary probability vectors(x_0, x_1). For this consider the following system of equations

$$x_0 B_{00} + x_1 B_{10} = x_0,$$
$$x_0 B_{01} + x_1 (A_1 + RA_2) = x_1.$$

From the second equation, we have

$$x_1 = x_0 B_{01}(I - A1 - RA_2)^{-1}.$$

Substituting this in the normalizing condition, $x_0 e + x_1(I - R)^{-1}e = 1$, one can solve for x_0 and x_1.

3.4   Marginal Probability Distributions

Let $P_i(v)$ be the probability that there are i jobs of in the queue $\mathcal{P}_v$ for $v = 1, 2, i \geq 0$

$$p_i(1) = x_i e,$$
$$p_i(2) = x_{0(i-1)} e(nm_2) + x_1(I - R)^{-1}\beta_i. \tag{4}$$

where β_i is an infinite column matrix whose $(inm_1 + 1)^{th}$ to $(i+1)nm_1^{th}$ entries are one and remaining values are zeros.

Let $E(v)$ be the expected number jobs in the queue $\mathcal{P}_v$ for $v = 1, 2$
Then

$$E(1) = 0x_0 e + 1x_1 e + 2x_2 e + \dots$$
$$= x_1(1 + 2R + 3R^2 + \dots)e$$
$$= x_1(I - R)^{-2} e$$
$$\text{and } E(2) = \sum iq_i(2)$$
$$= 1x_{01}e(nm_2) + 2x_{02}e(nm_2) + \dots$$
$$+ 1x_1(I - R)^{-1}\beta_1 + 2x_1(I - R)^{-1}\beta_2 + \dots$$
$$= x_0\gamma_1 + x_1(I - R)^{-1}\gamma_2.$$

Where $\gamma_1 = \begin{bmatrix} 0 \otimes (e)(n) \\ \phi \otimes e(nm_2) \end{bmatrix}$, $\gamma_2 = \begin{bmatrix} 0 \otimes e(nm_2) \\ \phi \otimes e(nm_1) \end{bmatrix}$ and $\phi = \begin{bmatrix} 1 \ 2 \ 3 \dots \end{bmatrix}$.

4   Non-preemptive Self Generated Interruption

Here we assume all the previous assumptions except that the arrival of $\mathcal{P}_1$ customer does not interrupt the service of $\mathcal{P}_2$ customer who is already in service. In addition to above notation, let $S(n)$ denote the status of server at an epoch n which takes the value 1 and 2 according as server serve $\mathcal{P}_2$ and $\mathcal{P}_2$ customer respectively. Then $\{(N_1(n), N_2(n), S(n), I(n), J(n)); n \geq 1\}$ is a discrete time quasi birth and death process with state space $\{(0, i); 1 \leq i \leq n\} \cup \{(0, n_2, i, j); n_2 \geq 1, 1 \leq i \leq n, 1 \leq j \leq m_2\} \cup \{(n_1, n_2, s, i, j); n_1 \geq 1, n_2 \geq 0, s = 1, 2 \& 1 \leq j \leq m_v\}$. The transition probability matrix P' describing this QBD is given by

$$P' = \begin{bmatrix} B_{00} & B_{01} & 0 & & \\ B_{10} & A_1 & A_0 & & \\ & A_2 & A_1 & A_0 & \\ & & \ddots & \ddots & \ddots \\ & & & \ddots & \ddots & \ddots \end{bmatrix},$$

where B_{00} is the same as in matrix P, A_0, A_1, and A_2 possess subsquare matrix as above but whose values block matrices given by

$$A_0^1 = \begin{bmatrix} D_1 \otimes S(1) & 0 \\ D_1 \otimes S^0(2)\beta(1) & D_1 \otimes S(2) \end{bmatrix}, \ A_1^0 = \begin{bmatrix} D_1 \otimes S^{02}(2)\beta(1) & 0 \\ 0 & 0 \end{bmatrix},$$

$$A_1^1 = \begin{bmatrix} (D_1 \otimes S^0(1)\beta(1) + D_0 S(1)) & 0 \\ D_0 \otimes S^0(2)\beta(1) & D_0 \otimes S(2) \end{bmatrix}, \ A_2^1 = \begin{bmatrix} D_0 \otimes S^0(1)\beta(1) & 0 \\ 0 & 0 \end{bmatrix},$$

$$A_2^0 = \begin{bmatrix} D_0 \otimes S^{02}(2)\beta(1) & 0 \\ 0 & 0 \end{bmatrix},$$

$$B_{01} = \begin{bmatrix} B_{01}^* & & & \\ B_{01}^{**} & 0 & & \\ & B_{01}^{**} & 0 & \\ & & \ddots & \ddots \end{bmatrix}, \ B_{10} = \begin{bmatrix} B_{10}^* & B_{10}^0 & & \\ & B_{10}^1 & B_{10}^0 & \\ & & B_{10}^1 & B_{10}^0 \\ & & & \ddots & \ddots \end{bmatrix},$$

$$B_{01}^* = \begin{bmatrix} D_1 \otimes \beta(1) & 0 \end{bmatrix}, \ B_{01}^{**} = \begin{bmatrix} D_1 \otimes S^0(2)\beta(1) & D_1 \otimes S(2) \end{bmatrix},$$

$$B_{10}^* = \begin{bmatrix} D_0 \otimes S^0(1) \\ 0 \end{bmatrix}, \ B_{10}^0 = \begin{bmatrix} D_0 \otimes S^{02}(1)\beta(2) \\ 0 \end{bmatrix}, \ B_{10}^1 = \begin{bmatrix} D_0 \otimes S^0(1)\beta(2) \\ 0 \end{bmatrix}.$$

4.1   Stability

The condition for stability is the same as that of the preemptive case. Hence the above QBD is stable if and only if

$$\lambda\beta(1)(I - S(1))^{-1}e + \lambda\beta(1)(I - S(1))^{-2}S^{02}(1)\beta(2)(I - S(2))^{-1}e < 1. \quad (5)$$

4.2   Computation of Rate Matrix and Steady-state Probability Vector

The rate matrix R possesses an upper triangular structure as in the preemptive case with the only difference is that each R_i is block lower triangular having order $n(m_1 + m_2)$ of the form

$$R_i = \begin{bmatrix} R_{11}^i & 0 \\ R_{21}^i & R_{22}^i \end{bmatrix}.$$

The $(i, j)^{th}$ entries of the rate matrix R is the expected number of visits into state $(k + 1, j)$, starting from the state (k, i), until the first return to level k, $k > 1$. Since the structure of A_1^0, A_2^0 has second column blocks zeros, we can conclude that

$$R_{22}^i = 0, \text{ for } i \geq 1.$$

This can also be verified recursively by substituting the values in the Eq. (3). Let $\mathbf{x} = (x_0, x_1, x_2, \dots)$ be the steady-state probability vector of P' where $x_i =$

$(x_{i0}(1), x_{i1}(1), x_{i1}(2), x_{i2}(1), x_{i2}(2), \dots)$ for $i \geq 1$ and x_0 as in above preemptive case. Then here also

$$x_{i+1} = x_i R, \text{ for } i \geq 1.$$

The marginal probability density functions are calculated as in the previous case with the exception that for $i = 0$, β_i is an infinite column matrix whose first nm_1 entries are one and the remaining entries are zeros and for $i \geq 1$, it is infinite column matrix whose $(nm_1 + (i-1)n(m_1 + m_2) + 1)^{th}$ to $(nm_1 + in(m_1 + m_2))^{th}$ entries are one and remaining entries are zeros.

5   Numerical Illustrations

For a given model, we consider both preemptive and non-preemptive case for the computation of the rate matrix R. One can observe that R_i, the entries of R tends to zero as $i \to \infty$.

Consider the parameters of the model as

$$D_0 = \begin{bmatrix} 0.4 & 0.4 \\ 0.3 & 0.4 \end{bmatrix}, \quad D_1 = \begin{bmatrix} 0.1 & 0.1 \\ 0.1 & 0.2 \end{bmatrix},$$

$$S(1) = \begin{bmatrix} 0.1 & 0.2 \\ 0.3 & 0.3 \end{bmatrix}, S^0(1) = \begin{bmatrix} 0.6 \\ 0.2 \end{bmatrix}, S^{02}(1) = \begin{bmatrix} 0.1 \\ 0.2 \end{bmatrix}, S(2) = \begin{bmatrix} 0.1 & 0.2 \\ 0.2 & 0.3 \end{bmatrix},$$

$$\beta_1 = \begin{bmatrix} 0.2 & 0.8 \end{bmatrix} \text{ and } \beta_2 = \begin{bmatrix} 0.2 & 0.8 \end{bmatrix}.$$

Computed values of R_i in Preemptive case

$$R_0 = \begin{bmatrix} 0.0217 & 0.0383 & 0.0238 & 0.0424 \\ 0.0515 & 0.0653 & 0.0555 & 0.0734 \\ 0.0272 & 0.0473 & 0.0409 & 0.0743 \\ 0.0619 & 0.0826 & 0.0987 & 0.1268 \end{bmatrix}, R_1 = \begin{bmatrix} 0.0030 & 0.0068 & 0.0037 & 0.0087 \\ 0.0056 & 0.0127 & 0.0069 & 0.0163 \\ 0.0047 & 0.0105 & 0.0058 & 0.0137 \\ 0.0087 & 0.0195 & 0.0108 & 0.0255 \end{bmatrix},$$

$$R_2 = \begin{bmatrix} 0.0009 & 0.0020 & 0.0011 & 0.0025 \\ 0.0017 & 0.0037 & 0.0020 & 0.0047 \\ 0.0014 & 0.0031 & 0.0017 & 0.0039 \\ 0.0026 & 0.0058 & 0.0031 & 0.0073 \end{bmatrix}, R_3 = \begin{bmatrix} 0.0003 & 0.0007 & 0.0004 & 0.0009 \\ 0.0006 & 0.0014 & 0.0007 & 0.0017 \\ 0.0005 & 0.0012 & 0.0006 & 0.0014 \\ 0.0010 & 0.0022 & 0.0012 & 0.0027 \end{bmatrix},$$

$$R_4 = \begin{bmatrix} 0.0001 & 0.0003 & 0.0002 & 0.0004 \\ 0.0003 & 0.0006 & 0.0003 & 0.0007 \\ 0.0002 & 0.0005 & 0.0003 & 0.0006 \\ 0.0004 & 0.0009 & 0.0005 & 0.0011 \end{bmatrix}, R_5 = \begin{bmatrix} 0.0001 & 0.0001 & 0.0001 & 0.0002 \\ 0.0001 & 0.0003 & 0.0001 & 0.0003 \\ 0.0001 & 0.0002 & 0.0001 & 0.0003 \\ 0.0002 & 0.0004 & 0.0002 & 0.0005 \end{bmatrix}.$$

Computed values of R_i in Non-Preemptive case

$$
R_0 = \begin{bmatrix}
0.02167 & 0.03830 & 0.02378 & 0.04241 & 0.00000 & 0.00000 & 0.00000 & 0.00000 \\
0.05151 & 0.06527 & 0.05548 & 0.07344 & 0.00000 & 0.00000 & 0.00000 & 0.00000 \\
0.02724 & 0.04725 & 0.04088 & 0.07431 & 0.00000 & 0.00000 & 0.00000 & 0.00000 \\
0.06186 & 0.08261 & 0.09874 & 0.12676 & 0.00000 & 0.00000 & 0.00000 & 0.00000 \\
0.05737 & 0.13296 & 0.06498 & 0.14874 & 0.01510 & 0.02819 & 0.01584 & 0.02938 \\
0.05477 & 0.12659 & 0.06260 & 0.14370 & 0.02819 & 0.04328 & 0.02938 & 0.04523 \\
0.07797 & 0.17003 & 0.10475 & 0.25234 & 0.01734 & 0.03178 & 0.02870 & 0.05397 \\
0.07609 & 0.16803 & 0.09896 & 0.23628 & 0.03178 & 0.04912 & 0.05397 & 0.08268
\end{bmatrix},
$$

$$
R_1 = \begin{bmatrix}
0.00302 & 0.00680 & 0.00372 & 0.00874 & 0.00000 & 0.00000 & 0.00000 & 0.00000 \\
0.00562 & 0.01266 & 0.00693 & 0.01628 & 0.00000 & 0.00000 & 0.00000 & 0.00000 \\
0.00467 & 0.01048 & 0.00580 & 0.01368 & 0.00000 & 0.00000 & 0.00000 & 0.00000 \\
0.00870 & 0.01955 & 0.01082 & 0.02550 & 0.00000 & 0.00000 & 0.00000 & 0.00000 \\
0.01242 & 0.02806 & 0.01518 & 0.03559 & 0.00000 & 0.00000 & 0.00000 & 0.00000 \\
0.01360 & 0.03077 & 0.01655 & 0.03876 & 0.00000 & 0.00000 & 0.00000 & 0.00000 \\
0.01907 & 0.04301 & 0.02344 & 0.05506 & 0.00000 & 0.00000 & 0.00000 & 0.00000 \\
0.02082 & 0.04706 & 0.02543 & 0.05962 & 0.00000 & 0.00000 & 0.00000 & 0.00000
\end{bmatrix},
$$

$$
R_2 = \begin{bmatrix}
0.00089 & 0.00201 & 0.00108 & 0.00252 & 0.00000 & 0.00000 & 0.00000 & 0.00000 \\
0.00165 & 0.00374 & 0.00201 & 0.00470 & 0.00000 & 0.00000 & 0.00000 & 0.00000 \\
0.00138 & 0.00313 & 0.00168 & 0.00393 & 0.00000 & 0.00000 & 0.00000 & 0.00000 \\
0.00258 & 0.00583 & 0.00313 & 0.00733 & 0.00000 & 0.00000 & 0.00000 & 0.00000 \\
0.00404 & 0.00917 & 0.00490 & 0.01145 & 0.00000 & 0.00000 & 0.00000 & 0.00000 \\
0.00464 & 0.01053 & 0.00561 & 0.01312 & 0.00000 & 0.00000 & 0.00000 & 0.00000 \\
0.00625 & 0.01416 & 0.00756 & 0.01768 & 0.00000 & 0.00000 & 0.00000 & 0.00000 \\
0.00715 & 0.01621 & 0.00864 & 0.02020 & 0.00000 & 0.00000 & 0.00000 & 0.00000
\end{bmatrix},
$$

$$
R_3 = \begin{bmatrix}
0.00033 & 0.00075 & 0.00040 & 0.00093 & 0.00000 & 0.00000 & 0.00000 & 0.00000 \\
0.00061 & 0.00139 & 0.00074 & 0.00173 & 0.00000 & 0.00000 & 0.00000 & 0.00000 \\
0.00051 & 0.00116 & 0.00062 & 0.00144 & 0.00000 & 0.00000 & 0.00000 & 0.00000 \\
0.00096 & 0.00217 & 0.00115 & 0.00269 & 0.00000 & 0.00000 & 0.00000 & 0.00000 \\
0.00158 & 0.00358 & 0.00190 & 0.00444 & 0.00000 & 0.00000 & 0.00000 & 0.00000 \\
0.00185 & 0.00420 & 0.00223 & 0.00520 & 0.00000 & 0.00000 & 0.00000 & 0.00000 \\
0.00244 & 0.00554 & 0.00294 & 0.00686 & 0.00000 & 0.00000 & 0.00000 & 0.00000 \\
0.00285 & 0.00648 & 0.00343 & 0.00801 & 0.00000 & 0.00000 & 0.00000 & 0.00000
\end{bmatrix},
$$

$$
R_4 = \begin{bmatrix}
0.00014 & 0.00031 & 0.00017 & 0.00039 & 0.00000 & 0.00000 & 0.00000 & 0.00000 \\
0.00026 & 0.00059 & 0.00031 & 0.00072 & 0.00000 & 0.00000 & 0.00000 & 0.00000 \\
0.00022 & 0.00049 & 0.00026 & 0.00060 & 0.00000 & 0.00000 & 0.00000 & 0.00000 \\
0.00040 & 0.00091 & 0.00048 & 0.00113 & 0.00000 & 0.00000 & 0.00000 & 0.00000 \\
0.00068 & 0.00155 & 0.00082 & 0.00191 & 0.00000 & 0.00000 & 0.00000 & 0.00000 \\
0.00081 & 0.00184 & 0.00097 & 0.00227 & 0.00000 & 0.00000 & 0.00000 & 0.00000 \\
0.00106 & 0.00240 & 0.00127 & 0.00296 & 0.00000 & 0.00000 & 0.00000 & 0.00000 \\
0.00125 & 0.00284 & 0.00150 & 0.00350 & 0.00000 & 0.00000 & 0.00000 & 0.00000
\end{bmatrix},
$$

$$R_5 = \begin{bmatrix} 0.00006 & 0.00014 & 0.00008 & 0.00018 & 0.00000 & 0.00000 & 0.00000 & 0.00000 \\ 0.00012 & 0.00027 & 0.00014 & 0.00033 & 0.00000 & 0.00000 & 0.00000 & 0.00000 \\ 0.00010 & 0.00022 & 0.00012 & 0.00027 & 0.00000 & 0.00000 & 0.00000 & 0.00000 \\ 0.00018 & 0.00042 & 0.00022 & 0.00051 & 0.00000 & 0.00000 & 0.00000 & 0.00000 \\ 0.00032 & 0.00072 & 0.00038 & 0.00088 & 0.00000 & 0.00000 & 0.00000 & 0.00000 \\ 0.00038 & 0.00086 & 0.00045 & 0.00106 & 0.00000 & 0.00000 & 0.00000 & 0.00000 \\ 0.00049 & 0.00111 & 0.00059 & 0.00137 & 0.00000 & 0.00000 & 0.00000 & 0.00000 \\ 0.00058 & 0.00132 & 0.00070 & 0.00163 & 0.00000 & 0.00000 & 0.00000 & 0.00000 \end{bmatrix}.$$

For these parameter values, the entries of R_i decrease with the increase of i in both preemptive and non-preemptive cases. The traffic intensity, which is the expression on the left side of Eq. (1), is 0.7957 and hence the system is stable. The entries of R_i become negligible as i becomes large. In this example, we can neglect R_i for $i \geq 6$. This truncation leads to the truncation of the lower priority queue and hence the rate matrix R will become as a finite matrix. Now the boundary probability x_0 and x_1 can be easily calculated using formulas in Sect. 3.3. Using the set of Eqs. (4), The marginal probability density functions are calculated for both preemptive and non-preemptive cases and are expressed graphically in the following Figs. 1 and 2 respectively.

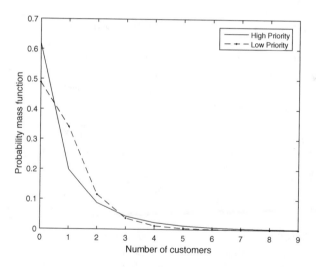

Fig. 1. Marginal probability distributions in preemptive discipline

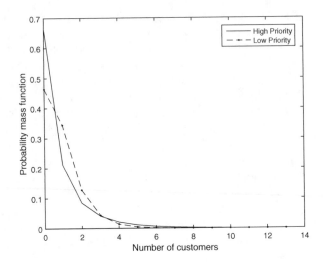

Fig. 2. Marginal probability distributions in non-preemptive discipline

6   Concluding Remarks

This paper looked at discrete-time, self-interrupting priority queues of MAP arrivals and service as the time till absorption. The absorbing Markov chain has two absorbing states through which a customer leaves the system after service or the interrupted customer moved to lower priority queue. The matrix-Analytic Method is used to analyze the model. The marginal probability distribution of queue length is discussed. For future studies, one can consider the chance of abandoning the service if interruption happened. A similar self-generated priority can be generated through a feedback queue of customers in which the customers after service may join to lower priority queue with some probability. Incorporation of inventory of items is also another interest of study.

References

1. Takagi, H.: Queueing Analysis: A Foundation of Performance Evaluation: Vacation and Priority Systems, Part 1, vol. 1. North-Holland, Amsterdam (1991)
2. Alfa, A.S.: Matrix-geometric solution of discrete time MAP/PH/1 priority queue. Naval Res. Logist. (NRL) **45**(1), 23–50 (1998)
3. Anilkumar, M.P., Jose, K.P.: A Geo/Geo/1 inventory priority queue with self induced interruption. Int. J. Appl. Comput. Math. **6**(4), 1–14 (2020)
4. Anilkumar, M.P., Jose, K.P.: Discrete time priority queue with self generated interruption. In: AIP Conference Proceedings, vol. 2261, p. 030031. AIP Publishing LLC (2020)
5. Brodal, G.S.: A survey on priority queues. In: Brodnik, A., López-Ortiz, A., Raman, V., Viola, A. (eds.) Space-Efficient Data Structures, Streams, and Algorithms. LNCS, vol. 8066, pp. 150–163. Springer, Heidelberg (2013). https://doi.org/10.1007/978-3-642-40273-9_11

6. Dudin, A., Jacob, V., Krishnamoorthy, A.: A multi-server queueing system with service interruption, partial protection and repetition of service. Ann. Oper. Res. **233**(1), 101–121 (2013). https://doi.org/10.1007/s10479-013-1318-3
7. Graham, A.: Kronecker Products and Matrix Calculus with Applications. Courier Dover Publications (2018)
8. Jacob, V., Chakravarthy, S.R., Krishnamoorthy, A.: On a customer-induced interruption in a service system. Stoch. Anal. Appl. **30**(6), 949–962 (2012)
9. Jaiswal, N.: Priority Queues. Academic Press, New York and London (1968)
10. Jaiswal, N.: Preemptive resume priority queue. Oper. Res. **9**(5), 732–742 (1961)
11. Krishnamoorthy, A., Babu, S., Narayanan, V.C.: The MAP/(PH/PH)/1 queue with self-generation of priorities and non-preemptive service. Eur. J. Oper. Res. **195**(1), 174–185 (2009)
12. Krishnamoorthy, A., Manjunath, A.: On priority queues generated through customer induced service interruption. Neural Parallel Sci. Comput. **23**, 459–486 (2015)
13. Krishnamoorthy, A., Pramod, P.K., Chakravarthy, S.R.: Queues with interruptions: a survey. Top **22**(1), 290–320 (2014)
14. Miller, D.R.: Computation of steady-state probabilities for M/M/1 priority queues. Oper. Res. **29**(5), 945–958 (1981)
15. Neuts, M.F.: Matrix-Geometric Solutions in Stochastic Models: An Algorithmic Approach. Courier Corporation (1994)
16. White, H., Christie, L.S.: Queuing with preemptive priorities or with breakdown. Oper. Res. **6**(1), 79–95 (1958)

On Comparison of Waiting Time Extremal Indexes in Queueing Systems with Weibull Service Times

Irina Peshkova[1,2(✉)] [ID], Evsey Morozov[1,2,3(✉)] [ID], and Maria Maltseva[2(✉)] [ID]

[1] Petrozavodsk State University, Lenin Str. 33, Petrozavodsk 185910, Russia
`iaminova@petrsu.ru`
[2] Institute of Applied Mathematical Research of the Karelian Research Centre of RAS, Pushkinskaya Str. 11, Petrozavodsk 185910, Russia
`emorozov@karelia.ru, masha.mariam.maltseva@mail.ru`
[3] Moscow Center for Fundamental and Applied Mathematics, Moscow State University, Moscow 119991, Russia

Abstract. The paper is devoted to the tail asymptotics analysis of the steady-state waiting times in the queuing systems in which service times have Weibull distributions. We deduce conditions under which the service times in two different queueing systems are stochastically ordered. Then we show that, under the same conditions, the normalizing sequences of the stationary waiting times and their extremal indexes are ordered. These results are then illustrated numerically for $GI/G/1$ queues with different shape parameters of the Weibull service times.

Keywords: Performance analysis · Queueing system · Extremal index · Weibull distribution

1 Introduction

The subexponential distributions form a subclass of the so-called *heavy-tailed* distributions, which arise in particular, in the insurance and various queueing applications. The tails of such distributions decrease more slowly than the exponential tails. It is established that the main reason why a sum of the subexponential random variables becomes 'large' is that one of the components of this sum is large. This property allows us to use the class heavy-tailed distributions to model and then simulate the processes which can take extremely large values with a probability that cannot be neglected [5]. For instance, it is known that in $GI/G/1$ queueing systems with heavy-tailed service time the tail waiting time asymptotics is defined by the service time distribution [1]. This property can be applied to derive the limiting distribution of the maximum of the waiting time based on the extreme value theory. To realize this idea, in this research we

The research has been prepared with the support of Russian Science Foundation according to the research project No. 21-71-10135.

A. Dudin et al. (Eds.): ITMM 2021, CCIS 1605, pp. 80–92, 2022.
https://doi.org/10.1007/978-3-031-09331-9_7

consider a queueing system with the Weibull service time distribution (denoted below $GI/Weibull/1$) with shape parameter $0 < \beta < 1$. In this case the distribution belongs to the subclass of *subexponential* distributions and it plays in particular a significant role in the reliability theory and in the survival analysis [5]. In this case we use the tail asymptotics of the waiting time given in [2] and then apply the regenerative approach [10] to estimate the extremal index.

The extreme events often occur in the clusters, and their prediction is an actual problem which however is highly difficult to be resolved in the most of cases [16]. We note that the extreme value theory in general is applied not only to describe the limiting distribution of a maximum but also to determine the size of the clusters and extreme's frequency via the so-called *extremal index*. This index, denoted by $\theta \in [0, 1]$, evaluates the reciprocal of the average cluster size and hence measures the degree of clustering of the extremes.

It is worth to mentioning that the idea of the predicting and mitigating extreme values of the performance measures in queueing systems is not a new one and has been studied, for example, in [1,3,7,8,15]. The limit theory of the Markov chains based on the extreme value theory is deeply analyzed in the fundamental paper [15]. The limit theorems for the maximum actual waiting time, maximum virtual waiting time in a $GI/G/1$ queueing system, for all (acceptable) values of the traffic intensity, are obtained in the paper [8]. The extremal properties of Markov chains and adapted algorithm for computing the extremal index in a stable $GI/G/1$ system is given in [7]. The comparison of performance measures of queueing systems with different distributions of input or service times based on stochastic ordering or failure rate ordering has been considered by the authors in the works [11,12]. The paper [13] considers comparison of the extremal indexes calculated for the stationary waiting time in $M/G/1$ queueing systems in which service times have Pareto distribution satisfying stochastic ordering.

In the case of the Poisson input process and exponential service times the extremal index is calculated explicitly in [3,7]. In some other cases it can be obtained iteratively, see for instance, [7]. Provided the service time distribution is subexponential, for example is Weibull with the shape parameter $\beta \in (0, 1)$ (for definition of β see (3) below), then we can use the tail asymptotic of the waiting time from [2] to derive the limiting distribution of the maximum waiting time. This analysis can be used to evaluate the extremal index of the strictly stationary waiting times and to compare the extremal index values in two $GI/G/1$ systems.

The purpose of this research is to provide that the stochastic ordering of the service times in two queueing systems allows to compare the normalizing sequences of the steady-state waiting times and their extremal indexes. We demonstrate this approach for the $GI/G/1$ systems with Weibull service times with the shape parameter $\beta \in (0, 1)$.

The paper is structured as follows. In Sect. 2, we describe the model and discuss the tail asymptotics of the waiting time distribution which is defined by so-called equilibrium distribution function of the remaining service time. The limiting distribution of the maximum waiting time (in a $GI/Weibull/1$ system), based on the tail asymptotics obtained in Sect. 2, is derived in Sect. 3 (see Lemma 1). In Sect. 4, we discuss the comparison of the extremal indexes of two random variables with Weibull distributions, in which shape parameters are properly ordered. This analysis is further applied to compare the extremal indexes of the stationary sequences of the waiting times in two queueing systems with the stochastically ordered service times (Theorem 1).

2   Model Description

We consider the $GI/G/1$ queueing system with a renewal input, and let $T_i = t_{i+1} - t_i$ be the independent, identically distributed (i.i.d.) interarrival times, $i \geq 0$. Denote by $\{S_i\}$ the i.i.d. service times. It is assumed that service discipline is FIFO (First-In-First-Out). Denote by $\lambda = 1/ET$ the input rate and by $\mu = 1/ES$ the service rate, and let $\rho = \lambda/\mu$ be the traffic intensity. (The serial index is omitted when we consider the generic element of an i.i.d sequence.) Let W_i be the waiting time of the i-th customer, and we recall that the sequence $\{W_i\}$ can be obtained by means of the *Lindley recursion*, which defines the accumulated work to be done at the instants $\{t_n^-\}$, that is, which a new arrival meets,

$$W_{n+1} = (W_n + S_n - T_n)^+, \quad n \geq 1, \tag{1}$$

where we assume that $W_1 = 0$ (zero-delayed process), and $(\cdot)^+ = \max(0, \cdot)$. If $\rho < 1$ and distribution of T is *non-lattice*, then there exists the stationary waiting time, that is, $W_n \Rightarrow W$, where $\Rightarrow$ denotes convergence in distribution (see, for instance, [1]). It is well-known that such a system *regenerates* (in the classic sense) when an arrival meets the system idle [10].

Denote by $B(x) = \mathsf{P}(S \leq x)$ the distribution function of the service times S. The distribution B is called *subexponential* if

$$\lim_{x \to \infty} \frac{\overline{B^{*n}}(x)}{n\overline{B}(x)} = 1 \text{ for all } n \geq 2,$$

where $\overline{B^{*n}}(x)$ is the tail of n-convolution of the distribution $B(x)$ with itself. In particular, $\overline{B}(x) = 1 - B(x)$ is the tail of B.

It is known that if the service times have subexponential distribution B with a finite mean, then the waiting time distribution in the $GI/G/1$ queueing system has the following tail asymptotic [2]:

$$\mathsf{P}(W > x) \sim \frac{\rho}{1 - \rho}\mathsf{P}(S_e > x), \quad x \to \infty, \tag{2}$$

if service time S and the stationary remaining renewal time S_e are subexponential. Note that S_e has the equilibrium density $\overline{B}(x)/ES$. (Relation $a \sim b$ in (2) means the asymptotic equivalence, that is $a/b \to 1$.)

Assume that the service time S has Weibull distribution

$$B(x) = 1 - e^{-x^\beta}, \quad \beta > 0, \ x \geq 0, \tag{3}$$

with the density function $f_B(x) = \beta x^\beta e^{-x^\beta}$. The equilibrium Weibull distribution function B_e is then defined by

$$B_e(x) = \frac{1}{ES} \int_0^x \overline{B}(t)dt = \frac{1}{\Gamma(1/\beta)} \int_0^{x^\beta} e^{-y} y^{1/\beta-1} dy = \frac{\gamma(1/\beta, x^\beta)}{\Gamma(1/\beta)}, \quad (4)$$

where

$$\Gamma(t) = \int_0^\infty e^{-y} y^{t-1} dy,$$

is the Gamma function and

$$\gamma(t, x) = \int_0^x e^{-y} y^{t-1} dy,$$

is the lower incomplete gamma function. Then it is easy to check that the corresponding tail satisfies the following asymptotic relation

$$\overline{B}_e(x) = \frac{\Gamma(1/\beta, x^\beta)}{\Gamma(1/\beta)} \sim \frac{x^{1-\beta} e^{-x^\beta}}{\Gamma(1/\beta)}, \quad \text{as } x \to \infty,$$

where

$$\Gamma(t, x) = \int_x^\infty e^{-y} y^{t-1} dy,$$

is the upper incomplete gamma function.

To check that the Weibull distribution B (of the service time S) and the corresponding equilibrium distribution B_e (with parameter $\beta \in (0, 1)$) both belong to the class of the subexponential distributions, it is enough to verify that B belongs to a special subclass $\mathcal{S}^*$ of the subexponential distributions. Namely, the distribution $B \in \mathcal{S}^*$ [5] if the service time S has a finite mean $1/\mu < \infty$ and moreover,

$$\lim_{x \to \infty} \int_0^x \frac{\overline{B}(x - y)}{\overline{B}(x)} \overline{B}(y) dy = \frac{2}{\mu}.$$

In practice, the following criteria for a distribution to belong $\mathcal{S}^*$ is often applied [5]. Denote the failure rate function

$$r(x) = \frac{f_B(x)}{\overline{B}(x)} \quad \text{and let } R(x) = -\log \overline{B}(x).$$

Suppose that

$$\lim_{x \to \infty} r(x) = 0 \quad \text{and} \quad \lim_{x \to \infty} x r(x) = \infty.$$

Then $B \in \mathcal{S}^*$ if, additionally, one of the following (incompatible) conditions holds [5]:

$$\text{a)} \quad \limsup_{x \to \infty} \frac{xr(x)}{R(x)} < 1; \tag{5}$$

$$\text{b)} \quad r \in \mathcal{R}(-\delta) \text{ for } \delta \in (0, 1]; \tag{6}$$

$$\text{c)} \quad R \in \mathcal{R}(\delta) \text{ for } \delta \in (0, 1), \tag{7}$$

where $\mathcal{R}(\delta)$ is the class of *regularly varying functions*, that is a function $g \in \mathcal{R}(\delta)$ if

$$\lim_{x \to \infty} \frac{g(tx)}{g(x)} = t^\delta, \text{ for all } t > 0.$$

It is easy to verify that the failure rate of Weibull distribution (with parameter $0 < \beta < 1$) decaying to zero, $xr(x) \to \infty$, and condition (5) holds. More exactly,

$$r(x) = \beta x^{\beta - 1} \to 0 \text{ as } x \to \infty;$$

$$xr(x) = \beta x^\beta \to \infty \text{ as } x \to \infty;$$

$$\frac{xr(x)}{R(x)} = \beta < 1 \text{ for all } x.$$

Hence the Weibull distribution B with parameter $\beta \in (0, 1)$ belongs to the subclass $\mathcal{S}^*$ of the subexponential distributions. Therefore B and the equilibrium distribution B_e both belong to the class of subexponential distributions, that is $B, B_e \in \mathcal{S}^*$. Actually, the subexponentiality of B_e follows from the following relations:

$$r_e(x) = \frac{f_{B_e}(x)}{\overline{B}_e(x)} = \frac{\beta e^{-x^\beta} \Gamma(1/\beta)}{\Gamma(1/\beta, x^\beta)} \sim \frac{\beta \Gamma(1/\beta)}{x^{1-\beta}} \to 0 \text{ as } x \to \infty;$$

$$xr_e(x) \sim \beta \Gamma(1/\beta) x^\beta \to \infty \text{ as } x \to \infty;$$

$$\frac{r_e(tx)}{r_e(x)} \to t^{\beta - 1} = t^{-\delta} \text{ as } x \to \infty, \ \delta = 1 - \beta < 1.$$

For the Weibull distribution (3), the traffic intensity is determined by the relation

$$\rho = \frac{\lambda \Gamma(1/\beta)}{\beta}. \tag{8}$$

It now follows from (2) and (4) that the waiting time tail distribution satisfies

$$P(W > x) \sim \frac{\lambda}{\beta - \lambda \Gamma(1/\beta)} x^{1-\beta} e^{-x^\beta} \text{ as } x \to \infty. \tag{9}$$

3   The Limiting Distribution of the Maximum Waiting Time

In this section we consider some basic concepts from the extreme value theory in order to apply them in the next section to the analysis of the limiting distribution of the stationary waiting time maximum in a $GI/G/1$ queueing system with Weibull service time.

Let $\{X_n,\ n \geq 1\}$ be a family of the i.i.d random variables (rv's) with a distribution function F. Then the distribution of maximum $M_n = \max(X_1, \ldots, X_n)$ satisfies

$$P(M_n \leq x) = F^n(x).$$

It is known [4,6,9,14] that if, for some sequences of the constants b_n, $a_n > 0$, $n \geq 1$, the normalized maximum $(M_n - b_n)/a_n$ has a non-degenerate limiting distribution function $G(x)$,

$$P((M_n - b_n)/a_n \leq x) \to G(x), \quad n \to \infty, \tag{10}$$

then $G(x)$ has one of the following forms:

$$\text{Type I:} \quad G(x) = \exp(-e^{-x}), \quad -\infty < x < \infty;$$

$$\text{Type II:} \quad G(x) = \begin{cases} 0, & x \leq 0; \\ \exp(-x^{-\eta}), & x > 0; \end{cases}$$

$$\text{Type III:} \quad G(x) = \begin{cases} \exp(-(-x)^{\eta})), & x \leq 0; \\ 1, & x > 0. \end{cases}$$

where parameter $\eta > 0$. Type I is called Gumbel distribution, Type II is Frechet distribution and Type III is called the reversed Weibull distribution.

Suppose that there exists a sequence of real constants $\{u_n,\ n \geq 1\}$ such that for some $0 \leq \tau \leq \infty$,

$$n\overline{F}(u_n) \to \tau \text{ as } n \to \infty. \tag{11}$$

Then it follows from [9] that

$$P(M_n \leq u_n) \to e^{-\tau} \text{ as } n \to \infty. \tag{12}$$

Conversely, if relation (12) holds for some $0 \leq \tau \leq \infty$ then the convergence (11) holds as well.

If condition (10) is satisfied, then convergence (12) is preserved for any linear normalizing sequence

$$u_n(x) = a_n x + b_n,\ n \geq 1,$$

where x takes real values and expression (12) becomes

$$P(M_n \leq u_n(x)) \to \tau(x),$$

where a concrete form of the function $\tau(x)$ depends on the type of the limiting distribution.

The sequence $\{u_n(x)\}$ for Weibull distribution (3) can be found in the following form:

$$u_n(x) = \frac{x(\log n)^{1/\beta-1}}{\beta} + (\log n)^{1/\beta}, \tag{13}$$

in which case the maximum M_n has Gumbel distribution (in the limit as $n \to \infty$), that follows from the asymptotics

$$n\overline{F}(u_n) = ne^{-\left(\frac{(\log n)^{1/\beta-1}}{\beta}x + (\log n)^{1/\beta}\right)^\beta}$$

$$= ne^{-\log n\left(1 + \frac{x}{\log n} + o(\frac{1}{\log n})\right)} \to e^{-x} \text{ as } n \to \infty. \tag{14}$$

The limit distribution of the maximum generated by the equilibrium Weibull distribution has a Gumbel form with the same normalizing sequence $\{u_n(x)\}$ defined by (13), since

$$nu_n(x)^{1-\beta}e^{-u_n(x)^\beta} = n(\log n)^{1/\beta}\log n^{-1}\left(\frac{x}{\beta\log n} + 1\right)\left(\frac{x}{\beta\log n} + 1\right)^{-\beta}$$

$$\times e^{-\log n\left(1 + \frac{x}{\log n} + o(1/\log n)\right)} \to \frac{e^{-x}}{\beta} \text{ as } n \to \infty.$$

To extend (12) to a non i.i.d. *strictly stationary* sequence $\{X_n\}$, an additional condition on the decay of the correlations is required (see condition $D(u_n)$ in [9]). In this case, instead of relation (12), we obtain

$$\mathsf{P}(M_n \le u_n(x)) \to e^{-\theta\tau(x)}, \text{ as } n \to \infty, \ 0 < \tau(x) < \infty, \tag{15}$$

where parameter $\theta \in [0, 1]$ defines the so-called *extremal index* of the sequence $\{X_n\}$. Moreover, like in the i.i.d. case, the same family of the extreme value distributions describes the maximum of the strictly stationary sequence.

Now we go back to the $GI/G/1$ system described in previous section. Denote the maximum waiting time by $W_n^* = \max(W_1, \ldots, W_n)$ for each $n \ge 1$. Then relations (9) and (14) lead to the following statement.

Lemma 1. *If the service time in a $GI/G/1$ queueing system has Weibull distribution (3) with parameter $\beta \in (0, 1)$ then the limiting distribution of W_n^* has the following Gumbel-type shape:*

$$P(W_n^* \le u_n(x)) \to e^{-\frac{\lambda}{(\beta - \lambda\Gamma(1/\beta)\beta}e^{-x}} \text{ as } n \to \infty, \tag{16}$$

with the normalized sequence

$$u_n(x) = a_n x + b_n = \frac{x(\log n)^{1/\beta-1}}{\beta} + (\log n)^{1/\beta}. \tag{17}$$

4   Comparison of the Waiting Time Extremal Indexes

In this section we return to the $GI/G/1$ queueing systems described in Sect. 2 and denote they by $\Sigma^{(1)}$ and $\Sigma^{(2)}$. Let $T^{(i)}$ be the generic interarrival time, $S^{(i)}$ the generic service time, and let $\mathsf{E}T^{(i)} = 1/\lambda_i$, $i = 1, 2$. (The superscript (i) relates to system i). Now we compare the steady-state waiting time processes in the systems $\Sigma^{(1)}$ and $\Sigma^{(2)}$. At the arrival instant of customer n in the system $\Sigma^{(i)}$, we denote by $\nu_n^{(i)}$ the number of customers, by $Q_n^{(i)}$ the *queue size* and by $W_n^{(i)}$ the (actual) *waiting time* of this customer, $n \geq 1$. Denote, when exists, the limits (in distribution)

$$Q_n^{(i)} \Rightarrow Q^{(i)}, \ \nu_n^{(i)} \Rightarrow \nu^{(i)}, \ W_n^{(i)} \Rightarrow W^{(i)}, \ n \to \infty, \ i = 1, 2.$$

These limits exists, in particular, when the interarrival times $T^{(i)}$, $i = 1, 2$ are *non-lattice* and $\rho_i = \lambda_i \mathsf{E}S^{(i)} < 1$ [1]. Assume that the following stochastic relations hold:

$$\nu_1^{(1)} = \nu_1^{(2)} = 0, \ T^{(1)} =_{st} T^{(2)}, \ S^{(1)} \leq_{st} S^{(2)}, \tag{18}$$

where, recall, the stochastic ordering $S^{(1)} \leq_{st} S^{(2)}$ means that the corresponding tail distributions satisfy $\overline{B}_{S^{(1)}}(x) \leq \overline{B}_{S^{(2)}}(x)$ for all x. Then it follows from [17] that

$$Q_n^{(1)} \leq_{st} Q_n^{(2)}, \ W_n^{(1)} \leq_{st} W_n^{(2)}, \ n \geq 1. \tag{19}$$

We assume that the systems $\Sigma^{(1)}$, $\Sigma^{(2)}$ have Weibull service times distributions in which parameters satisfy the inequalities $0 < \beta_1, \beta_2 < 1$. If moreover the assumption $\beta_1 \geq \beta_2$ holds then there exists a stochastic ordering between the service times, namely, $S^{(1)} \leq_{st} S^{(2)}$. Therefore, inequalities (19) hold and imply the stochastic ordering of the waiting times, $W_n^{(1)} \leq_{st} W_n^{(2)}$ [17]. In what follows we need the maximum waiting times which are defined as follows:

$$W_n^{(1)*} := \max(W_1^{(1)}, \ldots, W_n^{(1)}), \ W_n^{(2)*} := \max(W_1^{(2)}, \ldots, W_n^{(2)}), \ n \geq 1.$$

The following lemma allows us to obtain the main theoretical result of the research containing in Theorem 1 below.

Lemma 2. *Let $\{X_n\}$ and $\{Y_n\}$ be two stationary sequences with the corresponding generic rv's X and Y. Assume that X and Y have Weibull distributions with parameters β_1, β_2, respectively, $0 < \beta_2 \leq \beta_1 < 1$. Then the corresponding extremal indexes θ_X and θ_Y are ordered as*

$$\theta_X \geq \theta_Y. \tag{20}$$

The proof of this statement mainly follows the paper [13] in which we compare the extremal indexes of two stationary sequences $\{X_n\}$ and $\{Y_n\}$ with different Pareto distributions. We note that, to verify condition $u_n(x) \geq u_n'(x)$, it is enough to check that $(b_n' - b_n)/(a_n - a_n') \leq 0$ for all $x \geq 0$. Indeed, for $n \geq 3$,

$$\frac{b_n' - b_n}{a_n - a_n'} = \frac{\log n\big((\log n)^{1/\beta_2 - 1/\beta_1} - 1\big)}{\beta_1 \beta_2 \big(\beta_2 - \beta_1 (\log n)^{1/\beta_2 - 1/\beta_1}\big)} \leq 0. \tag{21}$$

Note that the limit distributions of $W_n^{(1)*}$ and $W_n^{(2)*}$ satisfy (16) with parameters β_1 and β_2, respectively. Lemma 2 and discussion above imply the following statement.

Theorem 1. *Assume that in the queueing systems under consideration the traffic intensities $\rho_i < 1$, $i = 1, 2$ and the parameters of Weibull service time distributions satisfy the inequalities*

$$1 > \beta_1 \geq \beta_2 > 0.$$

Then the extremal indexes of the stationary waiting times in these systems are ordered in the following way:

$$\theta_{W^{(1)}} \geq \theta_{W^{(2)}}. \tag{22}$$

5   Simulation Results

In this section, we discuss the extremal index estimation of the waiting times by the block method and the regenerative approach and present some numerical examples for described systems with Weibull service times.

The convergence (15) together with relation (11) imply the basic relation for extremal index

$$\theta = \lim_{n \to \infty} \frac{\log P(M_n \leq u_n)}{n \log F(u_n)}. \tag{23}$$

The main idea of the block method is to divide the sequence $X_1, \ldots, X_n$ into m blocks of the identical size h, where $n = m\,h$. After that, it is necessary to calculate the number of blocks with exceedances and the number of the exceedances of the threshold u_n by the sequence $X_1, \ldots, X_n$. Then the estimate of the extremal index is the ratio of these two quantities [4].

Regenerative approach can also be used to estimate the extremal index [8,15]. More exactly, consider a stationary regenerative sequence $\{Z_n\}$ with the regeneration instants β_k, and denote by M_{Y_k} its maximum in the kth regeneration cycle, that is,

$$M_{Y_k} = \sup_n \{Z_n, \, \beta_{k-1} \leq n < \beta_k\},$$

and also denote by $\mathsf{E}\alpha$ the mean cycle length. Then the stationary regenerative sequence $\{Z_n\}$ has the extremal index θ if and only if there exists the limit [15]

$$\theta = \lim_{n \to \infty} \frac{P(M_{Y_1} > u_n)/\mathsf{E}\alpha}{P(Z_1 > u_n)}, \tag{24}$$

for some normalizing sequence u_n satisfying relation (11).

By (24), the estimate of the extremal index based on the regeneration cycles can be constructed as follows (for n large enough):

$$\hat{\theta}_\alpha(n) := \hat{\theta}_\alpha = \frac{1}{\hat{\alpha}} \frac{\log\left(1 - \frac{\hat{m}(u_n)}{m}\right)}{\log\left(1 - \frac{\hat{N}(u_n)}{n}\right)}, \tag{25}$$

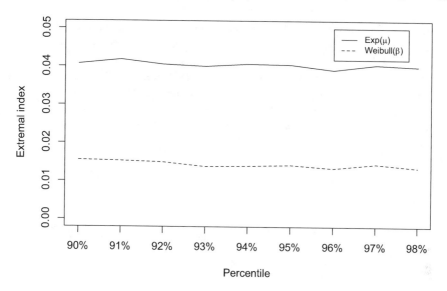

Fig. 1. The extremal indexes of the stationary waiting time in $M/M/1$ queue with input rate $\lambda = 0.4$, and service rate $\mu = 0.5$, and in $M/Weibull/1$ queue with $\lambda = 0.4$ and $\mu = 0.5$

where

$$\hat{N}(u_n) = \#(i \leq n : Z_i > u_n), \quad \hat{m}(u_n) = \#(i \leq m_\alpha : M_{Y_i} > u_n),$$

$\hat{m}(u_n)$ is the number of the exceedances within regeneration cycle, m_α is the number of regeneration points which are detected during simulation procedure, and $\hat{\alpha}$ is the sample mean of the cycle length.

If the input process is Poisson and the service times are exponential with parameters λ and μ, respectively, then the following explicit form of the extremal index is known [7]

$$\theta = (1 - \rho)^2, \tag{26}$$

where $\rho = \lambda/\mu$. This explicit form of the solution allows to compare it with the numerical results obtained by the simulation.

Figure 1 demonstrates the comparison of the estimates of the waiting time extremal indexes calculated, by the regenerative method, for the $M/M/1$ queueing system with input rate $\lambda_1 = 0.4$, and with service rate $\mu = 0.5$, and for the $M/G/1$ system with the same input rate $\lambda_2 = 0.4$ and Weibull service time with parameter $\beta = 0.5$. In both cases the traffic intensity turns out to be the same, $\rho_1 = \rho_2 = 0.8$.

The simulation results obtained by the regeneration method show that the extremal index of the waiting time in $M/M/1$ queueing system is close to exact value (26) which equals $\theta = 0.04$. We recall that the smaller extremal index is then the extreme values more often are. The extremal index in the 2nd (Weibull) system is close to 0.015, and it is at least in 2.5 times less than that in $M/M/1$ system. Thus we can interpret this result in such a way that the extreme values of the waiting times in the system with Weibull service time occur (approximately) in 2.5 times more frequent.

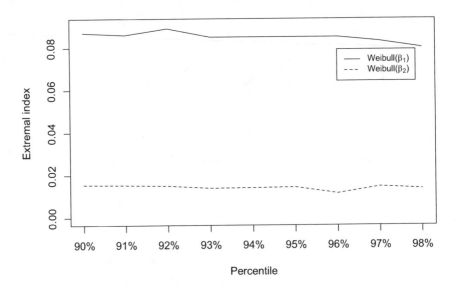

Fig. 2. The extremal indexes of the stationary waiting time in queueing systems $M/Weibull(\beta_i)/1$ with Weibull service times with input rates $\lambda_1 = \lambda_2 = 0.4$ and parameters $\beta_1 = 0,6$; $\beta_2 = 0.5$, respectively.

Now we analyze $M/Weibull/1$ queueing systems which are fed by the equivalent Poisson input processes but have different (Weibull) service times. Figure 2 demonstrates the values of the extremal indexes of the stationary waiting times in $M/Weibull(\beta_i)/1$ systems when the input rate equals $\lambda_1 = \lambda_2 = 0.4$, while the service times have Weibull distributions with parameters $\beta_1 = 0.6$ and $\beta_2 = 0.5$, respectively. These parameters guarantee that the systems are stationary because

$$\rho_1 = \frac{\lambda_1 \Gamma(1/\beta_1)}{\beta_1} \approx 0.6, \quad \rho_2 = \frac{\lambda_2 \Gamma(1/\beta_2)}{\beta_2} = 0.8,$$

(see (8)), and this implies the stochastic ordering $S^{(1)} \leq_{st} S^{(2)}$ between service times. Then, by Theorem 1, the extremal indexes of the corresponding waiting times satisfy the ordering:

$$\theta_{W^{(1)}} \geq \theta_{W^{(2)}},$$

and it is confirmed by the simulation as Figs. 1, 2 show. Note that $\beta_1 = 1.2\beta_2$, while the extremal indexes differ by a factor of four. These results show that the cluster size (extreme values) in the 2nd system occurs (about) four times more often.

6   Conclusion

In this research we study the extreme behaviour of the stationary performance indexes in $GI/Weibull(\beta)/1$ queueing systems with Weibull service times with shape parameter $0 < \beta < 1$. It is shown that if, in the two $GI/Weibull(\beta_i)/1$ systems having the same renewal inputs, the shape parameters of the service time distributions are ordered as $1 > \beta_1 \geq \beta_2 > 0$, then the corresponding (strictly stationary) waiting times have the extremal indexes which are ordered in the same way, that is $\theta_{W^{(1)}} \geq \theta_{W^{(2)}}$.

References

1. Asmussen, S.: Applied Probability and Queues. Stochastic Modelling and Applied Probability, 2nd edn. Springer, New York (2003). https://doi.org/10.1007/b97236
2. Asmussen, S., Kluppelberg, C., Sigman, K.: Sampling at subexponential times with queueing applications. Report TUM M9804 (1998)
3. Bertail, P., Clémençon, S., Tressou, J.: Regenerative block-bootstrap confidence intervals for the extremal index of Markov chains. In: Proceedings of the International Workshop in Applied Probability (2008)
4. Embrechts, P., Kluppelberg, C., Mikosch, T.: Modelling Extremal Events for Insurance and Finance. Applications of Mathematics, p. 660. Springer, Heidelberg (1997). https://doi.org/10.1007/978-3-642-33483-2
5. Goldie, C. M., Klüuppelberg, C.: Subexponential distributions. A Practical Guide to Heavy Tails: Statistical Techniques for Analysing Heavy Tails. Birkhauser, Basel (1997)
6. de Haan, L., Ferreira, A.: Extreme Value Theory: An Introduction. Springer Series in Operations Research and Financial Engineering, p. 491. Springer, Heidelberg (2006). https://doi.org/10.1007/0-387-34471-3
7. Hooghiemstra, G., Meester, L.E.: Computing the extremal index of special Markov chains and queues. Stochast. Process. Their Appl. **65**(2), 171–185 (1996). https://doi.org/10.1016/S0304-4149(96)00111-1
8. Iglehart, D.L.: Extreme values in GI/G/1 queue. Ann. Math. Stat. **43**(2), 627–635 (1972). https://doi.org/10.1214/aoms/1177692642
9. Leadbetter, M.R., Lindgren, G., Rootzin, H.: Extremes and Related Properties of Random Sequences and Processes. Springer, New York (1983). https://doi.org/10.1007/978-1-4612-5449-2

10. Morozov, E., Steyaert, B.: Stability Analysis of Regenerative Queueing Models. Springer, Heidelberg (2021). https://link.springer.com/book/10.1007/978-3-030-82438-9

11. Morozov, E., Peshkova, I., Rumyantsev, A.: On failure rate comparison of finite multiserver systems. In: Vishnevskiy, V.M., Samouylov, K.E., Kozyrev, D.V. (eds.) DCCN 2019. LNCS, vol. 11965, pp. 419–431. Springer, Cham (2019). https://doi.org/10.1007/978-3-030-36614-8_32

12. Peshkova, I., Morozov, E., Maltseva, M.: On comparison of multiserver systems with two-component mixture distributions. In: Vishnevskiy, V.M., Samouylov, K.E., Kozyrev, D.V. (eds.) DCCN 2020. CCIS, vol. 1337, pp. 340–352. Springer, Cham (2020). https://doi.org/10.1007/978-3-030-66242-4_27

13. Peshkova, I., Morozov, E., Maltseva, M.: On regenerative estimation of extremal index in queueing systems. In: Vishnevskiy, V.M., Samouylov, K.E., Kozyrev, D.V. (eds.) DCCN 2021. LNCS, vol. 13144, pp. 251–264. Springer, Cham (2021). https://doi.org/10.1007/978-3-030-92507-9_21

14. Resnick, S.: Extreme values, Regular Variation and Point Processes. Springer Series in Operations Research and Financial Engineering, 334 p. (2008). https://doi.org/10.1007/978-0-387-75953-1

15. Rootzen, H.: Maxima and exceedances of stationary Markov chains. Adv. Appl. Probabil. **20**(2), 371–390 (1998). https://doi.org/10.2307/1427395

16. Smith, L., Weissman, I.: Estimating the extremal index. J. Roy. Stat. Soc. Ser. B: Methodol. **56**(3), 515–528 (1994). https://doi.org/10.1111/J.2517-6161.1994.TB01997.X

17. Whitt, W.: Comparing counting processes and queues. Adv. Appl. Probab. **13**, 207–220 (1981)

On the Existence of the Stationary Distribution in a Cyclic Polling System with Autoregressive Poisson Inputs

Andrei V. Zorine$^{(\boxtimes)}$ (iD)

Lobachevsky University, Nizhni Novgorod 603022, Russian Federation
andrei.zorine@itmm.unn.ru

Abstract. In this paper we introduce a new queueing model with a special kind of input processes. It is assumed that the number of arrivals during consecutive time intervals makes an autoregressive sequence with conditional Poisson distributions. A single server serves input flows one by one in cyclic order with instantaneous switching. A d-limited policy is used. The mathematical model of the queueing process takes form of a multidimensional discrete Markov chain. The Markov chain keeps track of the server state, recent arrival numbers and queues' lengths. The necessary and sufficient condition for the existence of the stationary probability distribution is found. A possibility to give an explicit solution for the stationary equations for the probability generating functions is discussed.

Keywords: Autoregressive Poisson process · polling system · cyclic service · stationarity conditions · probability generating functions

Introduction

Studies of many real flows in telecommunication networks and vehicular control at junctions made it evident that a simple Poisson model or a renewal model [1] are often statistically inadequate. In the last five decades models with different kinds of dependence between some of the flow constituents. There are a least two options to add dependence to the mode. One can think of a random arrival rate. It leads to Cox's doubly stochastic flows [2], Markov-modulated flows of Neuts and Lucantoni [3]. On the other hand, dependence of the conditional probability distribution for inter-arrival time intervals on past arrivals can be introduced explicitely. On this way we come, for instance, to auto-regressive time series models formed by successive inter-arrival times (see [4]). In [6] following [5], a single-line queueing system with group arrivals is considered in which the group sizes make a certain Markov chain. Since any of the above-mentioned models watches after each single arrival time,

$$\tau_1' \leqslant \tau_2' \leqslant \dots, \tag{1}$$

© Springer Nature Switzerland AG 2022
A. Dudin et al. (Eds.): ITMM 2021, CCIS 1605, pp. 93–104, 2022.
https://doi.org/10.1007/978-3-031-09331-9_8

this approach can be called *local* [7].

In [7] a non-classical approach was proposed and started developing. According to this approach, the flow is observed only at special chosen epochs. At that, only a total random number of arrivals between two observation epochs becomes known. This approach is called *non-local*. Let us cite here an appropriate definition.

Definition 1. *Let* $0 = \tau_0^{(obs)} < \tau_1^{(obs)} < \ldots$ *be a point sequence on the axis* Ot *(here the superscript "obs" stands for "observation"), not coinciding with* (1), $\eta_i^{(obs)}$ *be a random number of requests from the flow* Π *during the time-interval* $\left(\tau_i^{(obs)}, \tau_{i+1}^{(obs)}\right]$, *and* $\nu_i^{(obd)}$ *be some characteristic(a mark) of those requests that arrive during the time-interval* $\left(\tau_i^{(obs)}, \tau_{i+1}^{(obs)}\right]$. *A random vector sequence*

$$\left\{ \left(\tau_i^{(obs)}, \eta_i^{(obs)}, \nu_i^{(obs)}\right); i = 0, 1, \ldots \right\}$$

is called a flow of non-homogeneous requests under its incomplete(non-local) description.

Informally speaking, our *non-local auto-regressive flow* is understood as a flow with a linear form $a\eta_{i-1}^{(obs)} + b$ for the regression equation of $\eta_i^{(obs)}$ onto $\eta_0^{(obs)}, \eta_1^{(obs)}, \ldots, \eta_{i-1}^{(obs)}$. For count time-series, this kind of stochastic processes was studied e.g. in [8].

The queueing system belongs to a class of polling systems [9]. Besides the inputs, it differs from classical polling systems by an assumption on the service process. Service time distributions are not known (in real queueing systems service times can be dependent and have different probability distributions), but the server's sojourn time distribution for each node is given together with the upper limit on the number of services customers. It models for example a roads intersection controlled by a fixed-cycle traffic-light, and data transmission nodes governed by a Round Robin algorithm.

We will demonstrate that even under simple assumptions on the queueing system structure the equation for the stationary probability distribution generating function is hard to solve. Still we will obtain conditions for the existence of the stationary probability distribution in the system using the iterative-dominating approach [10,11].

1   The Queueing System

Let us assume that all random variables and random elements in what follows are defined on a probability space $(\Omega, \mathfrak{F}, \mathbb{P})$. Then $\mathbb{E}(\cdot)$ denotes the mathematical expectation with respect to the probability measure $\mathbb{P}$. Set $\varphi(x; a) = a^x e^{-a}/x!$ for $a > 0$ and $x = 0, 1, \ldots$.

Consider a queueing system with $m < \infty$ input flows and a single server. Customers from the j-th flow join an infinite-capacity buffer O_j. Probability properties of the input flows will be defined later. The server spends a constant time $T > 0$ in front of each queue, and then an instant switch-over to the next queue occurs. After the last queue the first queue is visited. The server implements a d-limited policy: during its stay at the j-th queue the server can provide service to $d = \ell_j$ customers at most from that queue, no matter when exactly they arrived if they have arrived before the time T expired.

Let τ_0, $\tau_{i+1} = \tau_i + T = (i+1)T$, $i = 0, 1, \ldots$ be the time instants when the server switches to a next queue. Denote by $\Gamma^{(r)}$ the server state when is at the r-th queue, $i = 1, 2, \ldots, m$ and let $\Gamma = \{\Gamma^{(1)}, \Gamma^{(2)}, \ldots, \Gamma^{(m)}\}$ be the server state space. Let a random variable $\Gamma_i \in \Gamma$ be the server state during the time interval $(\tau_{i-1}, \tau_i]$ for $i = 1, 2, \ldots$, and $\Gamma_0 \in \Gamma$ be the random server state at time τ_0. Let $r \oplus 1 = r + 1$ for $r < m$ and $m \oplus 1 = 1$. Then $\Gamma_{i+1} = \Gamma_{i+1}(\omega) = \Gamma^{(r \oplus 1)}$ for all $\omega \in \Omega$ such that $\Gamma_i = \Gamma^{(r)}$.

Denote by $\eta_{j,i}$, $i = 1, 2, \ldots$ the random number of new customers arriving from the flow Π_j during the time interval $(\tau_i, \tau_{i+1}]$, $j = 1, 2, \ldots, m$. Let $\eta_{j,-1}$ be a non-negative integer-values random variable, $j = 1, 2, \ldots, m$. Let us assume that the conditional probability distribution of $\eta_{j,i+1}$ for any given $\eta_{j,-1} = x_{-1}$, $\eta_{j,0} = x_0, \ldots, \eta_{j,i} = x_i$ is the Poisson distribution with parameter $(a_j x_i + b_j)$ for some $a_j > 0$ and $b_j > 0$, so that the regression of $\eta_{j,i+1}$ on past numbers of arrivals equals

$$\mathbb{E}(\eta_{j,i+1} \mid \{\eta_{j,-1} = x_{-1}, \eta_{j,0} = x_0, \ldots, \eta_{j,i} = x_i\}) = a_j x_i + b_j.$$

We will call such an input flow an *autoregressive Poisson flow*. The previous number of arrivals, $\eta_{j,i-1}$, can be used as a mark of requests during the time-interval $(\tau_i, \tau_{i+1}]$. Then the non-local description of the autoregressive Poisson flow Π_j is a marked point process

$$\{(\tau_i, \eta_{j,i}, \eta_{j,i-1}); i = 0, 1, \ldots\}.$$

In particular, if the flow Π_j is a classical Poisson with intensity λ_j then we will have $a_j = 0$ and $b_j = \lambda_j T$. Further, let us assume that the stochastic sequences

$$\{\eta_{j,i}; i = -1, 0, \ldots\}, \qquad j = 1, 2, \ldots, m$$

are independent.

Denote by $\kappa_{j,i}$ the random number of customers in the queue O_j at time instant τ_i. Denote by $\xi_{j,i}$ the largest number of customers which can be serviced from O_j during the time interval $(\tau_i, \tau_{i+1}]$. Then the probability

$$\mathbb{P}(\{\xi_{1,i} = y_1, \xi_{2,i} = y_2, \ldots, \xi_{m,i} = y_m\} \mid \{\Gamma_i = \Gamma^{(r)}\})$$

equals 0 for $ty_j > 0$ and $y_k > 0$ for some $k \neq j$; it equals 1 for $y_{r \oplus 1} = \ell_j$. We have

$$\kappa_{j,i+1} = \max\{0, \kappa_{j,i} + \eta_{j,i} - \xi_{j,i}\}, \quad i = 0, 1, \ldots; j = 1, 2, \ldots, m. \qquad (2)$$

The recurrent equations and probability distributions given above prove the following claims.

Theorem 1. *For a given probability distribution of the vertor*

$$(\Gamma_0, \varkappa_{1,0}, \varkappa_{2,0}, \ldots, \varkappa_{m,0}, \eta_{1,-1}, \eta_{2,-1}, \ldots, \eta_{m,-1}),$$

random sequences

$$\{(\Gamma_i, \kappa_{1,i}, \kappa_{2,i}, \ldots, \kappa_{m,i}, \eta_{1,i-1}, \eta_{2,i-1}, \ldots, \eta_{m,i-1}); i = 0, 1, \ldots\},$$
$$\{(\Gamma_i, \varkappa_{j,i}, \eta_{j,i-1}); i = 0, 1, \ldots\}, \qquad j = 1, 2, \ldots, m$$

are irreducible periodic Markov chains.

2　Analysis of the Model

The main purpose of this section is to establish necessary and sufficient conditions for the existence of the stationary probability distribution of the Markov chain $\{(\Gamma_i, \varkappa_{j,i}, \eta_{j,i-1}); i = 0, 1, \ldots\}$ for $j = 1, 2, \ldots, m$, since it is easy to prove then, that the Markov chain

$$\{(\Gamma_i, \kappa_{1,i}, \kappa_{2,i}, \ldots, \kappa_{m,i}, \eta_{1,i-1}, \eta_{2,i-1}, \ldots, \eta_{m,i-1}); i = 0, 1, \ldots\}$$

has a stationary probability distribution if and only if each single

$$\{(\Gamma_i, \varkappa_{j,i}, \eta_{j,i-1}); i = 0, 1, \ldots\}, \qquad j = 1, 2, \ldots, m$$

does. In the remainder of this section the value of the index j is fixed.

In the first place, for the existence of the stationary distributions of the Markov chains, the inputs $\{\eta_{j,i}; i = 0, 1, \ldots\}$ need to have statioinary probability distribution. This is possible only if $0 < a_j < 1$ for all $j = 1, 2, \ldots, m$. We assume so in the rest of the section.

Let us define

$$Q_{j,i}(r, x, y) = \mathbb{P}(\{\Gamma_i = \Gamma^{(r)}, \kappa_{j,i} = x, \eta_{j,i-1} = y\}).$$

Let $I(\cdot)$ denote the indicator random variable for the event given in the parentheses. Let us introduce for $|z| \leqslant 1$, $|w| \leqslant 1$ and $i = 0, 1, \ldots$ the probability generating functions

$$\Psi_{j,i}(z, w; r) = \sum_{x=0}^{\infty} \sum_{y=0}^{\infty} z^x w^y Q_{j,i}(r, x, y) \mathbb{E}\left(z^{\kappa_{j,i}} w^{\eta_{j,i-1}} I(\{\Gamma_i = \Gamma^{(r)}\})\right).$$

Theorem 2. *The following recurrent equations with respect to* $i = 0, 1, \ldots$ *hold:*

$$\Psi_{j,i+1}(z, w; r \oplus 1) = e^{b_j(zw-1)} \Psi_{j,i}(z, e^{a_j(zw-1)}; r), \qquad r \oplus 1 \neq j;$$

$$\Psi_{j,i+1}(z, w; r \oplus 1) = z^{-\ell_j} e^{b_j(zw-1)} \Psi_{j,i}(z, e^{a_j(zw-1)}; r)$$
$$+ \sum_{x=0}^{\ell_j-1} \sum_{n=0}^{\ell_j-x-1} \left(\sum_{y=0}^{\infty} Q_{j,i}(r, x, y) \varphi(n; a_j y + b_j)\right)(1 - z^{x+n-\ell_j}) w^n$$

for $r \oplus 1 = j$.

Proof. Let $r \oplus 1 \neq j$. Then

$$\Psi_{j,i+1}(z, w; r \oplus 1) = \mathbb{E}[z^{\kappa_{j,i}+\eta_{j,i}} w^{\eta_{j,i}} I(\Gamma_i = \Gamma^{(r)})]$$
$$= \mathbb{E}\big(\mathbb{E}(z^{\kappa_{j,i}}(zw)^{\eta_{j,i}} I(\Gamma_i = \Gamma^{(r)}) \mid \kappa_{j,i}, \eta_{j,i-1}, \Gamma_i)\big)$$
$$= \mathbb{E}[z^{\kappa_{j,i}} e^{(a_j\eta_{j,i}+b_j)(zw-1)} I(\Gamma_i = \Gamma^{(r)})]$$
$$= e^{b(zw-1)}\Psi_{j,i}(z, e^{a(zw-1)}; r).$$

For $r \oplus 1 = j$,

$$\Psi_{j,i+1}(z, w; r \oplus 1) = \mathbb{E}[z^{\kappa_{j,i}+\eta_{j,i}-\ell_j} w^{\eta_{j,i}} I(\Gamma_i = \Gamma^{(r)})]$$
$$+ \mathbb{E}[(1 - z^{\kappa_{j,i}+\eta_{j,i}-\ell_j}) w^{\eta_{j,i}} I(\Gamma_i = \Gamma^{(r)}, \kappa_{j,i} + \eta_{j,i} < \ell_j)]$$
$$= z^{-\ell_j} e^{b_j(zw-1)} \Psi_{j,i}(z, e^{a_j(zw-1)}; r)$$
$$+ \mathbb{E}[(1 - z^{\kappa_{1,i}+\eta_{1,i}-\ell}) w^{\eta_{1,i}} I(\Gamma_i = \Gamma^{(m)}, \kappa_{1,i} + \eta_{1,i} < \ell)]$$
$$= z^{-\ell} e^{b(zw-1)} \Psi_1(z, e^{a(zw-1)}; m)$$
$$+ \sum_{x=0}^{\ell-1} \sum_{n=0}^{\ell-x-1} \Big(\sum_{y=0}^{\infty} Q_1(m, x, y) \frac{(ay+b)^n}{n!} e^{-(ay+b)}\Big)(1 - z^{x+n-\ell}) w^n.$$

Using methods from [10, 11] we get.

Theorem 3. *For the existence of the stationary probability distribution of the Markov chain $\{(\Gamma_i, \varkappa_{j,i}, \eta_{j,i-1}); i = 0, 1, \ldots\}$ it is necessary and sufficient that*

$$\frac{b_j}{1 - a_j} m < \ell_j. \tag{3}$$

The condition in the last theorem can be easily interpreted from a physical point of view because the quantity $mb_j(1 - a_j)^{-1}$ is the stationary expected number of arrivals from the flow Π_j during a complete cycle of the server.

In course of the proof of Theorem 3 the following Lemma is essential.

Lemma 1. *If $0 < a < 1$ then the equation $w = e^{a(wz-1)}$ has a unique solution*

$$w(z) = e^{-a} + \sum_{n=1}^{\infty} z^n \frac{(n+1)^{n-1} a^n e^{-(n+1)a}}{n!},$$

convergent in the open disk $|z| < a^{-1} e^{a-1}$, such that $w(1) = 1$, $|w(z)| < 1$ for $|z| < 1$.

Proof. Let's fix $|z| < 1$. We have an estimate from below for the magnitude of the complex quantity

$$|e^{a(zw-1)}| = e^{-a}|e^{azw}| \leqslant e^{-a} e^{a|w||z|} < e^{-a} e^{a|w|}.$$

So, on the circle $|w| = 1$ we have

$$|e^{a(zw-1)}| < e^{a(|w|-1)} = 1 = |w|. \tag{4}$$

By the classical Rouchè's theorem, for any such z there is a unique solution $w = w(z)$ of the equation $w = e^{a(wz-1)}$ such that $|w(z)| \leqslant 1$. It can be computed by evaluating the integral

$$w(z) = \frac{1}{2\pi i} \int\limits_{|w|=1} w \cdot \frac{F'_w(z,w)}{F(z,w)} \, dw = \frac{1}{2\pi i} \int\limits_{|w|=1} \frac{w - azwe^{a(zw-1)}}{w - e^{a(zw-1)}} \, dw,$$

where $i = \sqrt{-1}$ and $F(z,w) = w - we^{a(zw-1)}$. We only need to prove analyticity of $w(z)$ in the open unit disk.

Let $|w| = 1$ and $0 < r < 1$ be fixed. Let us consider the function $w(z)$ in a disk $|z| \leqslant r$. Since

$$|F(z,w)| \geqslant \left| |w| - |e^{a(wz-1)}| \right| \geqslant |w| - e^{a(|w||z|-1)} \geqslant 1 - e^{a(r-1)} > 0,$$

a function $wF'_w(z,w)/F(z,w)$ is analytic inside the open disk $|z| < r$ and with uniformly bounded absolute value as a ratio of two continuous functions in bath variables in a closed set $\{(z,w): |z| \leqslant r, |w| = 1\}$. A corollary from Vitali's theorem, the function $w(z)$ is an analytic function of z in the open disk $|z| < r$, and hence in the open disk $|z| < 1$.

From inequality (4) it follows that

$$\left| \frac{e^{a(wz-1)}}{w} \right| < 1,$$

so that the integral can be represented by a series:

$$\frac{1}{2\pi i} \int\limits_{|w|=1} \frac{w - azwe^{a(zw-1)}}{w - e^{a(zw-1)}} \, dw = \frac{1}{2\pi i} \int\limits_{|w|=1} \frac{1 - aze^{a(zw-1)}}{1 - \frac{e^{a(zw-1)}}{w}} \, dw$$

$$= \sum_{n=0}^{\infty} \frac{1}{2\pi i} \int\limits_{|w|=1} (1 - aze^{a(zw-1)}) \frac{e^{na(wz-1)}}{w^n} \, dw.$$

Using the Cauchy's Integral representation, we get

$$\frac{1}{2\pi i} \int\limits_{|w|=1} (1 - aze^{a(zw-1)}) \frac{e^{na(wz-1)}}{w^n} \, dw = 0 \qquad \text{for } n = 0,$$

$$\frac{1}{2\pi i} \int\limits_{|w|=1} (1 - aze^{a(zw-1)}) \frac{e^{na(wz-1)}}{w^n} \, dw$$

$$= \frac{1}{(n-1)!} \frac{d^{n-1}}{dw^{n-1}} (1 - aze^{a(zw-1)}) e^{na(wz-1)} \Big|_{w=0}$$

$$= \frac{1}{(n-1)!} ((naz)^{n-1} e^{-na} - az(a(n+1)z)^{n-1} e^{-(n+1)a}) \qquad \text{for } n \geqslant 1.$$

So,

$$\sum_{n=1}^{\infty} \frac{1}{(n-1)!}\left((naz)^{n-1}e^{-na} - az(a(n+1)z)^{n-1}e^{-(n+1)a}\right)$$

$$= e^{-a} + \sum_{n=1}^{\infty} z^n \left(\frac{(n+1)^n a^n e^{-(n+1)a}}{n!} - \frac{a^n (n+1)^{n-1} e^{-(n+1)a}}{(n-1)!}\right)$$

$$= e^{-a} + \sum_{n=1}^{\infty} z^n \frac{(n+1)^{n-1} a^n e^{-(n+1)a}}{n!}$$

The convergence radius R is found from

$$\frac{1}{R} = \lim_{n\to\infty} \sqrt[n]{\frac{(n+1)^{n-1}a^n e^{-(n+1)a}}{n!}} = \lim_{n\to\infty} \sqrt[n]{\frac{(n+1)^n a^n e^{-(n+1)a}}{(n+1)!}}$$

$$= \lim_{n\to\infty} \sqrt[n]{\frac{(n+1)^n a^n e^{-(n+1)a}}{\sqrt{2\pi(n+1)}(n+1)^{n+1}e^{-(n+1)}}} = ae^{1-a}.$$

Now let us prove that the series at $z = 1$ equals $w(1) = 1$. Any convergent series is a continuous function inside its disk of convergence. Here we focus on real values for z and $w > 0$. Then

$$z = \frac{1}{aw}(a + \ln w), \qquad \frac{dz}{dw} = \frac{1 - a - \ln w}{aw^2}.$$

In a neighborhood of $w = 1$ it is a continuous monotonously increasing function for $0 < w < e^{1-a}$ and it takes on value $z = 1$ at $w = 1$. Its inverse function takes on values $w < 1$ for $z < 1$, and it takes on value $w = 1$ for $z = 1$.

Proof (to Theorem 3). 1) Necessity. Let us assume that the stationary probability distribution exists. By substituting it in place of the initial probability distribution we guarantee the existence of limits

$$\lim_{i\to\infty} Q_{j,i}(r, x, y) = Q_j(r, x, y)$$

equal to the stationary probabilities. Let $r(j)$ be the solution to $r \oplus 1 = j$. To obtain equations for the time-stationary probability generating functions we can omit indices i and $i + 1$ in the equations in Theorem 3. Substituting there $w = w(z)$ from Lemma 1 where $a = a_j$ and $b = b_j$, and denoting

$$A_j(x, n) = \sum_{y=0}^{\infty} Q_j(r(j), x, y)\varphi(n; a_j y + b_j)$$

we get

$$\Psi_j(z, w(z); r \oplus 1) = e^{b_j(zw(z)-1)}\Psi_j(z, w(z); r), \qquad r \oplus 1 \neq j; \qquad (5)$$

$$\Psi_j(z, w(z); j) = z^{-\ell_j} e^{b_j(zw(z)-1)} \Psi_j(z, w(z); r(j))$$

$$+ \sum_{x=0}^{\ell_j-1} \sum_{n=0}^{\ell_j-x-1} A_j(x, n)(1 - z^{x+n-\ell_j})(w(z))^n. \tag{6}$$

Summation of Eqs. (5), (6) with respect to $r = 1, 2, \ldots, m$ results in

$$\sum_{r=1}^{m} \Psi_j(z, w(z); r) = \sum_{r \neq r(j)} e^{b_j(zw(z)-1)} \Psi_j(z, w(z); r) + z^{-\ell_j} e^{b_j(zw(z)-1)}$$

$$\times \Psi_j(z, w(z); r(j)) + \sum_{x=0}^{\ell_j-1} \sum_{n=0}^{\ell_j-x-1} A_j(x, n)(1 - z^{x+n-\ell_j})(w(z))^n. \tag{7}$$

In the left neighborhood of $z = 1$ (on the real axis) we have Taylor expansions

$$e^{b_j(zw(z)-1)} = 1 + (b_j + b_j w'(1))(z - 1) + o((z - 1))$$

$$= 1 + \frac{b_j(z - 1)}{1 - a_j} + o(z - 1),$$

$$z^{-\ell_j} e^{b_j(zw(z)-1)} = 1 + \left(\frac{b_j}{1 - a_j} - \ell_j\right)(z - 1) + o(z - 1),$$

$$(1 - z^{x+n-\ell_j})(w(z))^n = \left(\ell_j - x - n + \frac{na_j}{1 - a_j}\right)(z - 1) + o(z - 1).$$

There expansions substituted into (7), we get after collecting terms

$$0 = \sum_{r \neq r(j)} \frac{b_j(z - 1)}{1 - a_j} \Psi_j(z, w(z); r) + \left(\frac{b_j}{1 - a_j} - \ell_j\right)(z - 1)\Psi_j(z, w(z); r(j))$$

$$+ \sum_{x=0}^{\ell_j-1} \sum_{n=0}^{\ell_j-x-1} A_j(x, n)\left(\ell_j - x - n + \frac{na_j}{1 - a_j}\right)(z - 1) + o(z - 1). \tag{8}$$

Divide by $(z - 1)$ and send z to 1 from the left. We get

$$0 = \sum_{r \neq r(j)} \frac{b}{1 - a} \Psi_j(1, 1; r) + \left(\frac{b}{1 - a} - \ell_j\right)\Psi_j(1, 1; r(j))$$

$$+ \sum_{x=0}^{\ell-1} \sum_{n=0}^{\ell-x-1} A_j(x, n)\left(\ell_j - x - n + \frac{na}{1 - a}\right). \tag{9}$$

Substituting $z = 1$ into (5) and (6) leads to $\Psi_j(1, 1; r) = m^{-1}$ for all $r = 1, 2,$ $\ldots, m$. So, we finally come to

$$0 = \frac{b}{1 - a} - \frac{\ell_j}{m}$$

$$+ \sum_{x=0}^{\ell-1} \sum_{n=0}^{\ell-x-1} \left(\sum_{y=0}^{\infty} Q_1(m, x, y)\frac{(ay + b)^n}{n!} e^{-(ay+b)}\right)\left(\ell_j - x - n + \frac{na}{1 - a}\right). \tag{10}$$

Taking into account that $\ell_j - x - n + \frac{na}{1-a} > 0$ for those x and n which occur at summation, we draw the conclusion that for the existence of a stationary probability distribution it is necessary that

$$\frac{b}{1-a} - \frac{\ell_j}{m} < 0.$$

2) Sufficiency. Let us assume for now that Inequality (3) is true, but no stationary probability distribution exists. All the states of the Markov chain are essential and belong to a single class of communicating states, one must have

$$\lim_{i \to \infty} Q_{j,i}(x, y; r) = 0$$

for all x, y, and r, It follows then that the sequence of mathematical expectations $\mathbb{E}\kappa_{j,i}$ $i = 0, 1, \ldots$ unboundly grows. We claim that, on the contrary, the mathematical expectations are bounded if the condition from the theorem holds.

Let us setup the initial probability distribution so that the probability generating functions $\Psi_{j,0}(z, w; r)$ are analytic in $(z, w) \in \mathbb{C}^2$. Then all the next probability generating functions $\Psi_{j,i}(z, w; r)$, $i = 1, 2, \ldots$ can have analytical continuations onto whole $\mathbb{C}^2$. Consequently, the functions $\Psi_{j,i}(z, w(z); r)$ will be analytic in the disk $|z| < 1 + \varepsilon < 1/(ae^{1-a})$ (i.e. inside the disk of convergence of the series $w(z)$) and will satisfy equations

$$\Psi_{j,i+1}(z, w(z); r \oplus 1) = z^{-\ell_j} e^{b_j(zw(z)-1)} \Psi_{j,i}(z, w(z); r), \quad r \oplus 1 \neq j;$$

$$\Psi_{j,i+1}(z, w(z); j) = z^{-\ell_j} e^{b_j(zw(z)-1)} \Psi_{j,i}(z, w(z); r(j))$$
$$+ \sum_{x=0}^{\ell_j-1} \sum_{n=0}^{\ell_j-x-1} A_j(x, n)(1 - z^{x+n-\ell_j})(w(z))^n.$$

Let us fix a z, $1 < z < 1 + \varepsilon$ and let $r \oplus m = r = j$. The one has $(A_j(x, n) \leqslant 1)$:

$$\Psi_{j,i+m}(z, w(z); r \oplus m) \leqslant z^{-\ell_j} e^{b_j(zw(z)-1)} \Psi_{j,i+m-1}(z, w(z); r \oplus (m-1))$$
$$+ \sum_{x=0}^{\ell_j-1} \sum_{n=0}^{\ell_j-x-1} (1 - z^{x+n-\ell_j})(w(z))^n$$
$$= \left(z^{-\ell_j} e^{b_j(zw(z)-1)}\right)^2 \Psi_{j,i+m-2}(z, w(z); r \oplus (m-2))$$
$$+ \sum_{x=0}^{\ell_j-1} \sum_{n=0}^{\ell_j-x-1} (1 - z^{x+n-\ell_j})(w(z))^n = \ldots$$
$$= \left(z^{-\ell_j} e^{b_j(zw(z)-1)}\right)^m \Psi_{j,i}(z, w(z); r) + \sum_{x=0}^{\ell_j-1} \sum_{n=0}^{\ell_j-x-1} (1 - z^{x+n-\ell_j})(w(z))^n.$$

Since

$$\frac{d}{dz}\left(z^{-\ell_j} e^{b_j(zw(z)-1)}\right)^m \bigg|_{z=1} = -m\ell_j + \frac{mb_j}{1-a_j} < 0,$$

the sequence

$$\Psi_0^+ = \Psi_{j,0}(z, w(z); j), \quad \Psi_1^+ = \Psi_{j,1}(z, w(z); j), \quad \ldots, \Psi_{m-1}^+ = \Psi_{j,m-1}(z, w(z); j),$$

$$\Psi_{i+m}^+ = z^{-m\ell_j} e^{mb_j(zw(z)-1)} \Psi_i^+ + \sum_{x=0}^{\ell_j-1} \sum_{n=0}^{\ell_j-x-1} (1 - z^{x+n-\ell_j})(w(z))^n, \quad i = 0, 1, \ldots$$

converges, and hence is bounded. At the same time, for all $i = 0, 1, \ldots$ we have

$$\Psi_{j,i}(z, w(z); r) \leqslant \Psi_i^+.$$

It follows that all numbers (for this z) $\Psi_{j,i}(z, w(z); r)$, $r = 1, 2, \ldots, m$, and $i = 0, 1, \ldots$ are bounded by some constant $C > 0$. Then,

$$\mathbb{E}(\kappa_{j,i}) = \left| \frac{1}{2\pi \mathbf{i}} \int_{|\zeta-1|=\delta} \frac{\sum_{r=1}^m \Psi_{j,i}(z, 1; r)}{(z-1)^2} \, dz \right|$$

$$\leqslant \int_0^1 \frac{\sum_{r=1}^m \Psi_{j,i}(1+\delta, 1+\delta; r)}{\delta} \, du \leqslant \frac{mC}{\delta}.$$

This contradiction prove the claim.

To solve Eqs. (5), (6) for the functions $\Psi_j(z, w(z); r)$, $r = 1, 2, \ldots, m$, one needs to identify $\ell_j(\ell_j + 1)/2$ unknown constants $A(x, n)$, $0 \leqslant x + n < \ell_j$, n, x integers. We get

$$(z^{\ell_j} - e^{mb_j(zw(z)-1)})\Psi_j(z, w(z); r)$$

$$= \sum_{x=0}^{\ell_j-1} \sum_{n=0}^{\ell_j-x-1} A_j(x, n)(z^{\ell_j} - z^{x+n})(w(z))^n, \quad r \oplus 1 = j.$$

Case 1. If $\ell_j = 1$, then the only unknown constant is $A_j(0, 0)$. Recalling that $\Psi_j(z, z; r) = 1/m$ and expanding terms $z - e^{mb_j(zw(z)-1)}$, $(1 - z^{-1})$ in the left neighborhood of $z = 1$, we get

$$\left(1 - \frac{mb_j}{1 - a_j}\right) \frac{1}{m} = A_j(0, 0).$$

Case 2. If $\ell_j > 1$, we have $\ell_j(\ell_j + 1)/2 > 1$ unknown constants. Let us study the equation

$$z^{\ell_j} - e^{b_j(zw(z)-1)} = 0.$$

It follows from the modified Rouché theorem [12] and the Lemma below that it has exactly $\ell_j - 1$ zeros inside the unit disk $|z| < 1$ when the stationarity condition (3) is fulfilled.

Lemma 2. *If inequality (3) is fulfilled, then* $|e^{b_j(zw(z)-1)}| < 1$ *for all* $|z| = 1$, $z \neq 1$.

Proof. Let $z = e^{iu}$, $w(z) = Re^{i\varphi}$, $0 \leqslant u < 2\pi$, $0 \leqslant \varphi < 2\pi$. Then

$$Re^{i\varphi} = e^{a(e^{iu} \cdot Re^{i\varphi} - 1)}.$$

Its right-hand side equals $e^{a(Re^{i(u+\varphi)}-1)}$. By comparing moduli, we get

$$R = e^{aR\cos(u+\varphi)-a}.$$

We have $R = 1$ if and only if $a\cos(u + \varphi) - a = 0$, whence $\cos(u + \varphi) = 1$. But then $\sin(u + \varphi) = 0$ and it's the argument value φ of the complex number $Re^{i\varphi}$. Finally, from $1 = \cos(u + \varphi) = \cos u$ we get $u = 0$.

Denote these zeros by $\beta_1, \beta_2, \ldots, \beta_{\ell_j-1}$.

Theorem 4. *If inequality* (3) *is fulfilled then the following equations take place:*

$$\sum_{x=0}^{\ell_j-1} \sum_{n=0}^{\ell_j-x-1} A_j(x,n)(\ell_j - x - n) = \frac{\ell_j}{m} - \frac{b_j}{1 - a_j},$$

$$\sum_{x=0}^{\ell_j-1} \sum_{n=0}^{\ell_j-x-1} A_j(x,n)((\beta_k)^{\ell_j} - (\beta_k)^{x+n})(w(\beta_k))^n = 0, \quad k = 1, 2, \ldots \ell_j - 1.$$

The number of linear equations given by Theorem 4 is less than the number of unknown constants. Still, it was to be expected, since Eqs. (5) and (6) are not equivalent to equations of Theorem 2 and by substituting $w = w(z)$ there we lose evidently essential parts of information about the generating functions of interest. Moreover, once we obtain all $A_j(x,n)$, $0 \leqslant x + n < \ell_j$, we still need to solve a functional equation relating $\Psi_j(z, w; r \oplus 1)$ to $\Psi_j(z, e^{a_j(zw-1)}; r)$ in the polydisk $\{(z, w) : |z| \leqslant 1, |w| \leqslant 1\} \subset \mathbb{C}^2$.

Since the main functional transform (2) for a queue length is produces a random walk with reflection at zero, a many times studied (under a variety of assumptions) process, it is of interest to compare the assumptions on the input processes, such as input flows and service processes, in our work and in other classical works. Usually (c.f. [13]) it is assumed that the sequence (in our notation)

$$\{\eta_{j,i} - \xi_{j,i}; i = 0, 1, \ldots\}$$

is a stationary process. In our case, it would imply not only that the input sequence $\{\eta_{j,i}; i = -1, 0, 1, \ldots\}$ is stationary, but also that the initial server state, Γ_0, is random with the uniform probability distribution on Γ. Our exposition leaves more freedom for the input flow and the initial server state.

3 Conclusion

It was shown in this work that discrete-time models of queueing systems with auto-regressive input process may lead to a challenging problem in the domain of several complex variables in terms of multivariate probability generating functions. This problem still wait for its solution. Nethertheless, the necessary and sufficient conditions on the parameters of the queueing system which guarantee the existence of the stationary probability distribution can be found by careful analysis of these (yet unsolved) equations. For the polling queueing system under study, these conditions are easily verifiable and have natural physical interpretation in terms of mean values for the basic quantities like numbers of arrivals and saturation flow intensity.

References

1. Klimow, G.P.: Bedienungsprozesse. Springer, Basel (1979)
2. Grandell, J.: Doubly Stochastic Poisson Processes. Springer, Berlin, Heidelberg, New York (1976). https://doi.org/10.1007/BFb0077758
3. Lucantoni, D.M.: New results on the single server queue with a batch Markovian arrival process. Commun. Stat.: Stochast. Models **1**(7), 1–46 (1991)
4. Jagerman, D. L., Melamed, B., Willinger, W.: Stochastic modeling of traffic processes. In: Dshalalow, J.H. (ed.) Frontiers in Queueing: Models and Applications in Science and Engineering, pp. 271–320 (1997)
5. Hwang, G.U., Sohraby, K.: On the exact analysis of a discrete-time queueing system with autoregressive inputs. Queueing Syst. **1–2**(43), 29–41 (2003)
6. Leontyev, N.D., Ushakov, V.G.: Analysis of a queueing system with autoregressive arrivals. Inform. Appl. **3**(8), 39–44 (2014)
7. Fedotkin, M. A.: Incomplete description of flows of non-homogenous requests. In: Queueing Theory. Moscow, MGU-VNIISI, 113–118 (1981). (in Russian)
8. Heinen, A.: Modelling Time Series Count Data: An Autoregressive Conditional Poisson Model (2003). https://dx.doi.org/10.2139/ssrn.1117187
9. Vishnevskii, V.M., Semenova, O.V.: Mathematical methods to study the polling systems. Autom. Remote. Control. **2**(67), 173–220 (2006)
10. Zorine, A.: Stochastic model for communicating retrial queuing systems with cyclic control in random environment. Cybern. Syst. Anal. **6**(49), 890–897 (2013)
11. Zorine, A.V.: Study of a service process by a loop algorithm by means of a stopped random walk. Commun. Comput. Inf. Sci. **1109**, 121–135 (2019)
12. Klimenok, V.: On the modification of Rouche's theorem for the queueing theory problems. Queueing Syst. **38**, 431–434 (2001)
13. Borovkov, A.A.: Stochastic Processes in Queueing Theory. Springer, New York (1976). https://doi.org/10.1007/978-1-4612-9866-3

Evaluation of the New and Accepted Customers Blocking Probabilties in a Network of Resource Loss Systems

Eduard Sopin[1,3] , Vyacheslav Begishev[1(✉)] , Dmitri Moltchanov[2] ,
and Andrey Samuylov[1]

[1] Peoples' Friendship University of Russia (RUDN University),
Moscow, Russian Federation
{sopin-es,begishev-vo,samuylov-ak}@rudn.ru
[2] Tampere University, Tampere, Finland
dmitri.moltchanov@tuni.fi
[3] Institute of Informatics Problems, FRC CSC RAS, Moscow, Russian Federation

Abstract. The paper considers a network of resource loss systems (ReLS) with random resource requirements and two types of nodes. Customers initially arrive to the first type of nodes, where they receive service for exponentially distributed time. The service of customers can be interrupted. In this case, they are rerouted to the second type of nodes, where they receive service for an exponentially distributed time. Once the service is completed, they return back to the original node and continue its service. Customers require a random volume of limited resources. If there are not enough of unoccupied resources upon the arrival of a customer, then it is considered lost. Similarly, if an accepted customer is rerouted to another node and finds that there are not enough of resources to meet its requirements, then it is also lost. In this paper, we provide an approach to analyze the stationary behavior of the considered system, as well as establish expressions for the new customer loss probability and the accepted customer loss probability. The developed model has a wide range of applications in performance evaluation of fifth generation (5G) New Radio (NR) access networks. To this aim, we investigate the response of the considered service system in detail by revealing critical dependencies and trade-offs between input system parameters and performance measures of interest.

Keywords: Resource loss system · queuing network · loss probability · wireless networks · 5G NR · multiconnectivity

Introduction

The introduction of fifth generation (5G) systems promises to deliver not only extraordinary access rates but also the quality of service (QoS) guarantees to

The research was funded by the Russian Science Foundation, project no. 21-79-10139, https://rscf.ru/en/project/21-79-10139/.

the air interface. This will potentially enable applications requiring extremely high and constant bit rate (CBR) service such as augmented/virtual realities (AR/VR), holographic telepresence, 16/8K streamed video, and tactile Internet [11,14]. This is a principle paradigm shift as compared to fourth generation (4G) long-term evolution (LTE) systems that require appropriate mechanisms at the radio interface allowing to conceal the effects of wireless transmission medium.

In addition to much higher propagation losses in millimeter wave frequency (mmWave) band, 5G New Radio (NR) systems will be subject to outages caused by a dynamic blockage by human bodies [6] or buildings [1,5]. This behavior heavily affects the QoS characteristics provided to users and may even cause service interruptions. To alleviate the effect of blockage, the authors in [3] proposed to reserve a fraction of bandwidth at the serving base station (BS). However, this approach can only be utilized when users do not experience outage conditions in case of blockage. Alternatively, one may use recently standardized 3GPP multiconnectivity functionality [8,16]. According to it, user equipment (UE) is allowed to maintain more than a single link to nearby BSs and switch them in case of outage events by dynamically rerouting the traffic between locally available 5G NR BSs. Performance characterization of this mechanism as well as joint implementation to resource reservation and multiconnectivity naturally calls for queuing network formalism.

The first study that utilized a network of ReLS is [12], where a continuous-time ReLS has been applied to assess the performance of 5G mmWave NR deployments with multiconnectivity operation. Later on, a discrete variant of ReLS has been applied in [4]. Other applications of the ReLS in 5G NR and sixth generation (6G) systems operating in terahertz (THz) frequency band are detailed in [9]. In [10], the general approach for the analysis of networks of ReLS was described. In this paper, we analyze the network of ReLS that can be used to model muticonnectivity operation in 5G NR systems, define performance metrics, and provide an iterative approximate algorithm.

The rest of the paper is organized as follows. We introduce our model in Sect. 1. The analysis is performed in Sect. 2. Numerical results are provided and discussed in Sect. 3. Finally, conclusions are drawn in the last section.

1   Model Description

We consider a network of resource loss systems (ReLS) with two types of nodes. There are $N - 1$ first type nodes and one node of the second type. Each node has K_i servers and R_i resources, $i = 1, 2, ..., N$ (see Fig. 1). Customers arrive according to the Poisson process with intensities λ_1 and λ_2 to the first and second type nodes, respectively. The service times are exponentially distributed with parameters μ_1 and μ_2. Each customer requires not only a free server, but also a random discrete volume of resources, which are determined according to the probability distributions $\{f_{1,j}\}$ and $\{f_{2,j}\}$. Besides, each customer currently served at the first type of nodes is associated with a Poisson flow of signals with intensity α that causes rerouting of the customers to the node N. Rerouted customers stay at node N for exponentially distributed time with parameter β and return back to its original node, or leave the system if its service is completed.

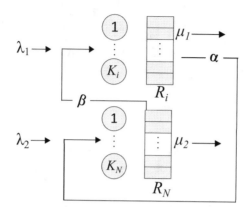

Fig. 1. Illustration of the considered queuing system: there are $N-1$ first type of nodes and one second type of node; the service intensity at first type of nodes is the same μ_1.

To analyze this model we follow the decomposition approach, which is a powerful methodology for queuing networks [7]. The core assumption here is that the service process at each BS in the network is independent of the service processes at other nodes. The relation between the service processes at the nodes of both types is incorporated into the numerical solution algorithm, where the characteristics of the entire network are calculated iteratively until the procedure converges. The stability properties of this class of models were analyzed in [2].

2   Model Analysis

In this section, we analyze the presented model. We start with an analysis of the service process at individual nodes of the first and second type and then proceed to derive the metrics of interest.

2.1   Service Process at the First Type Nodes

Consider the first type of nodes. Due to the memoryless property of the exponential distribution, the residual service time of returning sessions is also exponential with the same parameter μ_1. Let φ_i be the intensity of the returning session arrivals at node i, $i = 1, 2, ..., N-1$. Then the total session arrival intensity to node i is thus $\lambda_1 + \varphi_i$ and the total intensity of departures is $\mu_1 + \alpha$. The stochastic behavior of the node i can be described by the Markov process $X_i(t) = \{\xi_i(t), \delta_i(t)\}$, where $\xi_i(t)$ is the number of customers at node i at time

t and $\delta_i(t)$ is the total volume of occupied resources. Denote the stationary probabilities $q_{i,n}(r)$ as

$$q_{i,n}(r) = \lim_{t\to\infty} P\{\xi_i(t) = n, \delta_i(t) = r\}, \quad n = 0, 1, 2\ldots, K_i, \tag{1}$$

$$r = 0, 1, 2\ldots, R, \quad i = 1, 2, \ldots, N-1.$$

The process $X_i(t)$ describes the considered ReLS. According to [15], the stationary distribution (1) is given by

$$q_{i,0} = \left(1 + \sum_{n=1}^{K_i} \frac{\rho_i^n}{n!} \sum_{r=0}^{R_i} f_{1,r}^{(n)}\right)^{-1}, \tag{2}$$

$$q_{i,n}(r) = q_{i,0}\frac{\rho_i^n}{n!}f_{1,r}^{(n)}, \quad n = 1, 2, \ldots, K_i, \tag{3}$$

where $\rho_i = (\lambda_1 + \varphi_i)/(\mu_1 + \alpha)$ and $f_{1,j}^{(n)}, j \geq 0$ is the n-fold convolution of pmf $\{f_{1,j}\}, j \geq 0$. Note that the probability $f_{1,j}^{(n)}$ can be interpreted as the probability that n sessions on a first type node totally occupy j resources. Practically, the convolutions of discrete distributions may be evaluated using the following iterative procedure

$$f_{l,j}^{(n)} = \sum_{r=0}^{j} f_{l,r}f_{l,j-r}^{(n-1)}, \quad l = 1, 2, j \geq 0, n \geq 2, \tag{4}$$

where $f_{l,j}^{(1)} = f_{l,j}, j \geq 0$.

2.2   Service Process at the Second Type Node

The behavior of the second type of nodes (node N) can also be described in terms of the queuing systems with random resource requirements. As at nodes $1, 2, ..., N-1$, there are also two types of arrivals: customers that arrive initially to the node N with the intensity λ_2 and customers that are rerouted from the first type of nodes with intensity φ_N. However, in this case, the service times differ from each other: the service intensity of the initially arriving customers is μ_2, and for the rerouted customers it is $\mu_1 + \beta$. The arrival intensity φ_N for the rerouted customers is obtained by summing up all the rerouting intensities of the first type nodes, i.e.,

$$\varphi_N = \sum_{i=1}^{N-1} (\lambda_1 + \varphi_i)\frac{\alpha}{\mu_1 + \alpha}. \tag{5}$$

In (5), the term $\alpha/(\mu_1 + \alpha)$ refers to the probability that a customer from a first type node is rerouted to node N before its service completion. The intensities $\varphi_i, i = 1, 2, ..., N-1$, of customers returning back to their original node has the following form

$$\varphi_i = (\lambda_1 + \varphi_i)(1 - \pi_{i,1})\frac{\alpha}{\alpha + \mu_1}(1 - \pi_{N,1})\frac{\beta}{\beta + \mu_1}, \tag{6}$$

where $\pi_{i,1}$ is the loss probability of arriving customers at node i, $i = 1, 2, ..., N$ and $\beta/(\beta + \mu_1)$ is the probability that a rerouted customer at node N returns to its original node before service completion.

Observe that (6) implies that the flow of rerouted customers at node i, $i = 1, 2, ..., N-1$ equals to the fraction of the accepted flow, that was initially routed to the node N with the probability $\alpha/(\alpha + \mu_1)$, then accepted by the node N with the probability $1 - \pi_{N,1}$, and finally rerouted back with the probability $\beta/(\beta + \mu_1)$.

The expression for the stationary probabilities $q_{N,n_1,n_2}(r_1, r_2)$ that there n_1 first type of customers that totally occupy r_1 resources and n_2 second type of customers occupying r_2 resources also has the product form

$$q_{N,n_1,n_2}(r_1, r_2) = q_{N,0,0} \frac{\rho_{N,1}^{n_1}}{n_1!} \frac{\rho_{N,2}^{n_2}}{n_2!} f_{2,r_1}^{(n_1)} f_{2,r_2}^{(n_2)}, \tag{7}$$

$$q_{N,0,0} = \left(1 + \sum_{1 \leq n_1+n_2 \leq K_N} \frac{\rho_{N,1}^{n_1}}{n_1!} \frac{\rho_{N,2}^{n_2}}{n_2!} \sum_{0 \leq r_1+r_2 \leq R_N} f_{2,r_1}^{(n_1)} f_{2,r_2}^{(n_2)} \right)^{-1}, \tag{8}$$

where $\rho_{N,1} = \frac{\lambda_2}{\mu_2}$ and $\rho_{N,2} = \frac{\varphi_N}{\mu_1+\beta}$.

According to [15], the ReLS with two arrival flows can be analyzed similarly to ReLS with one aggregated arrival flow. Thus, the stationary probabilities $q_{N,n}(r)$ that there are n customers of any types in the system that totally occupy r resources have the following form

$$q_{N,n}(r) = q_{N,0} \frac{\rho_N^n}{n!} f_{2,r}^{(n)}, \tag{9}$$

$$q_{N,0} = \left(1 + \sum_{n=1}^{K_N} \frac{\rho_N^n}{n!} \sum_{r=0}^{R_N} f_{2,r}^{(n)} \right)^{-1}, \rho_N = \rho_{N,1} + \rho_{N,2}, \tag{10}$$

that can be evaluated numerically, see, e.g., [13].

2.3   Solution and Performance Metrics

Having obtained the stationary state probabilities for all the nodes, one may proceed with deriving the performance metrics. Recall that our solution is iterative in nature as one needs to add another layer of rerouted customers at each iteration until a parameter converges to its stable value with a given accuracy. The procedure is terminated once the required precision level is achieved. Particularly, at the first iteration, there are no rerouted customers, and thus $\varphi_i = 0$, $i = 1, 2, ..., N$. Then, the algorithm continues as follows:

1. Based on the system parameters, new customer loss probabilities $\pi_{i,1}$ at nodes $i = 1, 2, ..., N$ and arrival intensity of rerouted customers φ_N at node N are evaluated.

2. New values of φ_i, $i = 1, 2, ..., N - 1$ are calculated according to (6) by substituting their previous values into the right-hand side; if the difference between the new and the previous value meets the required precision, the algorithm proceeds with 3). Otherwise, it returns to 1).
3. When φ_i, $i = 1, 2, ..., N - 1$ converges to a stable value with the desired accuracy, all other performance metrics are evaluated.

The iterative solution outlined above requires new session drop probabilities. These can be evaluated as follows

$$\pi_{i,1} = 1 - q_{i,0} \sum_{n=0}^{K_i-1} \frac{\rho_i^n}{n!} \sum_{r=0}^{R_i} f_{1,r}^{(n+1)}, i = 1, 2, ..., N - 1 \tag{11}$$

$$\pi_{N,1} = 1 - q_{N,0} \sum_{n=0}^{K_N-1} \frac{\rho_N^n}{n!} \sum_{r=0}^{R_N} f_{2,r}^{(n+1)}. \tag{12}$$

Calculation of the probability that an accepted customer is lost is more involved process. Let us introduce the conditional probability Π_i, $i = 1, 2, ..., N - 1$ that a customer originally arriving and accepted at node i and is lost, given that it is rerouted, i.e.,

$$\Pi_i = \pi_{N,1} + (1 - \pi_{N,1})\frac{\beta}{\beta + \mu_1}\pi_{i,1}, \tag{13}$$

where the first term corresponds to the case of customer loss at entering the node N, while the second term is the probability that the rerouted customer is accepted at node N but then lost upon its return to the original node due to insufficient amount of available resources.

The average number of accepted customers lost as a result of rerouting during a time interval of length T is $\alpha \tilde{N}_i \Pi_i T$, where $\tilde{N}_i$ is the mean number of customers at node $i = 1, 2, ..., N - 1$. The mean number of customers that are accepted during the same time interval is $\lambda_1(1 - \pi_{i,1})T$. Hence, the probability that a customer, that was initially accepted at the node i, is eventually dropped is

$$\pi_{i,2} = \frac{\alpha \tilde{N}_i \Pi_i}{\lambda_1(1 - \pi_{i,1})}. \tag{14}$$

Finally, the average number of occupied resources b_i, $i = 1, 2, ..., N$ at node i has the following form

$$b_i = q_{i,0} \sum_{n=1}^{K_i} \frac{\rho_i^n}{(n-1)!} \sum_{r=0}^{R_i} f_{1,r}^{(n)}, \quad i = 1, 2, ..., N - 1 \tag{15}$$

$$b_N = q_{N,0} \sum_{n=1}^{K_N} \frac{\rho_N^n}{(n-1)!} \sum_{r=0}^{R_N} f_{2,r}^{(n)}. \tag{16}$$

3   Application to the Analysis of NR/LTE Deployment

In this section, we numerically investigate the performance response of the considered system. Specifically, we investigate the behavior of new and accepted customer drop probabilities as a function of the customer arrival intensities, the signal arrival intensities, and the resource requirements distribution parameters.

We consider two nodes each having its own customer arrival intensity, λ_1 and λ_2. Customers at the first node can be interrupted by the external signals with intensity α. Note that the scenario considered below can be interpreted as the service process of user sessions at dual mmWave NR (or alternatively THz) and LTE deployment, where sessions currently served at mmWave BS can be temporarily offloaded to LTE BS when the line-of-sight (LoS) path gets blocked to ensure session continuity. In this scenario, the service process of user sessions on the mmWave NR BS is modeled by the first type nodes, while the second type node models the service process at the LTE BS. The default system parameters are provided in Table 1.

Table 1. Parameters for numerical assessment.

Parameter	Value
Number of servers, $K_i, i = 1, 2, ..., N$	50
Number of resources $R_i, i = 1, 2, ..., N$	100
The intensity of the arrival of the first type of customers, λ_1	$[0.2, .., 0.65]$
The intensity of the arrival of the priority type of customers, λ_2	$[0.2, .., 0.65]$
Customer service intensity, $\mu_1 = \mu_2$	$1/30$
The intensity of signal, α	0.2
Interruption intensity, β	10

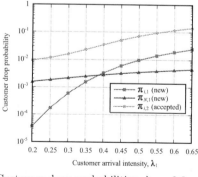

(a) Customer drop probabilities, $\lambda_2 = 0.3$

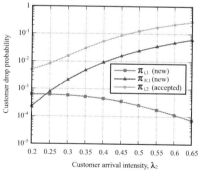

(b) Customer drop probabilities, $\lambda_1 = 0.3$

Fig. 2. New and accepted customer drop probabilities.

We start with new and accepted customer drop probabilities as a function of arrival intensity at NR and LTE BSs as illustrated in Fig. 2(a). By analyzing the presented data one may observe that for considered system parameters, the new customer drop probability at both systems increases with the increase in λ_1. Logically, the increase in λ_1 much heavily affects new and accepted customer drop probability at NR as compared to LTE. However, the new customer drop probability at LTE also increases and this is mainly caused by the temporal offloading of customers from NR system. Analyzing the trends dictated by the increase in λ_2, we observe slightly different behavior. First of all, expectedly, the new LTE customer drop probability increases. However, this effect causes the increase in the accepted NR customer drop probability. The latter phenomenon makes more resources available for newly arriving NR customers decreasing the corresponding probability as seen in Fig. 2(b). Thus, we may conclude that the session arriving process at the system utilized for temporal offloading (e.g., LTE) may drastically affect user session performance of the system subject of outages caused by blockage phenomenon (e.g., mmWave or THz).

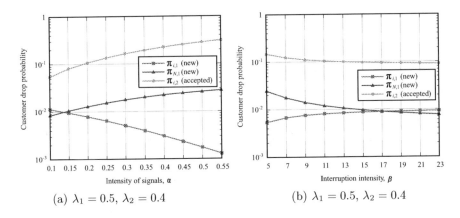

(a) $\lambda_1 = 0.5$, $\lambda_2 = 0.4$ (b) $\lambda_1 = 0.5$, $\lambda_2 = 0.4$

Fig. 3. New and accepted customer drop probabilities.

We now proceed considering the effect of the interruption process. To this aim, Fig. 3 illustrates the new and accepted customer drop probabilities as a function of both intensity of signals and interruption intensity. By analyzing the presented results, we observe that the accepted customer drop probability is heavily affected by the signal intensity. As these sessions are offloaded onto LTE, the corresponding new customer drop probability drastically increases. At the same time, the new customer drop probability at NR decreases. The overall effect is extremely negative from the QoS perspective – under high values of α most NR sessions are accepted for service and then eventually lost. The overall effect of the interruption intensity is also interesting. As β increases, the NR accepted customer drop probability decreases. The rationale is that it also results in a lower intensity of switching from one system to another and, thus, smaller chances to lose the session accepted for service. Logically, we also observe that this effectively lowers the new customer drop probability at LTE and increases the new customer drop probability at NR. However, we note that in real deployments we cannot affect the value of β as it is given by the environmental conditions, particularly, by the density of blockers and their movement patterns.

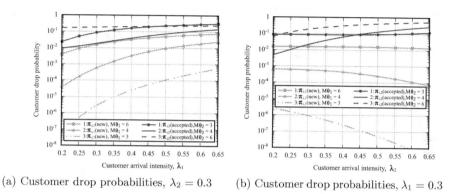

(a) Customer drop probabilities, $\lambda_2 = 0.3$ (b) Customer drop probabilities, $\lambda_1 = 0.3$

Fig. 4. New and accepted customer drop probabilities.

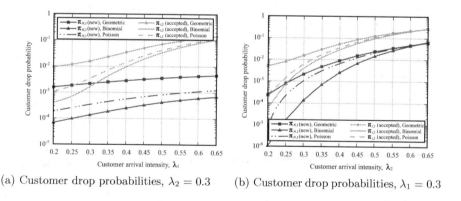

(a) Customer drop probabilities, $\lambda_2 = 0.3$ (b) Customer drop probabilities, $\lambda_1 = 0.3$

Fig. 5. New and accepted customer drop probabilities.

The amount of resources requested by a customer is one of the critical parameters for resource queuing systems. We now investigate it by analyzing the effect of the resource requirements parameters. Specifically, Fig. 4 shows the new and accepted customer drop probabilities for different values of the mean number of resources $E[\theta_1]$ and $E[\theta_2]$ required by NR and LTE customers as a function of arrival intensities λ_1 and λ_2. We consider three cases: (i) NR customers require two times more resources than LTE, (ii) an equal amount of resources is required by both LTE and NR customers, and (iii) NR customers require two times less resources. The presented results are logical – the increase in the amount of required resources at NR increases the NR new and accepted customer drop probabilities overall considered range of λ_1. However, as λ_2 increases, the situation becomes more complex as shown in Fig. 4(b). First of all, for a fixed NR arrival intensity and smaller NR resource requirements distribution, the new NR customer drop probability decreases. The rationale is that under this relation between the mean amount of required resources, offloaded sessions are rarely lost at LTE system increasing the new LTE customer drop probability. However, as the mean value of the required resources at NR system increases, the new NR customer drop probability increases as well.

Not only the mean value, but the type of the distribution may produce a significant effect on performance characteristics of the session service process. To this aim, we now consider the effect of the variance and distribution type jointly in Fig. 5 as a function of session arrival intensities, λ_1 and λ_2. Specifically, we consider: (i) geometric distribution with variance $\sigma^2 = 12$, (ii) binomial distribution with variance $\sigma^2 = 0.75$, and (iii) Poisson distribution with variance $\sigma^2 = 3$. Note that the distribution parameters are chosen so that the mean resource requirement is equal to 4 for all the considered distributions. The analysis of the results reveals that the best performance is produced by distributions having smaller values of variance. Furthermore, this effect can be quite substantial with the gap between new and accepted customer drop probabilities reaching orders of magnitude. More specifically, the considered binomial distribution with $\sigma^2 = 0.75$ leads to the new NR customer drop probability of approximately 10^{-8} for $\lambda_{NR} = 0.2$. For the same arrival intensity, geometric distribution with a much larger variance of $\sigma^2 = 12$ results in 10^{-4} new NR customer drop probability. The rationale is that higher variance leads to more variability in the customer sizes and thus more customers are accepted at the LTE system. The same observations can be made for other metrics of interest. This trend is best highlighted in Fig. 5(b), where these probabilities decrease as a function of λ_2 and there is a large gap between them.

Conclusion

In this paper, we considered a network of resource loss systems. Using the decomposition approach, we derived formulas for single node characteristics, and proposed an iterative algorithm to evaluate the performance measures of the whole system, including the new customer loss probability and the accepted customer loss probability. The proposed model can be utilized to investigate the performance of user sessions in joint mmWave/LTE deployments with UEs supporting inter radio access technology (RAT) multiconnectivity functionality. In these deployments blockage of the LoS path between mmWabe BS and UE may lead to outage conditions and LTE BS can be utilized to temporarily offload the served sessions. More complex deployment conditions and technologies such as mmWave and THz RATs can be considered, where a session may experience blockage conditions at both types of RATs.

The carried out numerical analysis allowed to make several critical observations. First of all, the session arriving process at the system utilized for temporal offloading (e.g., LTE) may drastically affect user session performance of the system subject of outages caused by blockage phenomenon (e.g., mmWave or THz). Higher intensity of signals negatively affects the new session drop probability at the system utilized for temporal offloading of sessions. The overall effect is extremely negative from the QoS perspective – under high values of α most NR sessions are accepted for service and then eventually lost. Further, the duration of interruption has a high impact on accepted session drop probability. The mean amount of required resources have a complex effect on performance metrics, specifically, the ratio between mean values of resources may heavily affect the new and accepted session drop probability. Finally, the variance of the resource request distribution may produce an extreme impact on session service performance with the gap between new and accepted session drop probabilities reaching orders of magnitude.

References

1. 3GPP: Study on channel model for frequencies from 0.5 to 100 GHz (Release 16). 3GPP TR 38.901 V16.1.0, July 2020
2. Ageev, K., Sopin, E., Shorgin, S., Chursin, A.: The probabilistic measures approximation of a resource queuing system with signals. In: Vishnevskiy, V.M., Samouylov, K.E., Kozyrev, D.V. (eds.) DCCN 2021. LNCS, vol. 13144, pp. 80–91. Springer, Cham (2021). https://doi.org/10.1007/978-3-030-92507-9_8
3. Begishev, V., et al.: Quantifying the impact of guard capacity on session continuity in 3GPP new radio systems. IEEE Trans. Veh. Technol. **68**(12), 12345–12359 (2019)
4. Begishev, V., et al.: Joint use of guard capacity and multiconnectivity for improved session continuity in millimeter-wave 5G NR systems. IEEE Trans. Veh. Technol. **70**(3), 2657–2672 (2021)
5. ITU-R: Propagation data and prediction methods required for the design of terrestrial broadband radio access systems operating in a frequency range from 3 to 60 GHz. Rec. ITU-R P.1410, February 2012
6. Jain, I.K., Kumar, R., Panwar, S.: Driven by capacity or blockage? A millimeter wave blockage analysis. In: 2018 30th International Teletraffic Congress (ITC 30), vol. 1, pp. 153–159. IEEE (2018)
7. Kuehn, P.: Approximate analysis of general queuing networks by decomposition. IEEE Trans. Commun. **27**(1), 113–126 (1979). https://doi.org/10.1109/TCOM.1979.1094270
8. Moltchanov, D., et al.: Performance characterization and traffic protection in street multi-band millimeter-wave and microwave deployments. IEEE Trans. Wirel. Commun. **21**, 163–178 (2021)
9. Moltchanov, D., Sopin, E., Begishev, V., Samuylov, A., Koucheryavy, Y., Samouylov, K.: A tutorial on mathematical modeling of millimeter wave and terahertz cellular systems. arXiv preprint arXiv:2109.08651 (2021)
10. Naumov, V.A., Samuilov, K.E.: Analysis of networks of the resource queuing systems. Autom. Remote. Control. **79**(5), 822–829 (2018)

11. Nikolich, P., et al.: Standards for 5G and beyond: their use cases and applications. IEEE 5G Tech Focus **1**(2) (2017)
12. Petrov, V., et al.: Dynamic multi-connectivity performance in ultra-dense urban mmWave deployments. IEEE J. Sel. Areas Commun. **35**(9), 2038–2055 (2017). https://doi.org/10.1109/JSAC.2017.2720482
13. Sopin, E.S., Ageev, K.A., Markova, E.V., Vikhrova, O.G., Gaidamaka, Y.V.: Performance analysis of M2M traffic in LTE network using queuing systems with random resource requirements. Autom. Control. Comput. Sci. **52**(5), 345–353 (2018). https://doi.org/10.3103/S0146411618050127
14. Ullah, H., Nair, N.G., Moore, A., Nugent, C., Muschamp, P., Cuevas, M.: 5G communication: an overview of vehicle-to-everything, drones, and healthcare usecases. IEEE Access **7**, 37251–37268 (2019)
15. Vikhrova, O.: About probability characteristics evaluation in queuing system with limited resources and random requirements. Discrete Continuous Models Appl. Comput. Sci. **25**(3), 209–216 (2017)
16. Wolf, A., Schulz, P., Dörpinghaus, M., Santos Filho, J.C.S., Fettweis, G.: How reliable and capable is multi-connectivity? IEEE Trans. Commun. **67**(2), 1506–1520 (2018)

Application of k-out-of-n:G System and Machine Learning Techniques on Reliability Analysis of Tethered Unmanned Aerial Vehicle

Nika Ivanova[1,2]($\boxtimes$) (iD) and Vladimir Vishnevsky[1] (iD)

[1] V.A. Trapeznikov Institute of Control Sciences of Russian Academy of Sciences, 65 Profsoyuznaya Street, 117997 Moscow, Russia
nm_ivanova@bk.ru, vishn@inbox.ru
[2] Peoples' Friendship University of Russia (RUDN University), 6 Miklukho-Maklaya St., 117198 Moscow, Russian Federation

Abstract. The purpose of the article is to investigate the reliability of an unmanned high-altitude module based on a mathematical model of the k-out-of-n:G system. An analytical model of the k-out-of-n:G system under two system failure scenarios is considered. In the first case, the system failure occurs after $(n - k + 1)$ elements failure. The second one examines the system failure depending on the location of the failed elements. The sensitivity analysis of system reliability characteristics to the shape of the lifetime distribution function of the components has been carried out. The impact of the coefficient of variation of the system elements lifetime on its operating probability without failure is investigated. Several machine learning methods are used to calculate reliability characteristics for arbitrary input data based on practically significant parameters. The accuracy of the trained models is expressed in terms of estimated mean values.

Keywords: Telecommunication high-altitude platform · tethered unmanned aerial vehicle · k-out-of-n:G system · system reliability · sensitivity analysis · coefficient of variation · simulation modeling · machine learning

1 Introduction and Motivation

Currently, telecommunication high-altitude platforms (THAP), which are implemented on autonomous unmanned aerial vehicles (UAV), are widely developed and used in various fields of human activity [1,2]. The main disadvantage of UAVs is the limited operating time associated with the short service life of UAV batteries equipped with electric motors or a limited supply of fuel for internal combustion engines. In this regard, such UAVs cannot be effectively used

Supported by the Russian Foundation for Basic Research, project no. 19-29-06043 and the RUDN University Strategic Academic Leadership Program.

in systems that require a long operating time. The long-term operation can be ensured by tethered THAP, in which the engines and payload equipment are powered from ground-based energy sources [3–5]. The ability to transmit high-power energy (10–15 kW) through a cable from the ground to the THAP's board allows lifting and holding at altitudes of 100–200 m of a payload telecommunication load for a long time, limited only by the reliability characteristics of the platform [6–9]. High reliability of the tethered unmanned module is achieved by the following ways: 1) choice of propulsion systems with a large meantime between failures; 2) redundancy of individual elements of the control system; 3) the usage of a multi-rotor architecture (for example, in a quadcopter, a failure of one engine leads to a complete cessation of operation, and in an eight-rotor version, in case of failure two motors, the copter may continue to run) and so on.

The reliability of such complex systems is effectively investigated using a mathematical model of the k-out-of-n system [10]. Such a system has broad practical applications in various industries: telecommunications and robotics [11,12], oil and gas [13], subsea pipeline monitoring systems [14], cryptography [15], etc. This model has been widely studied under many assumptions about the structure of such a model, for example, the dependence and independence of the system elements, the shape of life and repair times distributions, different recovery scenarios, and others. To study various k-out-of-n systems, both analytical methods based on multidimensional Markov processes and simulation are used [11,16–18].

Sensitivity analysis is a significant research stage, especially for redundancy systems like k-out-of-n system. In stochastic systems, stability is often understood as the insensitivity or low sensitivity of their output characteristics to the shape of some input distributions. The term "sensitivity" in other areas, for example, civil engineering, can be defined differently [19]. In queuing theory, the first results of sensitivity research are presented by Sevastyanov, Kovalenko, Gnedenko, Soloviev, and others. Some of the latest studies see in [18] and its references.

An additional research method considered in this paper is machine learning (ML). In queuing and reliability theories, ML methods are usually used for studying various probabilistic and time characteristics of complex systems. They are also useful in those cases when it is impossible to obtain results either analytically or using simulation [20]. The application of ML techniques for analyzing the reliability of an unmanned high-altitude module is due to the following factors.

1. From a practical point of view, the system service time is often estimated by its average value, while the shape of the lifetime distribution is unknown and can only be assumed based on some statistical data. ML model can operate based on the mean value without considering a specific distribution function of the lifetime of system elements.
2. Some parameters inside the system can significantly impact its reliability. However, from practice, this information may also be absent. The sensitivity analysis helps identify these weaknesses, after which they will be included in the ML model.

3. A model built and trained using ML techniques can predict the system reliability characteristics faster than a simulation model. In addition, it allows making accurate predictions on many data simultaneously, while simulation can only give a similar result after a lot of iterations.
4. Trained model can be useful and used by engineers at the development stage of such modules for many aims: to determine a highly reliable system architecture (parameters k, n), select the module components, the characteristics of which will support reliability and long-term operation of THAP (mean lifetime a and the coefficient of variation v), and also predict how long this unmanned module will operate with a satisfactory level of reliability.

There are many machine learning techniques. In the article, we will consider supervised learning for some types of regressions and neural networks using a Python programming language [21]. For this Scikit-learn [22] and TensorFlow [23] libraries will be used.

This paper continues studies related to reliability and sensitivity and considers a hot standby non-repairable system using analytical and simulation methods. The current paper aims to study the reliability of tethered THAP using the k-out-of-n:G system and ML methods, which make it possible to determine a satisfactory level of module reliability at different initial parameters with high accuracy.

The article is organized as follows. The next section introduces the problem setting and some notations. In Sect. 3, reliability function of homogeneous k-out-of-n:G system will study. Subsections 3.1 and 3.2 contain analytical results for a simple homogeneous k-out-of-n:G system and a homogeneous k^*-out-of-n:G system, the failure of which depends on the location of the failed elements. A numerical example and sensitivity analysis of the considered systems are presented in Subsects. 3.3. In Sect. 4, various ML techniques for predicting the level of reliability of unmanned module will discuss. Subsection 4.1 describes the methods and data used in this research, which are implemented in Subsects. 4.2 and 4.3. The paper ends with a conclusion and some problems descriptions.

2   Problem Setting

Due to the multi-rotor architecture of the high-altitude module, which consists of n identical engines, consider homogeneous k-out-of-n:G system. Such a system consists of n elements and remains operational iff at least k out of n elements are operational. Denote by A_i, $i = 1, 2, ...$, lifetimes of the system elements. Suppose that these random variables are independent and identically distributed, thus the corresponding cumulative density function is defined as $A(t) = \mathbf{P}\{A_i \leq t\}$. Suppose also that instantaneous failures are impossible and their mean times are finite:

$$A(0) = 0, \quad a = \int_0^\infty (1 - A(t))dt.$$

For the system study, introduce the random process $J = \{J(t),\ t \geq 0\}$ with

$$J(t) = \text{number of working components in time } t$$

with the set of states $E = \{j = \overline{0,k}\}$, where j is number of working units.

Denote also by T time to first system failure $T = \inf\{t : J(t) \in E_1\}$, where $E_1 = \{j = \overline{0, k-1}\}$ is the set of UP states. $E_0 = \{j = k\}$ is the set of DOWN states. Thus, we are interesting in calculation of reliability function

$$R(t) = \mathbf{P}\{T > t\},$$

and the mean time to system failure (MTTF)

$$m = \int_0^\infty R(t)dt.$$

3   Analytical Models and Sensitivity Analysis

3.1   Reliability Function of Homogeneous k-out-of-n:G System

Consider homogeneous k-out-of-n:G system, $A_i(t) = A(t)$ $(i = \overline{1,n})$. It is well known, the probability that exactly i elements of the system from n at time t are in a working state has the form

$$\mathbf{P}(t) = C_n^i (1 - A(t))^i A(t)^{n-i}.$$

Thus, the reliability function of such a system (the probability of the system operating for a certain time t without failure) is

$$R(t) = \mathbf{P}\{T > t\} = \sum_{i \geq k}^{n} C_n^i (1 - A(t))^i A(t)^{n-i}. \tag{1}$$

3.2   Reliability Function of Homogeneous k-out-of-n:G System Taking into Account the Location of the Failed Units

To investigate the reliability function of more complex homogeneous system, the failure of which depends on the location of its failed components, introduce a vector description of the state of the system $\mathbf{j} = (j_i, j_2, ..., j_n)$, where $j_i = 0$ if i-th component failed and $j_i = 1$ if it works. Then the probability of state j in time t equals to

$$p_{\mathbf{j}}(t) = \prod_{1 \leq i \leq n} (1 - A(t))^{j_i} A(t)^{1-j_i}.$$

The probabilities of the operable and failure states of the system at the time t take the forms

$$\mathbf{P}(UP) = \sum_{\mathbf{j} \in E_1} p_{\mathbf{j}}(t), \quad \mathbf{P}(DOWN) = \sum_{\mathbf{j} \in E_0} p_{\mathbf{j}}(t).$$

Thus, the system reliability function is

$$R(t) = \sum_{\mathbf{j} \in E_1} \prod_{1 \leq i \leq n} (1 - A(t))^{j_i} A(t)^{1-j_i}. \tag{2}$$

3.3   Numerical Examples and Sensitivity Analysis

As a numerical example consider the case of 4-out-of-6:G system. It is supposed that the lifetime of the system's units have the following distributions:

- Gamma $\left[\Gamma\left(1/v^2, av^2\right)\right]$;
- Gnedenko-Weibull $\left[GW\left(\mu, \frac{a}{\Gamma(1+1/\mu)}\right)\right]$;
- Log-normal $\left[LnN\left(\ln\frac{a}{\sqrt{1+v^2}}, \sqrt{\ln\left(1+v^2\right)}\right)\right]$,

where a is mean lifetime of the system components and v is its coefficient of variation. μ is the shape parameter of GW distribution and selected based on the value of v.

In our experiments we choose $a = 1$ and $v = [0.1, 0.5, 1, 5, 10]$. First, consider the simple case of homogeneous 4-out-of-6:G system.

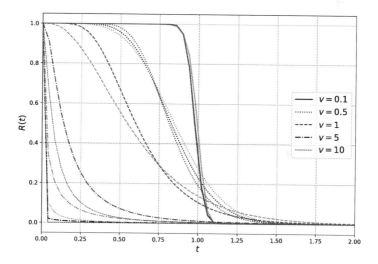

Fig. 1. Reliability function $R(t)$ of homogeneous 4-out-of-6:G system

Figure 1 shows the dependence of system reliability function on the time t calculated by formula (1). Black, red and blue colors correspond the Γ, GW, and LnN distributions, respectively. As it can be seen from the curves, the reliability function of the system is asymptotically insensitive to the form of the lifetime distribution at fixed mean and coefficient of variation $v \leq 1$. At the same time, with $v > 1$, this insensitivity disappears, and the system loses its reliability very quickly. We can conclude that the system behavior depends on the value v.

Further, look at the reliability of the 4-out-of-6:G system taking into account the location of the failed units. Denote such a system as a 4*-out-of-6:G system. Suppose that the system is operational as long as at least 4 out of 6 engines are running, and two failed motors should not be located next to each other. In other words, the system fails when two adjacent motors stop operate, or when any three engines fail.

Due to the complexity (time and computational) of calculating the reliability function using the formula (2) for arbitrary $A(t)$, k, and n, here we will apply simulation modeling to achieve our goals. The numerical example for the case of exponential distribution of system elements lifetime can be found in paper [11].

To build a simulator Python programming was chosen. The constructed simulation model is shown graphically as a process flowchart (Fig. 2). As a result of the algorithm, we can get the empirical reliability function $\hat{R}(t)$, and MTTF.

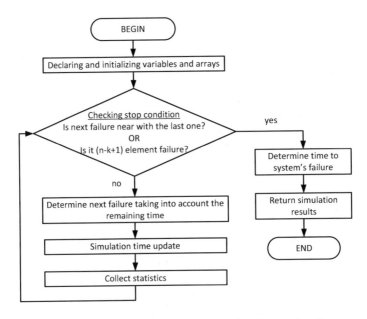

Fig. 2. Flowchart of the simulation model of a k^*-out-of-n:G system

Figure 3 shows evaluation result using simulation. The example of both the same system and parameters as before are used.

As can be seen from the numerical examples, the behavior of 4-out-of-6:G and 4*-out-of-6:G system reliability functions is very similar. To see the difference between them, consider corresponding MTTF (Table 1).

The results of the calculation of m confirm the conclusions of the sensitivity analysis. Moreover, the 4-out-of-6:G system, without dependence on the location of the failed elements, is efficient for a longer time than the other one.

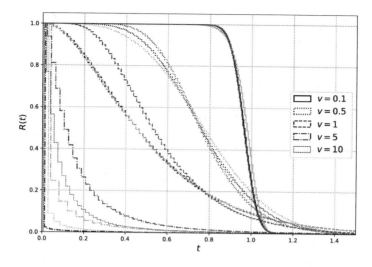

Fig. 3. Reliability function $R(t)$ of homogeneous 4^*-out-of-6:G system

Table 1. Mean lifetime m of 4-out-of-6:G/4^*-out-of-6:G systems

$A(t)$ \ v	0.1	0.5	1	5	10
Γ	0.9775 / 0.9608	0.8445 / 0.7747	0.6174 / 0.5154	0.0077 / 0.0041	$8*10^{-5}$ / $3*10^{-5}$
GW	0.9886 / 0.9693	0.8593 / 0.7777	0.6176 / 0.5177	0.0697 / 0.0503	0.0197 / 0.0136
LnN	0.9760 / 0.9598	0.8349 / 0.7743	0.6509 / 0.5714	0.2047 / 0.1621	0.1154 / 0.0874

4 Machine Learning Methods and Their Application to the Task

This section presents the results of prediction THAP reliability using ML methods.

4.1 Methods and Data

As ML methods [21], we will consider the followings from scikit-learn (for regressions) and TensorFlow (for neural network) libraries:

- Linear regression (LinReg),
- Polynomial regression (degree = 4) (PolyReg),
- K-nearest neighbors regression (n_neighbors = 5) (KNN),
- Multi-output regression with cross-validation (scoring = MSE) based on Ridge regression (MultiReg),

– Artificial neural network with three hidden layers (optimizer = RMSprop(1e-3), loss = MSE, batch size = 96) (ANN).

As it was noted in the introduction, the purpose of machine learning application is to predict the reliability and time characteristics of a tethered unmanned module. Therefore, the output parameters are R, t, m (Table 2). The set of parameters, as well as their ranges, is associated with the following. The previous section concludes some hidden parameters of the system, namely the coefficient of variation, have a significant impact on its behavior and performance. Moreover, the system is insensitive to the shape of the lifetime distribution with $v < 1$. In addition, from a practical point of view, we assume that the system is at a satisfactory level of reliability if $R(t) \geq 0.5$.

Table 2. Variables for machine learning models and their ranges

Type	Variables	Symbol	Range
Input	Total number of system's elements	n	4–10
	Needed number of elements in operating states	k	$2 - (n - 1)$
	Mean lifetime	a	0.1–1
	Coefficient of variation	v	0.01–1
Output	Reliability	R	0.5–1
	Time to system acceptable level of R	t	>0
	MTTF	m	>0

We have generated two datasets for training the models.

1. To train the model, which describes the behavior of THAP by homogeneous k-out-of-n:G system, the dataset was generated using formula (1), in which $A(t) \sim \Gamma$.
2. For the second case, in which system failure depends on the location of the failed elements, simulation results were used, here also $A(t) \sim \Gamma$. This data supposes that a system failure occurs either when 2 adjacent or any $(n-k+1)$ elements have failed.

The architecture of the selected ML models is different. Some can predict several outputs simultaneously, while others can operate with only one outcome. The whole process contains two phases – training and testing. Before training, we divide the initial dataset into train and test sets with a ratio of 70% and 30%, respectively. The learning process for LinReg, PolyReg, and KNN is structured as follows. The first step is to predict reliability R using parameters n, k, a, v, t. Next, the model is trained for prediction t on parameters n, k, a, v, R. The last cycle ends with a forecast of m based on the set n, k, a, v, R, t. After each round, the accuracy of the trained model is assessed, and testing begins on a new data sample. For MultiReg and ANN, there is one training cycle, in which the model predicts R, t, m simultaneously based on n, k, a, v. These models provide an additional phase for monitoring training, the so-called cross-validation. In this way, the initial set is divided into 70%, 20% and 10% for train, validation and final test.

4.2 Training and Testing Results for k-out-of-n:G System

Now move on to the results of ML techniques application for analyzing the reliability of a tethered unmanned high-altitude module. First, consider the k-out-of-n:G system. Table 3 shows the mean square error (MSE) for the predicted values on the training set. The table results show the smallest prediction error was achieved using PolyReg and KNN. The greatest error corresponds to Multi-Reg. The closest prediction in the training phase among all methods was made for MTTF m.

Table 3. Accuracy of training

		LinReg	PolyReg	KNN	MultiReg	ANN
MSE	R	0.0094	0.0028	10^{-4}	0.0313	0.0094
	t	0.0246	0.0149	0.0110	0.0322	0.0090
	m	0.0093	10^{-4}	$4 * 10^{-4}$	0.0123	10^{-4}

Table 4 demonstrates MSE, mean absolute error (MAE) as well as the coefficient variation (R^2) for the test set. Analyzing the results obtained, we can note that MSE estimate for all cases lies in an acceptable interval. MAE estimate shows the relative value of the prediction error. In our task, MAE ≥ 0.05 is considered unsatisfactory. Therefore, only the K-nearest neighbors regression shows the obtained accuracy result among all the considered cases. R^2 estimate

Table 4. Accuracy of testing

		LinReg	PolyReg	KNN	MultiReg	ANN
MSE	R	0.0094	0.0028	$2 * 10^{-4}$	0.0239	0.0131
	t	0.0177	0.0339	0.0117	0.0344	0.0322
	m	0.0107	0.0056	$2 * 10^{-6}$	0.0105	0.0043
MAE	R	0.0708	0.0365	0.0033	0.0761	0.0746
	t	0.1001	0.1125	0.0395	0.1395	0.1342
	m	0.0690	0.0121	$3 * 10^{-4}$	0.0687	0.0273
R^2	R	0.3804	0.8134	0.9894	0.1343	0.1370
	t	0.6824	0.4274	0.7891	0.3807	0.4249
	m	0.8934	0.9295	0.9999	0.8942	0.9571

indicates how well the constructed model adequately describes the initial data. The best result for this indicator is again shown by the KNN method. Note that all methods are suitable for predicting the meantime m. The estimates MSE and MAE are quite small, and R^2 is high, which confirms the high dependence between the input and output parameters.

Consider prediction results on the test set graphically. Figure 4, 5, 6, 7 and 8 shows the scatter diagrams for ML methods described above. For each of these figures, 500 samples were taken at random. In reality, the test sample contains about 200.000 values. LinReg and MultiReg demonstrate similar results for all predicted parameters, but their accuracy is quite low. PolyReg and ANN show acceptable prediction accuracy of m. For the other two, the prediction error is too high. These methods present insufficient prediction accuracy. It suggests that models do not reflect the relationship between input and output data. Predictions for R and m using KNN are close enough to their exact values. For t, this is not so much accurate. Nonetheless, the application of the KNN method obtains the most accurate prediction result for all metrics among the considered ML techniques.

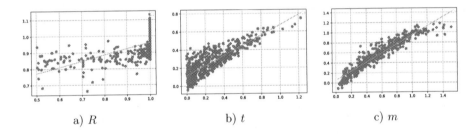

a) R b) t c) m

Fig. 4. Scatter plots for LinReg

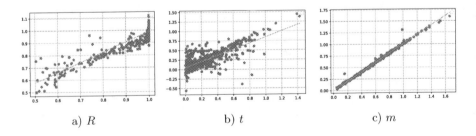

a) R b) t c) m

Fig. 5. Scatter plots for PolyReg

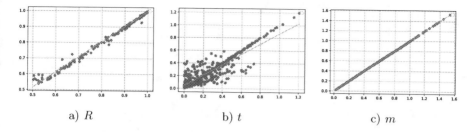

a) R b) t c) m

Fig. 6. Scatter plots for KNN

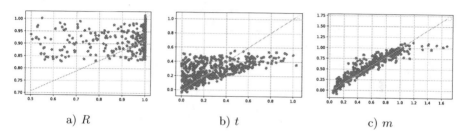

a) R b) t c) m

Fig. 7. Scatter plots for MultiReg

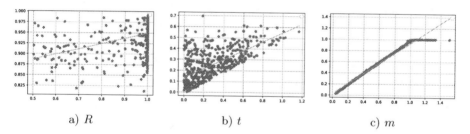

a) R b) t c) m

Fig. 8. Scatter plots for ANN

4.3 Training and Testing Results for k^*-out-of-n:G System

The application of machine learning techniques to the task at hand has shown that KNN most accurately predicts the reliability of an unmanned high-altitude module, the failure of which occurs after the failure of $(n - k + 1)$ its elements. Therefore, for the second case of dependence of the system failure on the location of the failed elements, we will consider only the KNN method. Consider the learning accuracy results (Table 5). MSE is small enough and takes the desired value.

Table 5. Accuracy of training (MSE)

	R	t	m
KNN	10^{-4}	$8.7 * 10^{-3}$	10^{-4}

The results on the test set are presented in Table 6 and Fig. 9. The results of the prediction accuracy take acceptable values. MSE and MAE are small enough, and the coefficient of determination R^2 is high.

Table 6. Accuracy of testing

	R	t	m
MSE	$1.6 * 10^{-4}$	$8.6 * 10^{-3}$	$1.9 * 10^{-6}$
MAE	$3.6 * 10^{-3}$	0.0492	$3.3 * 10^{-4}$
R^2	0.9904	0.7545	0.9999

The graphical results show similar prediction accuracy to the k-out-of-n:G system. The KNN model accurately reflects the dependence of R and m on the initial data, while the prediction of t is not so accurate, MAE $\approx 5\%$.

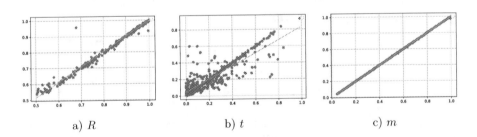

a) R b) t c) m

Fig. 9. Scatter plots for KNN

5   Conclusion

The paper investigates the reliability of an unmanned high-altitude module based on a mathematical model of the k-out-of-n system and machine learning methods. Two scenarios of the dependence of the system failure on the location of the failed elements were considered. Analytical results and sensitivity analysis demonstrated the dependence of the system reliability on the coefficient of variation of the lifetime for both scenarios. The application of machine learning methods showed that K-nearest neighbors regression describes the system reliability in the best way. As future research direction, we plan to improve chosen ML model to achieve more accurate predictions and consider other methods and models.

References

1. Mozaffari, M., Saad, W., Bennis, M., Nam, Y.-H., Debbah, M.: A tutorial on UAVs for wireless networks: applications, challenges, and open problems. IEEE Commun. Surv. Tutor. 410–438 (2019)
2. Khan, M.A., Hamila, R., Kiranyaz, M.S., Gabbou, A.M.: A novel UAV - aided network architecture using WiFi direct. IEEE Access **7**, 67305–67318 (2019)
3. Tognon, M., Franchi, A.: Theory and Applications for Control of Aerial Robots in Physical Interaction Through Tethers. Springer Tracts in Advanced Robotics. Springer, Heidelberg (2021). https://doi.org/10.1007/978-3-030-48659-4
4. Vishnevsky, V.M., Mikhailov, E.A., Tumchenok, D.A., et al.: Mathematical model of the operation of a tethered unmanned platform under wind loading. Math. Models Comput. Simul. **12**, 492–502 (2020). https://doi.org/10.1134/S2070048220040201
5. Kiribayashi, S., Yakushigawa, K., Nagatani, K.: Design and development of tether-powered multirotor micro unmanned aerial vehicle system for remote-controlled construction machine. In: Hutter, M., Siegwart, R. (eds.) Field and Service Robotics. SPAR, vol. 5, pp. 637–648. Springer, Cham (2018). https://doi.org/10.1007/978-3-319-67361-5_41
6. Ariram, S., Röning, J., Kowalczuk, Z.: Implementation of control system and tracking objects in a quadcopter system. In: Ronzhin, A., Rigoll, G., Meshcheryakov, R. (eds.) ICR 2019. LNCS (LNAI), vol. 11659, pp. 19–29. Springer, Cham (2019). https://doi.org/10.1007/978-3-030-26118-4_3
7. Wang, G., Samarathunga, W., Wang, S.: Uninterruptible power supply design for heavy payload tethered hexaroters. Int. J. Emerg. Eng. Res. Technol. **4**(2), 16–21 (2016)
8. Vishnevsky, V., Tereschenko, B., Tumchenok, D., Shirvanyan, A.: Optimal method for uplink transfer of power and the design of high-voltage cable for tethered high-altitude unmanned telecommunication platforms. In: Vishnevskiy, V.M., Samouylov, K.E., Kozyrev, D.V. (eds.) DCCN 2017. CCIS, vol. 700, pp. 240–247. Springer, Cham (2017). https://doi.org/10.1007/978-3-319-66836-9_20
9. Kozyrev, D.V., Phuong, N.D., Houankpo, H.G.K., Sokolov, A.: Reliability evaluation of a hexacopter-based flight module of a tethered unmanned high-altitude platform. In: Vishnevskiy, V.M., Samouylov, K.E., Kozyrev, D.V. (eds.) DCCN 2019. CCIS, vol. 1141, pp. 646–656. Springer, Cham (2019). https://doi.org/10.1007/978-3-030-36625-4_52
10. Shepherd, D.K.: k-out-of-n systems. In: Encyclopedia of Statistics in Quality and Reliability. Wiley, New York (2008)
11. Vishnevsky, V.M., Kozyrev, D.V., Rykov, V.V., Nguyen, Z.F.: Reliability modeling of an unmanned high-altitude module of a tethered telecommunication platform. Inf. Technol. Comput. Syst. **4**, 26–38 (2020). (In Russian)

12. Aslansefat, K., Marques, F., Mendonça, R., Barata, J.: A Markov process-based approach for reliability evaluation of the propulsion system in multi-rotor drones. In: Camarinha-Matos, L.M., Almeida, R., Oliveira, J. (eds.) DoCEIS 2019. IAICT, vol. 553, pp. 91–98. Springer, Cham (2019). https://doi.org/10.1007/978-3-030-17771-3_8

13. Rykov, V.V., Sukharev, M.G., Itkin, V.Y.: Investigations of k-out-of-n systems application possibilities to objects of oil and gas industry. J. Marine Sci. Eng. 8(11), 928 (2020). https://doi.org/10.3390/jmse8110928

14. Rykov, V., Kochueva, O., Farkhadov, M.: Preventive maintenance of a k-out-of-n system with applications in subsea pipeline monitoring. J. Marine Sci. Eng. 9, 85 (2021). https://doi.org/10.3390/jmse9010085

15. Yang, C.-N., Lin, Y.-C., Li, P.: Cheating immune k-out-of-n block-based progressive visual cryptography. J. Inf. Secur. Appl. 55, Article ID 102660 (2020). https://doi.org/10.1016/j.jisa.2020.102660

16. Zhang, Yu., Wu, W., Tang, Y.: Analysis of a k-out-of-n: G system with repairman's single vacation and shut off rule. Oper. Res. Perspect. 4, 29–38 (2017). https://doi.org/10.1016/j.orp.2017.02.002

17. Ivanova, N.: Modeling and simulation of reliability function of a k-out-of-n:F system. In: Vishnevskiy, V.M., Samouylov, K.E., Kozyrev, D.V. (eds.) DCCN 2020. CCIS, vol. 1337, pp. 271–285. Springer, Cham (2020). https://doi.org/10.1007/978-3-030-66242-4_22

18. Rykov, V.V., Ivanova, N.M., Kozyrev, D.V.: Sensitivity analysis of a k-out-of-n:F system characteristics to shapes of input distribution. In: Vishnevskiy, V.M., Samouylov, K.E., Kozyrev, D.V. (eds.) DCCN 2020. LNCS, vol. 12563, pp. 485–496. Springer, Cham (2020). https://doi.org/10.1007/978-3-030-66471-8_37

19. Kala, Z.: Probability based global sensitivity analysis of fatigue reliability of steel structures. In: IOP Conference on Series: Material Science and Engineering, vol. 668, p. 012015 (2019). https://doi.org/10.1088/1757-899X/668/1/012015

20. Gorbunova, A.V., Vishnevsky, V.M.: Estimating the response time of a cloud computing system with the help of neural networks. Adv. Syst. Sci. Appl. 20(3), 105–112 (2020)

21. Bonetto, R.: Computing in communication networks. Mach. Learn. 135-167 (2021). https://doi.org/10.1016/B978-0-12-820488-7.00021-9

22. Scikit-learn, Machine Learning in Python. https://scikit-learn.org/stable/index.html. Accessed 16 Dec 2021

23. TensorFlow: A System for Large-Scale Machine Learning. https://www.tensorflow.org/. Accessed 16 Dec 2021

Mathematical Model of the Tandem Retrial Queue M | GI | 1 | M | 1 with a Common Orbit

Svetlana Paul[1]([⊠]) [iD], Anatoly Nazarov[1] [iD], Tuan Phung-Duc[2] [iD],
and Mariya Morozova[1] [iD]

[1] National Research Tomsk State University, 36 Lenina Avenue,
634050 Tomsk, Russia
paulsv82@mail.ru

[2] University of Tsukuba, 1-1-1 Tennodai, Tsukuba, Ibaraki 305-8573, Japan
tuan@sk.tsukuba.ac.jp

Abstract. This paper considers a retrial tandem queue with single orbit, Poisson arrivals of incoming calls and without intermediate buffer. The first server provides services for incoming calls for an arbitrary random time, while the second server does for an exponentially distributed random time. Blocked customers at either the first server or the second server join the orbit and stay there for an exponentially distributed time before retrying to enter the first server again. Under an asymptotic condition when the mean of retrial intervals is extremely large, we derive a diffusion limit, which is further utilized to obtain an approximation to the number of customers in the orbit in stationary regime.

Keywords: tandem queue · retrial queue · diffusion limit

1 Introduction

The new feature of retrial queues in comparison with the conventional ones is that blocked customers that cannot find an idle server upon arrival join the orbit and retry for service after some random time. These models have been extensively studied in the literature; see the books [1,2] and survey papers [3,4]. The paper [4] summarizes major analytical results on retrial queues up to 1990 for both single server and multiserver models. Reference [3] presents a careful survey on single server retrial models with and without impatient customers. Furthermore, a survey of recent results for retrial queues is presented in [5].

The analysis of retrial queues is more difficult in comparison with that of counterparts with infinite buffer because each orbiting customer independently retries leading to a total retrial rate that is proportional to the number of customers in the orbit.

Tandem queues are simple networks of queues connected in a line topology are widely used in many applications such as computer communication, manufacturing and service systems. For example, in call centers, customers first connect

A. Dudin et al. (Eds.): ITMM 2021, CCIS 1605, pp. 131–143, 2022.
https://doi.org/10.1007/978-3-031-09331-9_11

to IVR (Interactive Voice Response) unit and then to operators [6]. Some other applications can be found in transmitting multimedia information [7], and in [8] for modelling a multi-agent robotic system, etc.

To our knowledge, only a little attention was paid the study of on tandem queues with retrials due to the complex of these models. In [9], the authors consider a tandem system of two sequentially connected servers without an intermediate buffer. In this system the blocking phenomenon occurs at the first server when a customer finishes the service a the first server but sees the second server busy. Customers that cannot enter the first server because the server is busy or blocked join the orbit and retry to enter the first server according to a constant retrial rate policy. Furthermore, [10] presents an approximate analysis for a tandem queue with a common orbit and constant retrial rate.

As a closely related paper, Phung-Duc [11] obtained an explicit solution for a simple model where only blocked customers the first server joins the orbit while blocked customers at the second server are lost. In this line, [12] presented a matrix-analytic solution for a model with Batch Markovian Arrival Process (BMAP) and general service time distribution at the first server and customers from the first server are lost if the second server is busy.

Furthermore, in our recent papers, we obtained the approximation of the stationary probability distribution of the number of calls in the orbit by methods of asymptotic analysis [13] and asymptotic diffusion analysis [14] for a special case with exponential distributions for service times in both servers. Further related papers can be found in [15,16]. In [16] a fixed point approximation is proposed for a tandem retrial queue. Pourbabai [15] investigates the tandem behavior in telecommunication systems with finite buffer and with repeated calls of constant retrial time. In [15], an approximation method is proposed.

In this paper, we study the two-phase tandem retrial queue system with one orbit and arbitrary service time distribution at the first server by the method of asymptotic diffusion analysis under the condition when the delay of calls in the orbit is extremely large. To the best of our knowledge, this is the first work dealing with a tandem retrial queue with classical (linear) retrial rate and arbitrary service time distribution at the first server, where blocked customers at the first or the second server enters orbit.

The remaining parts of the paper are organized as follows. In Sect. 2, we present the description of the model in detail. In Sect. 3, we write down the set of Kolmogorov differential equations while Sects. 4 and 5 show to the first order analysis (fluid limit) and the second order analysis (diffusion limit). Section 6 shows the use of the diffusion limit to approximate queue-length distribution in the orbit in the steady-state. Section 7 demonstrates some numerical examples.

2   Analytical Model

We consider a tandem retrial queue with two sequentially connected servers where customers arrive at the server according to a Poisson process with rate λ (see the Fig. 1). In this paper, customers and calls are interchangeably used.

If the server is idle upon the arrival of a call, the call occupies it immediately for a random time with the distribution function $B(x)$ and then moves to the second server. In the case that the second server is free, the call occupies it for a random time exponentially distributed with mean $1/\mu$. On the other hand, if the first server is busy upon arrival of a customer, this customer immediately goes to the orbit staying there for a period of time which is exponentially distributed with parameter σ and then tries to enter the first server again. Upon the service completion at the first server, if the second server is busy, the call immediately goes to the same orbit, staying there for a random period of time which is exponentially distributed mean $1/\sigma$ and trying to enter the first server for service again. This process is repeated until the call successfully receives services from both servers and leave the system.

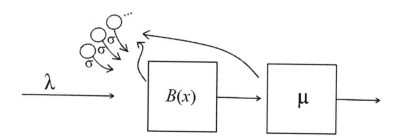

Fig. 1. The model

We define the following notations for further analysis.

The process $k(t)$ - the state of servers at time t: 0, if both servers are free; 1, if the first server is busy and the second one is free; 2, if the first server is free and the second one is busy; 3, if both servers are busy;

The process $z(t)$ - the remainder of service at the first server at time t;

The process $i(t)$ - number of retrial customers in the orbit at time t.

The purpose of the study is twofold: 1) to obtain the fluid and diffusion limit of $i(t)$ and 2) based on the diffusion limit, to build an approximation to the steady-state distribution of $i(t)$.

3   Kolmogorov Backward Equations

We define probabilities

$$P_k(i,t) = P\{k(t) = k, i(t) = i\}, k = 0, 2,$$

$$P_k(i,z,t) = P\{k(t) = k, i(t) = i, z(t) < z\}, k = 1, 3. \qquad (1)$$

The process $\{k(t), i(t)\}$, $k = 0, 2$, $\{k(t), i(t), z(t)\}$, $k = 1, 3$ is a Markov chain. Kolmogorov backward equations for (1) are given as follows.

$$\frac{\partial P_0(i,t)}{\partial t} = -(\lambda + i\sigma)P_0(i,t) + \mu P_2(i,t),$$

$$\frac{\partial P_1(i,z,t)}{\partial t} = \frac{\partial P_1(i,z,t)}{\partial z} - \frac{\partial P_1(i,0,t)}{\partial z} - \lambda P_1(i,z,t)$$
$$+ (i+1)\sigma B(z)P_0(i+1,t) + \lambda P_1(i-1,z,t)$$
$$+ \lambda B(z)P_0(i,t) + P_3(i,z,t)\mu,$$

$$\frac{\partial P_2(i,t)}{\partial t} = \frac{\partial P_1(i,0,t)}{\partial z} + \frac{\partial P_3(i-1,0,t)}{\partial z} - (\lambda + \mu + i\sigma)P_2(i,t),$$

$$\frac{\partial P_3(i,z,t)}{\partial t} = \frac{\partial P_3(i,z,t)}{\partial z} + \frac{\partial P_3(i,0,t)}{\partial z} - (\lambda + \mu)P_3(i,z,t)$$
$$+ \lambda P_3(i-1,z,t) + \lambda B(z)P_2(i,t) + (i+1)\sigma B(z)P_2(i+1,t). \tag{2}$$

We define partial characteristic functions, using $j = \sqrt{-1}$

$$H_k(u,t) = \sum_{i=0}^{\infty} e^{jui} P_k(i,t), k = 0, 2.$$

$$H_k(u,z,t) = \sum_{i=0}^{\infty} e^{jui} P_k(i,z,t), k = 1, 3. \tag{3}$$

We rewrite (2) using $H_k(u,t), k = 0, 2, H_k(u,z,t), k = 1, 3$ and add all the resulted equations with $z \to \infty$. We obtain following equations for further research in next sections.

$$\frac{\partial H_0(u,t)}{\partial t} = -\lambda H_0(u,t) + j\sigma \frac{\partial H_0(u,t)}{\partial u} + \mu H_2(u,t),$$

$$\frac{\partial H_1(u,z,t)}{\partial t} = \frac{\partial H_1(u,z,t)}{\partial z} - \frac{\partial H_1(u,0,t)}{\partial z} - j\sigma e^{-ju}\frac{\partial H_0(u,t)}{\partial u}B(z)$$
$$+ \lambda(e^{ju} - 1)H_1(u,z,t) + \lambda B(z)H_0(u,t) + \mu H_3(u,z,t),$$

$$\frac{\partial H_2(u,t)}{\partial t} = \frac{\partial H_1(u,0,t)}{\partial z} + e^{ju}\frac{\partial H_3(u,0,t)}{\partial z} + j\sigma \frac{\partial H_2(u,t)}{\partial u} - (\lambda + \mu)H_2(u,t),$$

$$\frac{\partial H_3(u,z,t)}{\partial t} = \frac{\partial H_3(u,z,t)}{\partial z} - \frac{\partial H_3(u,0,t)}{\partial z} - j\sigma e^{-ju}B(z)\frac{\partial H_2(u,t)}{\partial u}$$
$$+ (\lambda(e^{ju} - 1) - \mu)H_3(u,z,t) + \lambda B(z)H_2(u,t),$$

$$\frac{\partial H(u,t)}{\partial t} = (e^{ju} - 1)\left\{ j\sigma e^{-ju}\left(\frac{\partial H_0(u,t)}{\partial u} + \frac{\partial H_2(u,t)}{\partial u}\right) \right.$$
$$\left. + \lambda(H_1(u,t) + H_3(u,t)) + \frac{\partial H_3(u,0,t)}{\partial z} \right\}. \tag{4}$$

We are going to solve (4) under $\sigma \to 0$.

4 Fluid Limit

By denoting $\sigma = \varepsilon$ and performing substitution in (4)

$$\tau = t\varepsilon, u = \varepsilon w, H_k(u,t) = F_k(w, \tau, \varepsilon,),$$

$$H_k(u, z, t) = F_k(w, z, \tau, \varepsilon), \tag{5}$$

we obtain

$$\varepsilon \frac{\partial F_0(w, \tau, \varepsilon)}{\partial \tau} = -\lambda F_0(w, \tau, \varepsilon) + j \frac{\partial F_0(w, \tau, \varepsilon)}{\partial w} + \mu F_2(w, \tau, \varepsilon),$$

$$\varepsilon \frac{\partial F_1(w, z, \tau, \varepsilon)}{\partial \tau} = \frac{\partial F_1(w, z, \tau, \varepsilon)}{\partial z} - \frac{\partial F_1(w, 0, \tau, \varepsilon)}{\partial z} - je^{-jw\varepsilon} \frac{\partial F_0(w, \tau, \varepsilon)}{\partial w} B(z)$$

$$+ \lambda(e^{jw\varepsilon} - 1)F_1(w, z, \tau, \varepsilon) + \lambda B(z)F_0(w, \tau, \varepsilon) + \mu F_3(w, z, \tau, \varepsilon),$$

$$\varepsilon \frac{\partial F_2(w, \tau, \varepsilon)}{\partial \tau} = \frac{\partial F_1(w, 0, \tau, \varepsilon)}{\partial z} + e^{jw\varepsilon} \frac{\partial F_3(w, 0, \tau, \varepsilon)}{\partial z}$$

$$+ j \frac{\partial F_2(w, \tau, \varepsilon)}{\partial w} - (\lambda + \mu)F_2(w, \tau, \varepsilon),$$

$$\varepsilon \frac{\partial F_3(w, z, \tau, \varepsilon)}{\partial \tau} = \frac{\partial F_3(w, z, \tau, \varepsilon)}{\partial z} - \frac{\partial F_3(w, 0, \tau, \varepsilon)}{\partial z} - je^{-jw\varepsilon} B(z) \frac{\partial F_2(w, \tau, \varepsilon)}{\partial w}$$

$$+ (\lambda(e^{jw\varepsilon} - 1) - \mu)F_3(w, z, \tau, \varepsilon) + \lambda B(z)F_2(w, \tau, \varepsilon),$$

$$\varepsilon \frac{\partial F(w, \tau, \varepsilon)}{\partial \tau} = (e^{jw\varepsilon} - 1) \left\{ je^{-jw\varepsilon} \left(\frac{\partial F_0(w, \tau, \varepsilon)}{\partial w} + \frac{\partial F_2(w, \tau, \varepsilon)}{\partial w} \right) \right.$$

$$\left. + \lambda(F_1(w, \tau, \varepsilon) + F_3(w, \tau, \varepsilon)) + \frac{\partial F_3(w, 0, \tau, \varepsilon)}{\partial z} \right\}, \tag{6}$$

which we will solve under the assumption that functions $F_k(w, \tau, \varepsilon)$, $F_k(w, z, \tau, \varepsilon)$ and their derivatives have limits as $\varepsilon \to 0$.

Theorem 1. *We have*

$$\lim_{\sigma \to 0} Me^{jw\sigma i\left(\frac{\tau}{\sigma}\right)} = e^{jwx(\tau)}, \tag{7}$$

where $x = x(\tau)$ satisfies

$$x'(\tau) = (1 + b_1(\lambda + x))^{-1} \left(\lambda b_1(\lambda + x) - x + B^*(\mu) \frac{(\lambda + x)^2}{\mu + \lambda + x} \right), \tag{8}$$

and where $b_1 = \int_0^\infty x dB(x)$ and $B^*(\mu) = \int_0^\infty e^{-\mu x} dB(x)$.

Proof. We take the limit $\varepsilon \to 0$ in (6)

$$-\lambda F_0(w, \tau) + j\frac{\partial F_0(w, \tau)}{\partial w} + \mu F_2(w, \tau) = 0,$$

$$\frac{\partial F_1(w, z, \tau)}{\partial z} - \frac{\partial F_1(w, 0, \tau)}{\partial z} - j\frac{\partial F_0(w, \tau)}{\partial w}B(z)$$
$$+ \lambda B(z)F_0(w, \tau) + \mu F_3(w, z, \tau) = 0,$$

$$\frac{\partial F_1(w, 0, \tau)}{\partial z} + \frac{\partial F_3(w, 0, \tau)}{\partial z} + j\frac{\partial F_2(w, \tau)}{\partial w} - (\lambda + \mu)F_2(w, \tau) = 0,$$

$$\frac{\partial F_3(w, z, \tau)}{\partial z} - \frac{\partial F_3(w, 0, \tau)}{\partial z} - jB(z)\frac{\partial F_2(w, \tau)}{\partial w}$$
$$+ (\lambda - \mu)F_3(w, z, \tau) + \lambda B(z)F_2(w, \tau) = 0,$$

$$\frac{\partial F(w, \tau)}{\partial \tau} = jw\left\{ j\left(\frac{\partial F_0(w, \tau)}{\partial w} + \frac{\partial F_2(w, \tau)}{\partial w}\right)\right.$$

$$\left. + \lambda(F_1(w, \tau) + F_3(w, \tau)) + \frac{\partial F_3(w, 0, \tau)}{\partial z}\right\}. \tag{9}$$

We assume that (9) has a solution in the form

$$F_k(w, \tau) = r(x)e^{jwx(\tau)}, k = 0, 2, \quad F_k(w, z, \tau) = r(z, x)e^{jwx(\tau)}, k = 1, 3, \tag{10}$$

where $x = x(\tau)$ expresses $\lim_{\sigma \to 0} \sigma i(\tau/\sigma)$. Substituting (10) into (9), we obtain

$$-(\lambda + x)r_0(x) + \mu r_2(x) = 0,$$

$$\frac{\partial r_1(z, x)}{\partial z} - \frac{\partial r_1(0, x)}{\partial z} + (\lambda + x)B(z)r_0(x) + +\mu r_3(z, x) = 0,$$

$$\frac{\partial r_1(0, x)}{\partial z} + \frac{\partial r_3(0, x)}{\partial z} - (\lambda + \mu + x)r_2(x) = 0,$$

$$\frac{\partial r_3(z, x)}{\partial z} - \frac{\partial r_3(0, x)}{\partial z} - \mu r_3(z, x) + (\lambda + \mu)B(z)r_2(x) = 0, \tag{11}$$

$$x'(\tau) = \lambda(r_1(x) + r_3(x)) - x(r_0(x) + r_2(x)) + \frac{\partial r_3(0, x)}{\partial z}. \tag{12}$$

Summing up the first equation with the third, the second equation with the fourth of (11), we have

$$\frac{\partial r_1(z, x)}{\partial z} + \frac{\partial r_3(0, x)}{\partial z} = (\lambda + x)(r_0(x) + r_2(x))$$

$$\frac{\partial r_3(z, x)}{\partial z} - \frac{\partial r_3(0, x)}{\partial z} + \frac{\partial r_1(z, x)}{\partial z} - \frac{\partial r_1(0, x)}{\partial z}$$
$$+ (\lambda + x)B(z)(r_0(x) + r_2(x)) = 0. \tag{13}$$

We denote

$$r_{02}(x) = r_0(x) + r_2(x),$$

$$r_{31}(z, x) = r_1(z, x) + r_3(z, x),$$

$$r_{31}(0, x) = r_1(0, x) + r_3(0, x).$$

Then from (13) we obtain

$$r_{31}(z, x) = (\lambda + x)r_{02}(x) \int_0^z (1 - B(s))ds.$$

Letting $z \to \infty$ and denoting $r_k(\infty, x) = r_k(x)$, $k = 1, 3$, we have

$$r_1(x) + r_3(x) = (\lambda + x)b_1(r_0(x) + r_2(x)),$$

where $b_1 = \int_0^\infty x dB(x)$. Because $r_0(x) + r_1(x) + r_2(x) + r_3(x) = 1$, from the last equality we obtain

$$r_1(x) + r_3(x) = \frac{(\lambda + x)b_1}{1 + (\lambda + x)b_1},$$

$$r_0(x) + r_2(x) = \frac{1}{1 + (\lambda + x)b_1}.$$

Taking into account the first equation of (11), we write

$$r_0(x) = \frac{\mu}{\lambda + x}r_2(x).$$

We write the solution of the fourth differential equation of system (11) in the form

$$r_3(z, x) = e^{\mu z} \int_0^z e^{-\mu s} \left(\frac{\partial r_3(0, x)}{\partial z} - (\lambda + x)B(s)r_2(x) \right) ds. \qquad (14)$$

Let us send $z \to \infty$ in this equation to have

$$\mu \int_0^\infty e^{-\mu s} \left(\frac{\partial r_3(0, x)}{\partial z} - (\lambda + x)B(s)\dot{r}_2(x) \right) ds = 0.$$

The integrand satisfies the condition

$$\frac{\partial r_3(0, x)}{\partial z} = (\lambda + x)r_2(x)B^*(\mu), \qquad (15)$$

where $B^*(\mu) = \int_0^\infty e^{-\mu x} dB(x)$. Solution (14), taking into account (15), we rewrite under $z \to \infty$ in the form

$$r_3(x) = (\lambda + x)r_2(x)(1 - B^*(\mu)).$$

We obtain equations for the stationary probability distribution $r_k(x)$, $k = \overline{0,3}$ of the states of servers

$$r_0(x) = \frac{\mu}{\mu + \lambda + x}(1 + b_1(\lambda + x))^{-1},$$

$$r_1(x) = (1 + b_1(\lambda + x))^{-1}(\lambda + x)\left(b_1 - \frac{1}{\mu}\frac{\lambda + x}{\mu + \lambda + x}(1 - B^*(\mu))\right),$$

$$r_2(x) = \frac{\lambda + x}{\mu + \lambda + x}(1 + b_1(\lambda + x))^{-1},$$

$$r_3(x) = \frac{1}{\mu}\frac{(\lambda + x)^2}{\mu + \lambda + x}(1 - B^*(\mu))(1 + b_1(\lambda + x))^{-1}. \tag{16}$$

Let us substitute $r_k(x)$ into (12) in order to obtain

$$x'(\tau) = (1 + b_1(\lambda + x))^{-1}\left(\lambda b_1(\lambda + x) - x + B^*(\mu)\frac{(\lambda + x)^2}{\mu + \lambda + x}\right), \tag{17}$$

which coincides with (8).

Since $x(\tau)$ represents the asymptotic value ($\varepsilon \to 0$) of $\sigma i(\tau/\sigma)$, (7) holds. So, Theorem 1 is proved.

Let us denote

$$a(x) = x'(\tau) = (1 + b_1(\lambda + x))^{-1}\left(\lambda b_1(\lambda + x) - x + B^*(\mu)\frac{(\lambda + x)^2}{\mu + \lambda + x}\right). \tag{18}$$

$a(x)$ plays an important role for our analysis. First, as it is shown in Theorem 1, $a(x)$ represents the dynamic of $x(\tau)$, which is the limit under $\sigma \to 0$ for $\sigma i(\tau/\sigma)$. Second, as it will be shown, $a(x)$ expresses the drift coefficient for the diffusion process that represents a scaled version of $i(t)$.

5   Diffusion Limit

We carry out the following substitution in (4)

$$H_k(u, t) = e^{j\frac{u}{\sigma}x(\sigma t)}H_k^{(1)}(u, t), k = 0, 2$$

$$H_k(u, z, t) = e^{j\frac{u}{\sigma}x(\sigma t)}H_k^{(1)}(u, z, t), k = 1, 3. \tag{19}$$

For $H_k^{(1)}(u, t)$ and $H_k^{(1)}(u, z, t)$, $k = \overline{0,3}$, considering (18), we obtain

$$\frac{\partial H_0^{(1)}(u, t)}{\partial t} = -(\lambda + jua(x) + x)H_0^{(1)}(u, t)$$

$$+ j\sigma\frac{\partial H_0^{(1)}(u, t)}{\partial u} + \mu H_2^{(1)}(u, t),$$

$$\frac{\partial H_1^{(1)}(u,z,t)}{\partial t} = \frac{\partial H_1^{(1)}(u,z,t)}{\partial z} - \frac{\partial H_1^{(1)}(u,0,t)}{\partial z} - j\sigma e^{-ju}\frac{\partial H_0^{(1)}(u,t)}{\partial u}B(z)$$

$$+ (\lambda(e^{ju}-1) - jua(x))H_1^{(1)}(u,z,t) + (\lambda + xe^{-ju})B(z)H_0^{(1)}(u,t) + \mu H_3^{(1)}(u,z,t),$$

$$\frac{\partial H_2^{(1)}(u,t)}{\partial t} = \frac{\partial H_1^{(1)}(u,0,t)}{\partial z} + e^{ju}\frac{\partial H_3^{(1)}(u,0,t)}{\partial z} + j\sigma\frac{\partial H_2^{(1)}(u,t)}{\partial u}$$

$$- (\lambda + \mu + jua(x) + x)H_2^{(1)}(u,t),$$

$$\frac{\partial H_3^{(1)}(u,z,t)}{\partial t} = \frac{\partial H_3^{(1)}(u,z,t)}{\partial z} - \frac{\partial H_3^{(1)}(u,0,t)}{\partial z} - j\sigma e^{-ju}B(z)\frac{\partial H_2^{(1)}(u,t)}{\partial u}$$

$$+ (\lambda(e^{ju}-1) - \mu - jua(x))H_3^{(1)}(u,z,t) + (\lambda + xe^{-ju})B(z)H_2^{(1)}(u,t),$$

$$\frac{\partial H^{(1)}(u,t)}{\partial t} + jua(x)H^{(1)}(u,t)$$

$$= (e^{ju}-1)\left\{ j\sigma e^{-ju}\left(\frac{\partial H_0^{(1)}(u,t)}{\partial u} + \frac{\partial H_2^{(1)}(u,t)}{\partial u} \right) \right.$$

$$- xe^{-ju}(H_0^{(1)}(u,t) + H_2^{(1)}(u,t))$$

$$\left. + \lambda(H_1^{(1)}(u,t) + H_3^{(1)}(u,t)) + \frac{\partial H_3^{(1)}(u,0,t)}{\partial z} \right\}. \tag{20}$$

Because $H^{(1)}(u,t)$ is the characteristic function of $i(t) - \frac{1}{\sigma}x(\sigma t)$, we make the substitutions as follows.

By defining $\sigma = \varepsilon^2$ in (20) and substituting

$$\tau = t\varepsilon^2, \ u = w\varepsilon, \ H_k^{(1)}(u,t) = F_k^{(1)}(w,\tau,\varepsilon), \ k = 0,2,$$

$$H_k^{(1)}(u,z,t) = F_k^{(1)}(w,z,\tau,\varepsilon), \ k = 1,3, \tag{21}$$

we obtain

$$\varepsilon^2\frac{\partial F_0^{(1)}(w,\tau,\varepsilon)}{\partial \tau} = -(\lambda + j\varepsilon wa(x) + x)F_0^{(1)}(w,z,\tau,\varepsilon)$$

$$+ j\varepsilon\frac{\partial F_0^{(1)}(w,\tau,\varepsilon)}{\partial w} + \mu F_2^{(1)}(w,\tau,\varepsilon),$$

$$\varepsilon^2\frac{\partial F_1^{(1)}(w,z,\tau,\varepsilon)}{\partial \tau} = \frac{\partial F_1^{(1)}(w,z,\tau,\varepsilon)}{\partial z} - \frac{\partial F_1^{(1)}(w,0,\tau,\varepsilon)}{\partial z}$$

$$- j\varepsilon e^{-jw\varepsilon}\frac{\partial F_0^{(1)}(w,\tau,\varepsilon)}{\partial w}B(z) + (\lambda(e^{jw\varepsilon}-1) - j\varepsilon wa(x))F_1^{(1)}(w,z,\tau,\varepsilon)$$

$$+ (\lambda + xe^{-jw\varepsilon})B(z)F_0^{(1)}(w,\tau,\varepsilon) + \mu F_3^{(1)}(w,z,\tau,\varepsilon),$$

$$\varepsilon^2\frac{\partial F_2^{(1)}(w,\tau,\varepsilon)}{\partial \tau} = \frac{\partial F_1^{(1)}(w,0,\tau,\varepsilon)}{\partial z} + e^{jw\varepsilon}\frac{\partial F_3^{(1)}(w,0,\tau,\varepsilon)}{\partial z}$$

$$+ j\varepsilon\frac{\partial F_2^{(1)}(w,\tau,\varepsilon)}{\partial w} - (\lambda + \mu + j\varepsilon wa(x) + x)F_2^{(1)}(w,\tau,\varepsilon),$$

$$\varepsilon^2\frac{\partial F_3^{(1)}(w,z,\tau,\varepsilon)}{\partial \tau} = \frac{\partial F_3^{(1)}(w,z,\tau,\varepsilon)}{\partial z} - \frac{\partial F_3^{(1)}(w,0,\tau,\varepsilon)}{\partial z}$$

$$- j\varepsilon e^{-jw\varepsilon}B(z)\frac{\partial F_2^{(1)}(w,\tau,\varepsilon)}{\partial w} + (\lambda(e^{jw\varepsilon} - 1) - \mu - j\varepsilon wa(x))F_3^{(1)}(w,z,\tau,\varepsilon)$$

$$+ (\lambda + xe^{-jw\varepsilon})B(z)F_2^{(1)}(w,\tau,\varepsilon),$$

$$\varepsilon^2\frac{\partial F^{(1)}(w,\tau,\varepsilon)}{\partial \tau} + j\varepsilon wa(x)F^{(1)}(w,\tau,\varepsilon)$$

$$= (e^{jw\varepsilon} - 1)\left\{ j\varepsilon e^{-jw\varepsilon}\left(\frac{\partial F_0^{(1)}(w,\tau,\varepsilon)}{\partial w} + \frac{\partial F_2^{(1)}(w,\tau,\varepsilon)}{\partial w}\right)\right.$$

$$- xe^{-jw\varepsilon}(F_1^{(1)}(w,\tau,\varepsilon) + F_2^{(1)}(w,\tau,\varepsilon))$$

$$\left. + \lambda(F_1^{(1)}(w,\tau,\varepsilon) + F_3^{(1)}(w,\tau,\varepsilon)) + \frac{\partial F_3^{(1)}(w,0,\varepsilon)}{\partial z}\right\}. \tag{22}$$

which we will solve under the assumption that $F_k^{(1)}(w,\tau,\varepsilon)$, $F_k^{(1)}(w,z,\tau,\varepsilon)$ and their derivatives have limits as $\varepsilon \to 0$.

Theorem 2. $F_k^{(1)}(w,\tau)$ *is given by*

$$F_k^{(1)}(w,\tau) = \Phi(w,\tau)r_k(x), k = \overline{0,3} \tag{23}$$

where $\Phi(w,\tau)$ *satisfies*

$$\frac{\partial \Phi(w,\tau)}{\partial \tau} = a'(x)w\frac{\partial \Phi(w,\tau)}{\partial w} + b(x)\frac{(jw)^2}{2}\Phi(w,\tau) \tag{24}$$

and $r_k(x)$ *is defined in (16).* $a(x)$ *is defined by (18) and* $b(x)$ *is given by*

$$b(x) = a(x) + 2(\lambda(g_1(x) + g_3(x)) + g_3'(0,x) - x(g_0(x) + g_2(x) - r_0(x) - r_2(x))), \tag{25}$$

where

$$g_3'(0,x) = (\lambda + x)B^*(\mu)g_2(x) + ((a(x) - \lambda)(\lambda + x)B^{*\prime}(\mu) - xB^*(\mu)) \tag{26}$$

and $g_k(x)$, $k = \overline{0,3}$ *are defined by*

$$-(\lambda + x)g_0(x) + \mu g_2(x) = a(x)r_0(x),$$

$$(\lambda + x)g_0(x) + ((\lambda + x)(B^*(\mu) - 1) + \mu)g_2(x) + \mu g_3(x)$$

$$= xr_0(x) + (a(x) - \lambda)r_1(x) - ((a(x) - \lambda)(\lambda + x)B^{*\prime}(\mu) - a(x) + \lambda B^*(\mu))r_2(x),$$

$$g_1(x) + g_3(x) - (\lambda + x)b_1(g_2(x) + g_0(x))$$

$$= \left((\lambda - a(x))(\lambda + x)\frac{b_2}{2} - xb_1\right)(r_0(x) + r_2(x)),$$

$$g_0(x) + g_1(x) + g_2(x) + g_3(x) = 0, \tag{27}$$

and where $b_2 = \int_0^\infty x^2 dB(x)$.

Proof. The methodology of the proof is similar to that used in paper [14] before.

As it will be shown $a(x)$ in (18) and $b(x)$ in (25) are coefficients of a diffusion process. Later we will show their role in the approximation of the stationary distribution of $i(t)$.

Remark 1. The results in Theorem 2 show that in the heavy traffic regime ($\sigma \rightarrow 0$) $i(t)$ and the state of the servers are independent as their joint characteristic function is decomposed as a product of the orbit part and the server part.

6 Approximation of the Stationary Distribution Based on Diffusion Limit

In this section, we apply the diffusion limit to find the probability distribution of $i(t)$ under $\sigma \rightarrow 0$ in our system. This general method is also used other related work e.g. [14].

Lemma 1. *Under $\sigma \rightarrow 0$*

$$y(\tau) = \lim_{\sigma \rightarrow 0} \sqrt{\sigma} \left\{ i(\tau/\sigma) - \frac{1}{\sigma} x(\tau) \right\}, \tag{28}$$

is the solution of

$$dy(\tau) = a'(x)y d\tau + \sqrt{b(x)} dw(\tau). \tag{29}$$

We consider

$$l(\tau) = x(\tau) + \varepsilon y(\tau),$$

where $\varepsilon = \sqrt{\sigma}$ as before.

Lemma 2. *The process $l(\tau)$ is the solution of*

$$dl(\tau) = a(l)d\tau + \sqrt{\sigma b(l)} dw(\tau) \tag{30}$$

up to an infinitesimal of order ϵ^2.

Under the steady-state regime, we consider $l(\tau)$

$$s(l, \tau) = s(l) = \frac{\partial P\{l(\tau) < l\}}{\partial l}. \tag{31}$$

Theorem 3. *The density $s(l)$ of $l(\tau)$ is given by*

$$s(l) = \frac{C}{b(l)} \exp \left\{ \frac{2}{\sigma} \int_0^l \frac{a(x)}{b(x)} dx \right\}, \tag{32}$$

where C is some constant that satisfies the normalization condition.

7   Numerical Examples

Let us consider $G(i)$ in the form

$$G(i) = \frac{C}{b(\sigma i)} \exp\left\{\frac{2}{\sigma}\int_0^{\sigma i}\frac{a(x)}{b(x)}dx\right\}, \tag{33}$$

and define $P(i)$ as

$$P(i) = \frac{G(i)}{\sum\limits_{i=0}^{\infty} G(i)}. \tag{34}$$

We use $P(i)$ to approximate $P\{i(t) = i\}$.

We consider a particular case of $B(x)$ as a Gamma distribution with parameters of shape $\alpha = 2$ and of scale $\beta = 2$. We consider $\lambda = 0.5$ and $\mu = 1$.

Figure 2 presents the approximation of the probability distribution of the $i(t)$ with different values of calls' delay time in the orbit: P1 - the approximation with $\sigma = 0.5$, P2 - the approximation with $\sigma = 0.3$, P3 - the approximation with $\sigma = 0.1$.

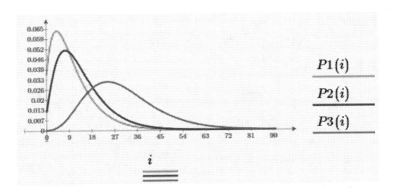

Fig. 2. The probability distribution $i(t)$

This figure shows the feasibility of our proposed approach.

8   Conclusion

In this paper, we have investigated the tandem retrial queue with two connected servers and without intermediate buffer. The first server provides services for calls for an arbitrary random time, while the second does for an exponentially distributed random time. Under the condition that $\sigma \to 0$, we have obtained diffusion limit of a scaled version of $i(t)$. The stationary probability density distribution of this diffusion process is used to approximate the stationary distribution of $i(t)$.

In further research, we plan to compare our approximate results with simulation

References

1. Falin, G., Templeton, J.G.: Retrial Queues, vol. 75. CRC Press, Boca Raton (1997)
2. Artalejo, J.R., Gómez-Corral, A.: Retrial Queueing Systems. Springer, Heidelberg (2008). https://doi.org/10.1007/978-3-540-78725-9
3. Yang, T., Templeton, J.: A survey on retrial queues. Queueing Syst. **2**, 203–233 (1987)
4. Falin, G.: A survey of retrial queues. Queueing Syst. **7**, 127–168 (1987)
5. Phung-Duc, T.: Retrial queueing models: a survey on theory and applications. In: Stochastic Operations Research in Business and Industry. Springer (2007)
6. Kumar, B.K., Sankar, R., Krishnan, R.N., Rukmani, R.: Performance analysis of multi-processor two-stage tandem call center retrial queues with non-reliable processors. Methodol. Comput. Appl. Probab. **24**, 1–48 (2021)
7. Vishnevsky, V.M., Larionov, A.A., Semyonova, O.V.: Evaluating the performance of a high-speed wireless tandem network using centimeter and millimeter-wave radio channels in road safety management systems. Manage. Probl. (4) (2013)
8. Kuznetsov, N.A., Myasnikov, D.V., Semenikhin, K.V.: Optimal control of data transmission in a mobile two-agent robotic system. J. Commun. Technol. Electron. **61**(12), 1456–1465 (2016). https://doi.org/10.1134/S1064226916120159
9. Moutzoukis, E., Langaris, C.: Two queues in tandem with retrial customers. Probab. Eng. Inf. Sci. **15**(3), 311–325 (2001)
10. Avrachenkov, K., Yechiali, U.: On tandem blocking queues with a common retrial queue. Comput. Oper. Res. **37**(7), 1174–1180 (2010)
11. Phung-Duc, T.: An explicit solution for a tandem queue with retrials and losses. Oper. Res. Int. J. **12**(2), 189–207 (2012)
12. Kim, C.S., Park, S.H., Dudin, A., Klimenok, V., Tsarenkov, G.: Investigation of the bmap/g/1→ ·/ph/1/m tandem queue with retrials and losses. Appl. Math. Model. **34**(10), 2926–2940 (2010)
13. Nazarov, A., Paul, S., Phung-Duc, T., Morozova, M.: Scaling limits of a tandem retrial queue with common orbit and poisson arrival process. In: Vishnevskiy, V.M., Samouylov, K.E., Kozyrev, D.V. (eds.) DCCN 2021. LNCS, vol. 13144, pp. 240–250. Springer, Cham (2021). https://doi.org/10.1007/978-3-030-92507-9_20
14. Nazarov, A., Paul, S., Phung-Duc, T., Morozova, M.: Analysis of tandem retrial queue with common orbit and poisson arrival process. In: Ballarini, P., Castel, H., Dimitriou, I., Iacono, M., Phung-Duc, T., Walraevens, J. (eds.) EPEW/ASMTA-2021. LNCS, vol. 13104, pp. 441–456. Springer, Cham (2021). https://doi.org/10.1007/978-3-030-91825-5_27
15. Pourbabai, B.: Tandem behavior of a telecommunication system with finite buffers and repeated calls. Queueing Syst. **6**, 89–108 (1990)
16. Takahara, G.: Fixed point approximations for retrial networks. Probab. Eng. Inf. Sci. **10**, 243–259 (1996)

Analyzing the Effect of Catastrophic Breakdowns with Retrial Queues in a Two-Way Communication System

Attila Kuki$^{(\boxtimes)}$ ⓘ, Tamás Bérczes ⓘ, and János Sztrik ⓘ

University of Debrecen, Debrecen, Hungary
{kuki.attila,berczes.tamas,sztrik.janos}@inf.unideb.hu

Abstract. A two-way communication system is modeled in this paper. A retrial queueing system with a finite and an infinite sources is used in the model. The system has two sources. The first source is finite, the second source is infinite. Jobs from the first source are the primary jobs (requests). They can be called as first order job, as well. Jobs from the second source are the secondary jobs. They can be called as second order job, as well. In case of an idle server, the second order customers are called for service. This situation is said as a special search for customers.

The non-reliable server is subject to random breakdowns. Two types of breakdowns are considered: the regular breakdown, when the first or second order customer under service is sent back to the orbit or the infinite source, respectively, and the catastrophic breakdown, when all of the requests at the server and in the orbit are sent back to the corresponding sources. The novelty of this paper is to investigate the effect of catastrophic breakdown in a two-way communication environment. The goal is to determine the steady-state probabilities and the system characteristics. The system balance equations are formulated for different cases, but the analytic solution is very difficult. A software tool is used instead. Figures illustrate the effect of the system parameters on the performance measures in scenarios of regular and catastrophic breakdowns.

Keywords: two-way communication · catastrophic breakdown · retrial queues

Introduction

For modeling different types of infocommunication and computer sciences, the retrial queueing systems are a useful and effective tool. Results can be found in various publications [1–6]. Several models assume finite sources. It means, that a finite number of population is in connection with the system. In some

The research work was supported by the Austro-Hungarian Cooperation Grant No 106öu4, 2020.

cases, these finite source models are more realistic and give a better description of the considered application [2,7]. In addition, the considered real-life systems are unfortunately non-reliable, that is the server can lose its efficiency or may break down. There is a large literature on this types of system subject to random breakdowns [1–6].

Sometimes the customers can not spend long time in the queue or in the orbit. They leave the system. In order not to lose these customers, a two-way communication system is built up. Here the customers can sign up for a special service. The system in an idle period will call these customers for the service. This field also has a literature. For the most interesting results see, [8–15].

This two-way communication principle can be considered as a special search for customers, as well. With this outgoing call, the organization (business, commercial etc.) can look for customers, send advertisement, and call them a personal encounter in case of interest. This way the efficiency of the system also can be increased, the ratio of the idle periods can be optimized. Results can be found, e.g. in [5,16–20].

In this paper, a special two-way communication system is considered. It is called searching for customers. Two types of customers are in the system. The organization has a small or large number of basic customers. They are called the primary customers. They make calls (request) towards the business entity. It can be considered as a server in the system. The principle of the retrial queueing system is applied for these customers. In case of an occupied server, the customer is not lost, it can wait and retry its request for the service. The customer keep retrying until a successful service. The first (finite) source contains these customers. During the idle periods of the server, outgoing calls are performed towards the secondary clients. They are in the second (infinite) source. These customers are the secondary customers, and they are reaching the system for some special reasons. For example, answer some promotion or check some personal data. They are called at an idle period of the system, but in the time period until arriving, a primary customer might be arrived. In this case the secondary customer finds the system occupied. A called customer can not be sent away, so it is placed in a priority (non-preemptive) buffer. For the next outgoing call this buffer will be applied.

The non-reliable server might break down. The main interest of this paper is to investigate the effect of a disaster event, which is called as a catastrophic event or breakdown. The most characteristic property of this type of breakdown is, that in case of the disaster event (or negative customer arrival) all of the services are interrupted, the customers from the servers, orbit, priority buffer are sent back to the corresponding source. There is a large literature on this phenomenon. For example, these types of investigations are very effective for describing the behavior of bank teller equipment. Here, several different disaster event can be imagined. For example, some mechanical malfunction, loss of power etc. Sometimes, the disaster is represented as the presence or arrival of a so called negative client. In this case, all of the service is stopped, the system is blocked, and all of the clients are removed from the system. Results on disaster and negative customers can be seen in [21–25], and reference therein.

The chapters of the paper are organized in the following way. In Sect. 1 the element of the system is given. It contains the stochastic description of the system (Markov analysis), the working modes, scenarios, parameters etc. Section 2 contains the system balance equations. The system probabilities can be calculated from those equation. For calculation the Mosel-2 software is used. From system probabilities the performance measures (mean waiting times, size of orbit etc.) can be obtained. The paper ends with a summary and conclusion.

1   Description of the Model

The system is modeled by a single-server retrial queueing system with a finite and an infinite sources. The functionality of the model is displayed in Fig. 1.

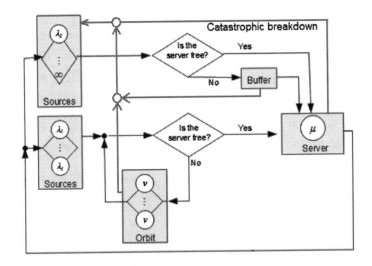

Fig. 1. The system model

The system has two sources. The first one is finite with the first order customers, the number of customers is N. These customers generate a job towards the server using the exponential law with parameter λ_1. For the first order customers, there is no queue at the server. After the service, the job goes back to the source can generate a new request again. The service time is also exponentially distributed with parameter μ_1. When the server is busy, the incoming job is transferred to the orbit. The size of the orbit is N. From the orbit the jobs after an exponential random time interval with parameter ν retry their request to the server until they are served.

The model has second order customers in an infinite number of sources. These customers generate triggered requests only. The idle server makes outgoing calls towards this infinite source, and the second order customers generate a request to be served. The generation is also exponential with parameter λ_2. The distribution of service times is also exponential with parameter μ_2. During the generation time interval of the second order customer, a new customer might arrive to the server. Thus, the called and incoming second order customer may find the server busy. In this case, the following scenarios can be investigated:

- The second order customer is sent back to the infinite source,
- The second order customer is transferred into a priority buffer. In the case of an idle server, a second order customer is called from this buffer. The size of the buffer is one because in the case of an idle server, there is no outgoing call when a customer is in the buffer.

In this model the single server is unreliable, it may be subject to breakdown. Here the regular and the catastrophic breakdowns are considered. In regular breakdown, the server stops working. The breakdown parameters are γ_0 and γ_1 for idle and busy servers, respectively. γ_2 is the parameter of the repair. The behavior of the customers at the time of breakdown is described below. The considered times are exponentially distributed. During the breakdown period, the sources can be blocked (they are not able to generate requests) or non-blocked. In this paper, the non-blocked case is considered. The other breakdown mode is the catastrophic breakdown. This is the situation when a disaster event removes all of the customers from the system (from the orbit, from the buffer, and from the server after interrupting the service). The repair of the system starts immediately. The breakdown parameters are γ_0 and γ_1 for idle and busy servers, respectively. γ_2 is the parameter of the repair. The considered times are exponentially distributed. During the breakdown period, the sources are blocked, they are not able to generate requests.

In case of a regular breakdown, a primary client can find the server down. In this case, it is sent to the orbit. In this server state, a secondary client might reach the system, as well. For this, the unoccupied server performs an outgoing call. The infinite source has an inter arrival time (request generation). This time has an exponential distribution. The parameter of the distribution is λ_2. During the time interval of the generation, the time between the call and the arrival of the secondary client, the server can go wrong. Two different scenarios can be considered here.

- The secondary client is transferred back to the infinite source,
- The secondary client is transferred into a priority buffer. In the case of an idle server, a second order customer is called from this buffer.

In case of a catastrophic breakdown, the sources are blocked. No new request is generated.

When the server is busy, a first order or a second order customer is under service. A breakdown can occur in the busy sate, as well. In case of regular breakdown, the behavior of the first order customers can be the followings:

- The primary client is transferred to the waiting facility (orbit),

- The primary client leaves the system and goes back to the source,
- The primary client remains at the system, namely, at the server. The service of the client will be restarting or resuming when the server is up again. (because of the memory-less property of the exponential distribution of the service times, restarting or continuing the service has no difference).

In case of catastrophic breakdown, all of the customers in the system, thus the customer under service are sent beck to the sources.

In case of regular breakdown, the behavior of the second order customers there are also some cases to be investigated.

- The secondary client remains at the server. The service will start again when the repair is finished.
- The secondary client leaves the system and goes back to the second source.

In case of catastrophic breakdown, all of the customers in the system, thus the customer under service are sent beck to the sources.

Let us denote $O(t)$ and $S(t)$ the number of requests in the orbit and the state of the server at a given time point of t:

$$
S(t) = \begin{cases}
0, & \text{when no job is at the server} \\
1, & \text{when the server is working} \\
 & \text{with a primary client} \\
2, & \text{when the server is working} \\
 & \text{with a secondary client} \\
3, & \text{when the server is down}
\end{cases}
$$

The state-space of the underlying Markovian-process $(S(t), O(t))$ can be described as a set of $\{0, 1, 2, 3\} \times \{0, 1, 2, ..., N\}$ elements. Although the system has an infinite source, the maximum number of the customers in the system is $(N + 1)$ (N in the orbit and one second order customer under service), there is no stability problems regarding the system. The state space is finite.

For buffered and non-buffered models the system balance equations can be formulated. For example, in the non-buffered case when a customer under service is sent back to the corresponding source at a breakdown the equations are the following.

$$
p_{i,j} = \lim_{t \to \infty} P(S(t) = i, O(t) = j), i = 0, 1, 2, 3 \text{ and } j = 0, 1, ..N \tag{1}
$$

$$
[(N - j)\lambda_1 + \lambda_2 + j\nu + \gamma_0]\, p_{0,j} = \mu p_{1,j} + \mu p_{2,j} + \gamma_2 p_{3,j} \tag{2}
$$

$$
[(N - j - 1)\lambda_1 + \mu + \gamma_1]\, p_{1,j} = (N - j)\lambda_1 p_{0,j} + (j + 1)\nu p_{0,j+1} \tag{3}
$$

$$
[(N - j)\lambda_1 + \mu + \gamma_1]\, p_{2,j} = \lambda_2 p_{0,j} \tag{4}
$$

$$[(N - j)\lambda_1] p_{3,j} = \gamma_0 p_{0,j} + \gamma_1 p_{1,j-1} + \gamma_1 p_{2,j} \tag{5}$$

with $p_{1,-1} = p_{0,N+1} = 0$.

The system balance equations in the case of a catastrophic breakdown look like this:

$$[(N - j)\lambda_1 + \lambda_2 + j\nu + \gamma_0] p_{0,j} = \mu p_{1,j} + \mu p_{2,j} \tag{6}$$

$$[(N - j - 1)\lambda_1 + \mu + \gamma_1] p_{1,j} = (N - j)\lambda_1 p_{0,j} + (j + 1)\nu p_{0,j+1} \tag{7}$$

$$[(N - j)\lambda_1 + \mu + \gamma_1] p_{2,j} = \lambda_2 p_{0,j} \tag{8}$$

$$\gamma_2 p_{3,0} = \gamma_0 p_{0,j} + \gamma_1 p_{1,j} + \gamma_1 p_{2,j} \tag{9}$$

with $p_{1,-1} = p_{0,N+1} = 0$.

The manual solution of the Kolmogorov-equations is very hard. An alternative method has to be found. For calculating the steady-state probabilities, the MOSEL-2 software has been used. Based on the calculated system probabilities the usual performance measures are provided by the software. Using the system probabilities, these measures can be calculated by the following formulas, as well.

– *Utilization 1*

$$U_1 = \sum_{o=0}^{N} P(1, o) \tag{10}$$

– *Utilization 2*

$$U_2 = \sum_{o=0}^{N} P(2, o) \tag{11}$$

– *Mean number of customers in the orbit*

$$\overline{O} = \sum_{s=0}^{3} \sum_{o=0}^{N} o P(s, o) \tag{12}$$

– *Mean number of active primary customers*

$$\overline{M} = N - \overline{O} - U_1 \tag{13}$$

– *Mean generation rate of primary customers*

$$\overline{\lambda_1} = \lambda_1 \overline{M} \tag{14}$$

– *Mean time spent in orbit by using Little-formula*

$$\overline{W} = \frac{\overline{O}}{\overline{\lambda_1}} \tag{15}$$

2   Numerical Results

The most important goal of these types of stochastic systems is to obtain the performance measures and system characteristics. Usually, the throughput, utilization, response times, waiting times, queue length are considered. Here the utilization and response time are focused.

The manual solution of the Kolmogorov-equations is very hard. There exist alternative methods for performing this task. For calculating the steady-state probabilities, several software tools can be applied. Based on the calculated system probabilities the usual performance measures are usually provided by the software. Because solving directly the balance equations is rather difficult, the MOSEL-2 tool is used. The system equations are solved by the SPNP (Stochastic Petri Net Program). The following figures illustrate the most interesting numerical results. The numerical values of the applied parameters in the model are listed in Table 1.

Table 1. Numerical values of model parameters

			Case studies				
No.	N	λ_1	λ_2	μ	ν	γ_0	γ_2
Fig. 2	100	$x - axes$	2	3	0.05	0.1	1
Fig. 3	100	$x - axes$	2	3	0.05	0.01	1
Fig. 4	100	$x - axes$	2	3	0.05	0.2, 0.5	1
Fig. 5	100	$x - axes$	2	3	0.05	0.2, 0.5	1
Fig. 6	100	$x - axes$	2	3	0.05	0.2	1
Fig. 7	100	$x - axes$	2	3	0.05	0.2, 0.5	1
Fig. 8	100	$x - axes$	2	$special$	0.05	0.2	1

The table contains only the idle time breakdown parameters. For calculations, the same values are used for the busy time breakdown parameters. The first two figures compare two different cases for the regular breakdown:

– In case of busy state breakdown, the service of the first order and second order customers are interrupted. The first or second order customer under the interrupted service is sent back to the orbit or to the infinite source, respectively.
– The service of both types of customers is interrupted. The customers are left at the server. When the server is up again, their service will continue or restart. Because of the exponentially distributed service time, this difference - restart or continue - has no effect on the system characteristics.

On Fig. 2 displays the difference of the scenarios mentioned above (leave or remain). The failure rate here is rather high, thus the difference between the scenarios is significant. The interruption is more frequent and the first order

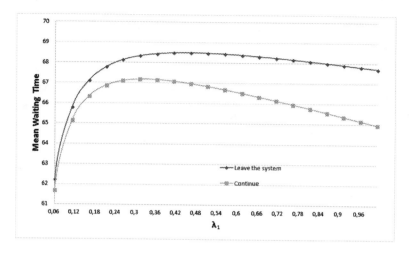

Fig. 2. Mean Waiting Time vs. λ_1

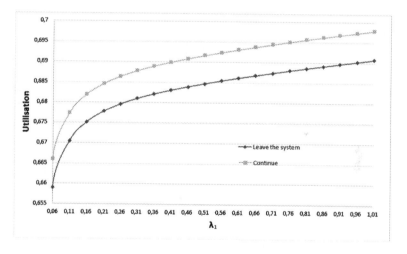

Fig. 3. Utilization vs. λ_1

customers are sent back to the orbit more frequently, which results higher waiting times. The waiting time of the 'leave the system' case is greater because the first order jobs go to the orbit and they have to try again.

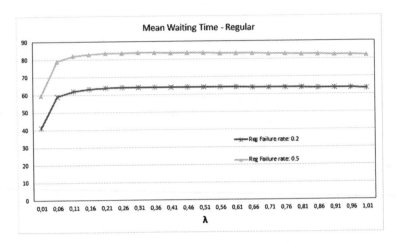

Fig. 4. Mean Waiting Time vs. λ_1

Figure 3 displays the utilization in function of the first order generation rate. When the failure rate is small, the difference between the two scenarios is not significant. This figure shows the situation when the failure rate is high, thus the differences in utilization are more significant. The utilization is greater for the 'Continue' scenario because after the repair the server state will be immediately busy. While for the other scenario the server will be idle, and a retrial, first or second order generation with an exponentially distributed time interval will take place.

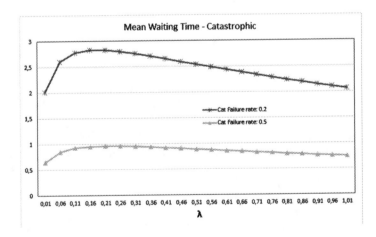

Fig. 5. Mean Waiting Time vs. λ_1

In Figs. 4 and 5 the running parameter is the generation rate for first order customers, λ_1, and the considered performance measure is the mean waiting time in the orbit. Here only the first order customers can take place. In Fig. 4 the case of regular breakdown is displayed for two different failure rates. Comparing the mean waiting times, a reverse effect can be observed. In case of regular breakdown (Fig. 4) for a higher failure rate, the waiting time is higher. The reason is that the customers spend more time in the orbit or at the server in down periods. While in case of catastrophic breakdown (Fig. 5) the waiting time is less for higher breakdown rates. Customers are sent to the sources in case of a breakdown.

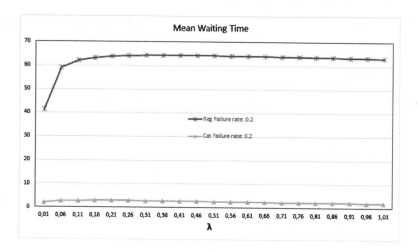

Fig. 6. Mean Waiting Time vs. λ_1

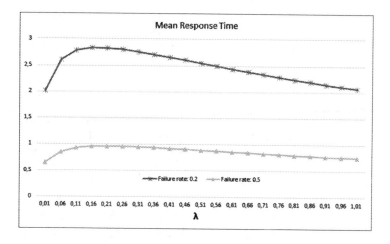

Fig. 7. Mean Response Time vs. λ_1

Figure 6 compares the waiting time between the regular and catastrophic breakdown at a given failure rate, $\gamma_0 = \gamma_1 = 0.2$. As it was expected, the waiting time is much higher for the regular breakdown.

In Fig. 7 and 8 the catastrophic breakdown is applied. The running parameter is $\lambda = \lambda_1$, the first order generation rate. In Fig. 7 the mean response time can be seen for two different failure rates ($\gamma_0 = \gamma_1$ in this figure). For a higher failure rate, lower response time can be observed because the jobs are more often kicked off to the source.

Figure 8 displays the server utilization. Here, the service rates for first (μ_1) and second order (μ_2) customers are different. $\mu_1 = 4$ and $\mu_2 = 2$. This is the reason, that the utilization is higher in the catastrophic case than in the normal breakdown case.

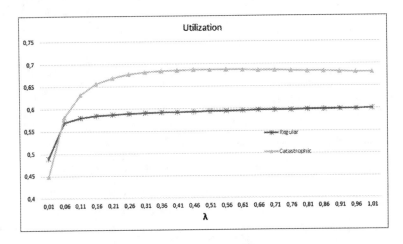

Fig. 8. Server utilization vs. λ_1

3   Conclusion

A special two-way communication system was investigated here. First order customers come from a finite source, while in the case of an idle server, second order customers can reach the system via a direct call. Different cases can be considered. Failure rates are set to be equal for idle server, for server with first order customer, and for server with second order customer. Two different cases were considered. The system is subject to regular breakdown and catastrophic breakdown. In case of regular breakdown, the "leave the system" and the "remain at server" scenarios were compared. Based on the numeric result, it can be stated, the second scenario is more effective regarding the response times, waiting times, and utilization. When the two breakdown models were compared with different failure rates, the expected reverse effect of the breakdown parameters can be observed.

References

1. Falin, G.I.: Waiting time in a single-channel queuing system with repeated calls. Mosc. Univ. Comput. Math. Cybern. **4**, 66–69 (1977)
2. Artalejo, J.R.: Retrial queues with a finite number of sources. J. Korean Math. Soc. **35**, 503–525 (1998)
3. Falin, G.I., Templeton, J.G.C.: Retrial Queues. Chapman and Hall, London (1997)
4. Templeton, J.G.C.: Retrial queues. Top **7**(2), 351–353 (1999)
5. Gans, N., Koole, G., Mandelbaum, A.: Telephone call centers: tutorial, review, and research prospects. Manuf. Serv. Oper. Manage. **5**(2), 79–141 (2003)
6. Artalejo, J.R., Gomez Corral, A.: Retrial Queueing Systems: A Computational Approach. Springer, Heidelberg (2008). https://doi.org/10.1007/978-3-540-78725-9
7. Falin, G.I., Artalejo, J.R.: A finite source retrial queue. Eur. J. Oper. Res. **108**, 409–424 (1998)
8. Falin, G.I.: Model of coupled switching in presence of recurrent calls. Eng. Cybern. **17**(1), 53–59 (1979)
9. Dimitriou, I.: A retrial queue to model a two-relay cooperative wireless system with simultaneous packet reception. In: Wittevrongel, S., Phung-Duc, T. (eds.) ASMTA 2016. LNCS, vol. 9845, pp. 123–139. Springer, Cham (2016). https://doi.org/10.1007/978-3-319-43904-4_9
10. Dragieva, V., Phung-Duc, T.: Two-way communication M/M/1 retrial queue with server-orbit interaction. In: Proceedings of the 11th International Conference on Queueing Theory and Network Applications, p. 11. ACM (2016)
11. Nazarov, A., Phung-Duc, T., Paul, S.: Heavy outgoing call asymptotics for MMPP/M/1/1 retrial queue with two-way communication. In: Dudin, A., Nazarov, A., Kirpichnikov, A. (eds.) ITMM 2017. CCIS, vol. 800, pp. 28–41. Springer, Cham (2017). https://doi.org/10.1007/978-3-319-68069-9_3
12. Nazarov, A.A., Paul, S., Gudkova, I., et al.: Asymptotic analysis of Markovian retrial queue with two-way communication under low rate of retrials condition. In: Proceedings 31st European Conference on Modelling and Simulation (2017)
13. Phung-Duc, T., Rogiest, W.: Two way communication retrial queues with balanced call blending. In: Al-Begain, K., Fiems, D., Vincent, J.-M. (eds.) ASMTA 2012. LNCS, vol. 7314, pp. 16–31. Springer, Heidelberg (2012). https://doi.org/10.1007/978-3-642-30782-9_2
14. Sakurai, H., Phung-Duc, T.: Two-way communication retrial queues with multiple types of outgoing calls. TOP **23**(2), 466–492 (2014). https://doi.org/10.1007/s11750-014-0349-5
15. Sakurai, H., Phung-Duc, T.: Scaling limits for single server retrial queues with two-way communication. Ann. Oper. Res. **247**(1), 229–256 (2015). https://doi.org/10.1007/s10479-015-1874-9
16. Aguir, S., Karaesmen, F., Akşin, O.Z., Chauvet, F.: The impact of retrials on call center performance. OR Spectr. **26**(3), 353–376 (2004)
17. Aksin, Z., Armony, M., Mehrotra, V.: The modern call center: a multi-disciplinary perspective on operations management research. Prod. Oper. Manag. **16**(6), 665–688 (2007)
18. Brown, L., et al.: Statistical analysis of a telephone call center: a queueing-science perspective. J. Am. Stat. Assoc. **100**(469), 36–50 (2005)
19. Pustova, S.V.: Investigation of call centers as retrial queuing systems. Cybern. Syst. Anal. **46**(3), 494–499 (2010)

20. Wolf, T.: System and method for improving call center communications. US Patent App. 15/604,068, 30 November 2017
21. Subramanian, S., et al.: A stochastic model for automated teller machines subject to catastrophic failures and repairs. Queueing Models Serv. Manage. 1(1), 75–94 (2018)
22. Thilaka, B., Poorani, B., Udayabaskaran, S.: Performance analysis for queueing systems with close down periods subject to catastrophe. Int. J. Pure Appl. Math. 119(7), 39–57 (2018)
23. Gupta, U.C., Kumar, N., Barbhuiya, F.P.: A queueing system with batch renewal input and negative arrivals. In: Joshua, V.C., Varadhan, S.R.S., Vishnevsky, V.M. (eds.) Applied Probability and Stochastic Processes. ISFS, pp. 143–157. Springer, Singapore (2020). https://doi.org/10.1007/978-981-15-5951-8_10
24. Piriadarshani, D., Narasimhan, S., Maheswari, M., James, B., et al.: A retrial queuing system operating in a random environment subject to catastrophes. Eur. J. Mol. Clin. Med. 7(2), 5029–5032 (2020)
25. Ammar, S.I., Zeifman, A., Satin, Y., Kiseleva, K., Korolev, V.: On limiting characteristics for a non-stationary two-processor heterogeneous system with catastrophes, server failures and repairs. J. Ind. Manage. Optim. 17(3), 1057 (2021)

Queueing System with Two Types of Customers and Limited Processor Sharing

Valentina Klimenok$^{(\boxtimes)}$ ⓘ, Alexander Dudin ⓘ, and Viktor Boksha

Department of Applied Mathematics and Computer Science,
Belarusian State University, 220030 Minsk, Belarus
{klimenok,dudin,fpm.boksha}@bsu.by

Abstract. In this paper a processor-sharing queueing system is investigated. Two types of customers enter the system according a marked Markovian arrival process. It is assumed that the number of customers of each type simultaneously being serviced is limited. The service times of customers have a phase type distribution the parameters of which depend both on the type of a customer and on the number of customers of this type in the system. The operation of the system is described in terms of a multi-dimensional Markov chain. We calculate the stationary probabilities, the main performance characteristics of the system and derive the Laplace–Stieltjes transform of the sojourn time distribution. We also present illustrative numerical examples to show the behavior of the performance measures of the system and to solve numerically an optimization problem.

Keywords: Processor sharing · Marked Markovian arrival process · Phase type distribution · Stationary distribution · Performance measures · Sojourn time

1 Introduction

Processor sharing technology is very popular in computer systems and telecommunications networks. It can be found a number of examples of real processor sharing systems and their mathematical models in the literature , see, e.g. the papers [1–8]. Most often, it is assumed that the processor can be used by an unlimited number of users, the input flow is stationary Poisson, and the service times are distributed exponentially. More general systems have been considered in the papers [9,10] where it was assumed that customers arrive into the system according to Markovian arrival process (MAP) and service times have a phase type distribution. In these papers, homogeneous traffic is assumed, which is not always suitable for describing next-generation wireless communication networks, implying, in particular, the use of the Internet of Things and the presence of interaction between H2H users and M2M devices, see, e.g. [6–8]. The presence of heterogeneous requests gives rise to the need to develop new mechanisms

© Springer Nature Switzerland AG 2022
A. Dudin et al. (Eds.): ITMM 2021, CCIS 1605, pp. 157–171, 2022.
https://doi.org/10.1007/978-3-031-09331-9_13

to maintain the specified quality of service parameters for both H2H users and M2M devices. At the same time, with an increase in the intensity of the proposed load, the planners at the base station of the LTE network must determine the optimal strategy for the allocation of radio resources based on the established restrictions, for example, the probability of loss of requests from H2H users and the average transmission time of data blocks from M2M devices.

The queueing system considered in this paper significantly expands the capabilities of modeling real systems with processor sharing. We believe that there are restrictions on the number of users of different types simultaneously in service, and we do not introduce restrictive assumptions such as the homogeneity and uncorrelated nature of the customers flow, as well as the exponential distribution of service times for customers of different types. We assume that the input flow to the system is correlated and described by the marked Markov arrival process $(MMAP)$ introduced in the paper [11]. For a more adequate description of the service process, we use a phase type distribution (PH) which is successfully used to approximate an arbitrary distribution.

Thus, in this paper we consider a queueing system with processor sharing which receives two types of customers arriving according to a $MMAP$. If at the moment of a customer arrival the number of customers of this type on the server is greater than a predetermined threshold, then the customer leaves the system un-handled, it is considered lost. Otherwise, the customer takes up part of the throughput of the channel and is serviced for a period of time having a PH distribution, the parameters of which differ for customers of different types.

2   Mathematical Model

We consider a queueing system with two type of customers and processor sharing. Customers of different types arrive into the system according to the $MMAP$ under control of the irreducible Markov chain ν_t, $t \geq 0$, which takes values in the set $\{0, 1, 2, \ldots, W\}$ and is called as an underlying process of the $MMAP$. The transitions of the underlying process accompanied by an arrival of a customer of type k are stored as entries of the matrix D_k, $k = 1, 2$, of order $\bar{W} \times \bar{W}$ where $\bar{W} = W + 1$ and idle transitions of this process are described by the matrix D_0.

The arrival rate of customers of type k in the $MMAP$ is given by $\lambda_k = \boldsymbol{\theta} D_k \mathbf{e}$, $k = 1, 2$, where the vector $\boldsymbol{\theta}$, is defined as the unique solution of the system $\boldsymbol{\theta} D(1) = \mathbf{0}, \boldsymbol{\theta}\mathbf{e} = 1$. The total arrival rate is $\lambda = \lambda_1 + \lambda_2$.

The variance of inter-arrival times of customers of type k is calculated by the formula

$$v_k = \frac{2\boldsymbol{\theta}(-D_0 - \sum_{l=1, l \neq k}^{2} D_l)^{-1}\mathbf{e}}{\lambda_k} - \left(\frac{1}{\lambda_k}\right)^2, \quad k = 1, 2.$$

The coefficient of correlation of the lengths of two adjacent intervals between the arrivals of customers of type k is calculated by

$$c_{cor}^{(k)} = \left[\frac{\boldsymbol{\theta}(D_0 + \sum\limits_{l=1,l\neq k}^{2} D_l)^{-1}}{\lambda_k} D_k (D_0 + \sum\limits_{l=1,l\neq k}^{2} D_l)^{-1}\mathbf{e} - \left(\frac{1}{\lambda_k}\right)^2 \right] v_k^{-1}, \ k = 1, 2.$$

A detailed description of the $MMAP$ can be found, for example, in [11].

In this paper we assume that the server can simultaneously serve up to N customers of type 1 and up to R customers of type 2. If only one customer of the kth type is serviced on the server, then its service time has the PH distribution given by the irreducible representation $(\boldsymbol{\beta}_k, S_k)$ and the underlying process $m_t^{(k)}$, $t \geq 0$, with the state space $\{1, \ldots, M_k, M_k + 1\}$, where the state $M_k + 1$ is absorbing. The intensities of transitions to the absorbing state are determined by the column vector $\boldsymbol{S}_0^{(k)} = -S_k\mathbf{e}$. The service rate of a customer of type k is calculated as $\mu_k = (\boldsymbol{\beta}_k(-S_k)^{-1}\mathbf{e})^{-1}$.

The customers of each type divide the throughput of the server allocated to them equally. If the server simultaneously serves n_k customers of the kth type, then the service time of any of these customers has the PH distribution given by the irreducible representation $(\boldsymbol{\beta}_k, \frac{1}{n_k}S_k)$ and the underlying process $m_t^{(k)}$, $t \geq 0$, with the state space $\{1, \ldots, M_k, M_k + 1\}$, where the state $M_k + 1$ is absorbing. The intensities of transitions to the absorbing state are determined by the column vector $\frac{1}{n_k}\boldsymbol{S}_0^{(k)}$.

If an incoming customer of type 1 finds $n < N$ customers on the server, then it is sent for service. In this case, the throughput of the server allocated to customers of the 1st type is divided equally between $n+1$ customers. Otherwise, the customer leaves the system un-handled, it is considered lost. Similarly, if a customer of the 2nd type finds $r < R$ customers on the server, then it is sent for service. The throughput of the server allocated to customers of type 2 is divided equally between $r + 1$ customers. Otherwise, the customer is lost.

3   Process of the System States

The operation of the system is described by the regular irreducible Markov chain

$$\xi_t = \{n_t, r_t, \eta_t^{(1)}, \eta_t^{(2)}, \ldots, \eta_t^{(M_1)}, \tau_t^{(1)}, \tau_t^{(2)}, \ldots, \tau_t^{(M_2)}, \nu_t\},$$

where at the moment t

- n_t is the number of customers of type 1 on the server, $n_t = \overline{0, N}$;
- r_t is the number of customers of type 2 on the server, $r_t = \overline{0, R}$;
- $\eta_t^{(m^{(1)})}$ is the number of customers of type 1 that are served in the phase $m^{(1)}$, $\eta_t^{(m^{(1)})} = \overline{0, n_t}$, $m^{(1)} = \overline{1, M_1}$;
- $\tau_t^{(m^{(2)})}$ is the number of customers of type 2 that are served in the phase $m^{(2)}$, $\tau_t^{(m^{(2)})} = \overline{0, r_t}$, $m^{(2)} = \overline{1, M_2}$;
- ν_t is the state of underlying process of the $MMAP$, $\nu_t = \overline{0, W}$,

In the following we will also use the processes

$$\mathbf{u}_t^{(1)} = \{\eta_t^{(1)}, \eta_t^{(2)}, \ldots, \eta_t^{(M_1)}\}; \ \mathbf{u}_t^{(2)} = \{\tau_t^{(1)}, \tau_t^{(2)}, \ldots, \tau_t^{(M_2)}\}.$$

Let us arrange the states of the considered Markov chain $\xi_t, t \geq 0$, as follows. We enumerate the components n_t, r_t in the direct lexicographic order and, for fixed values of these components, we renumber the states of the processes $\mathbf{u}_t^{(1)}$ and $\mathbf{u}_t^{(2)}$ in the reverse lexicographic order.

To further describe the transition rates of the chain, we need the matrices $P_i(\cdot)$, $A_i(\cdot, \cdot)$, and $L_i(\cdot, \cdot)$, which have the following probabilistic sense: the matrix $L_l(n, \tilde{S}_k)$ contains the transition rates of the process $\mathbf{u}_t^{(k)}$, leading to the end of servicing of one of $n - l$ customers of the kth type; the matrix $P_n(\beta_k)$ contains the transition probabilities of the process $\mathbf{u}_t^{(k)}$ leading to an increase in the number of customers of the kth type on the server from n to $n + 1$; the matrix $A_n(l, \tilde{S}_k)$ contains the transition rates of the process $\mathbf{u}_t^{(k)}$ in its state space without increasing or decreasing the number of customers of the kth type. Here $\tilde{S}_k = \begin{pmatrix} 0 & O \\ S_0^{(k)} & S_l \end{pmatrix}$, $k = 1, 2$. Algorithm for calculating matrices $P_i(\cdot)$, $A_i(\cdot, \cdot)$, and $L_i(\cdot, \cdot)$ follows from the results of V. Ramaswami and D. Lucantoni published in the papers [12, 13].

Let us introduce the notation $Q_{n,n'}$ for the matrices of transition rates of the chain from the states corresponding to the value n of the first component to the states corresponding to the value n' of this component, $n, n' = \overline{0, N}$. We also introduce the following notation:

- $C_n^m = \begin{pmatrix} n \\ m \end{pmatrix} = \frac{n!}{m!(n-m)!}$;
- $diag\{a_1, a_2, \ldots, a_n\}$ is a block diagonal matrix in which the diagonal blocks are equal to the elements listed in brackets, and the other blocks are zero;
- $diag^+\{a_1, a_2, \ldots, a_n\}$ ($diag^-\{a_1, a_2, \ldots, a_n\}$) is a square block matrix in which the off-diagonal (below-diagonal) blocks are equal to the elements listed in brackets, and the other blocks are zero.

Lemma 1. *The infinitesimal generator Q of a Markov chain $\xi_t, t \geq 0$, has the block three-diagonal structure*

$$Q = \begin{pmatrix} Q_{0,0} & Q_{0,1} & O & \cdots & O & O \\ Q_{1,0} & Q_{1,1} & Q_{1,2} & \cdots & O & O \\ O & Q_{2,1} & Q_{2,2} & \cdots & O & O \\ \vdots & \vdots & \vdots & \ddots & \vdots & \vdots \\ O & O & O & \cdots & Q_{N-1,N-1} & Q_{N-1,N} \\ O & O & O & \cdots & Q_{N,N-1} & Q_{N,N} \end{pmatrix},$$

where

$$Q_{0,0} = diag^- \{\tfrac{1}{r} L_{R-r}(R, \tilde{S}_2), r = \overline{1, R}\} \otimes I_{\bar{W}}$$

$$+ diag\{0, \tfrac{1}{r} A_r(R, S_2), r = \overline{1, R}\} \oplus D_0 + diag\{O_{R-1 \atop \sum\limits_{r=0} C_{r+M_2-1}^{M_2-1}}, I_{C_{R+M_2-1}^{M_2-1}}\} \otimes D_2$$

$$+ diag^+ \{P_r(\beta_2), r = \overline{0, R-1}\} \otimes D_2 + \Delta_0;$$

$$Q_{n,n+1} = P_n(\beta_1) \otimes I_{R \atop \sum\limits_{r=0} C_{r+M_2-1}^{M_2-1}} \otimes D_1, \ 0 \leq n \leq N-1;$$

$$Q_{n,n-1} = \tfrac{1}{n} L_{N-n}(N, \tilde{S}_1) \otimes I_{R \atop \sum\limits_{r=0} C_{r+M_2-1}^{M_2-1}} \otimes I_{\bar{W}}, \ 1 \leq n \leq N;$$

$$Q_{n,n} = I_{C_{n+M_1-1}^{M_1-1}} \otimes diag^- \{\tfrac{1}{r} L_{R-r}(R, \tilde{S}_2), r = \overline{1, R}\} \otimes I_{\bar{W}}$$

$$+ \tfrac{1}{n} A_n(N, \tilde{S}_1) \oplus diag\{0, \tfrac{1}{r} A_r(R, S_2), r = \overline{1, R}\} \oplus (D_0 + \delta_{n,N} D_1)$$

$$+ I_{C_{n+M_1-1}^{M_1-1}} \otimes diag\{O_{R-1 \atop \sum\limits_{r=0} C_{r+M_2-1}^{M_2-1}}, I_{C_{R+M_2-1}^{M_2-1}}\} \otimes D_2$$

$$+ I_{C_{n+M_1-1}^{M_1-1}} \otimes diag^+ \{P_r(\beta_2), r = \overline{0, R-1}\} \otimes D_2 + \Delta_n, \ 1 \leq n \leq N,$$

where $\otimes(\oplus)$ *denotes the Kronecker product (sum) of matrices,* $\delta_{n,N}$ *is the Kronecker symbol,* $\Delta_n, n = \overline{0, N}$, *are diagonal matrices, which are constructed so that the equality* $Q\mathbf{e} = \mathbf{0}^T$ *holds.*

Proof. The generator block $Q_{0,0}$ contains the transition rates in the set of states corresponding to the absence of customers of type 1. The corresponding transitions occur when

a) one of the customers of type 2 finishes the service. The corresponding rates are given by the matrix $diag^- \{\tfrac{1}{r} L_{R-r}(R, \tilde{S}_2), r = \overline{1, R}\} \otimes I_{\bar{W}}$;
b) the number of customers of type 2 that are in a certain phase of servicing is changed or the $MMAP$ underlying process makes an idle transition. The corresponding rates are given by the matrix $diag\{0, \tfrac{1}{r} A_r(R, S_2), r = \overline{1, R}\} \oplus D_0$;
c) a customer of type 2 arrives and take place on the server (the matrix $diag^+ \{P_r(\beta_2), r = \overline{0, R-1}\} \otimes D_2$) or, if all places for customers of this type are occupied, the customer leaves the system un-handled (the matrix $diag\{O_{R-1 \atop \sum\limits_{r=0} C_{r+M_2-1}^{M_2-1}}, I_{C_{R+M_2-1}^{M_2-1}}\} \otimes D_2$.

Block $Q_{n,n}, n = \overline{1, N}$, contains the transition rates in the set of states corresponding to the presence of n customers of type 1 in the system. The expression for this block differs from the expression for the block $Q_{0,0}$ only in the second term, which in this case specifies the transition rates of the processes of servicing customers of types 1 and 2 in their sets of states without changing their numbers or the $MMAP$ idle transition, or the loss of customer of type 1.

Block $Q_{n,n+1}, n = \overline{0, N-1}$, contains the rates of transitions accompanied by the arrival of a customer of type 1 which takes up place on the server.

Block $Q_{n,n-1}, n = \overline{1, N}$, contains the rates of transitions accompanied by the departure of the serviced customer of type 1 from the system.

All other blocks of the generator are zero matrces, since they consist of the rates of two or more transitions of the considered Markov chain on an infinitely small time interval.

4   Stationary Distribution. Performance Measures

In accordance with the described ordering of the states of the Markov chain ξ_t, we form the row vectors $\mathbf{p}_n, n = \overline{0, N}$, of the stationary probabilities of the states of the chain corresponding to the value n of the first component n_t. Let $\mathbf{p} = (\mathbf{p}_0, \mathbf{p}_1, \ldots, \mathbf{p}_N)$ be the vector of steady state probabilities of the chain. This vector is the unique solution to the system of linear algebraic equation $\mathbf{p}Q = \mathbf{0}, \mathbf{p}\mathbf{e} = 1$. If the dimension of this system is large, the solution can be calculated using the algorithm developed in [14].

Based on the stationary distribution, we can obtain formulas for calculating a number of stationary performance characteristics of the system. Below we present some of them.

- Joint distribution of the number of type 1 customers on the server, the number of type 1 customers in different service phases, and the states of the $MMAP$

$$\mathbf{p}_n^* = \mathbf{p}_n(I_{C_{n+M_1-1}^{M_1-1}} \otimes \mathbf{e}_{\underset{r=0}{\overset{R}{\sum}} C_{r+M_2-1}^{M_2-1}} \otimes I_{\bar{W}}), \ n = \overline{0, N}.$$

- Distribution of the number of customers of type 1 in the system $p_n = \mathbf{p}_n^* \mathbf{e}, \ n = \overline{0, N}$.
- Joint distribution of the number of type 2 customers on the server, the number of type 2 customers in different service phases, and the states of the $MMAP$

$$\mathbf{q}_r^* = \sum_{n=0}^{N} \mathbf{p}_n(I^{(n,r)} \otimes I_{\bar{W}}), r = \overline{0, R},$$

where

$$I^{(n,r)} = \begin{pmatrix} O \\ C_{n+M_1-1}^{M_1-1} \overset{r-1}{\underset{m=0}{\sum}} C_{m+M_2-1}^{M_2-1} \times C_{r+M_2-1}^{M_2-1} \\ \mathbf{e}_{C_{n+M_1-1}^{M_1-1}} \otimes I_{C_{r+M_2-1}^{M_2-1}} \\ O \\ C_{n+M_1-1}^{M_1-1} \overset{R}{\underset{m=r+1}{\sum}} C_{m+M_2-1}^{M_2-1} \times C_{r+M_2-1}^{M_2-1} \end{pmatrix}.$$

- Distribution of the number of customers of type 2 in the system $q_r = \mathbf{q}_r^* \mathbf{e}, r = \overline{0, R}$.
- The probability of losing a customer of the kth type

$$P_{loss,k} = \frac{\lambda_k - \varphi_k}{\lambda_k}, \ k = 1, 2,$$

where λ_k is the arrival rate of customers of kth type, φ_k is the output rate of customers of kth type. The value of φ_k is calculated as

$$\varphi_1 = \sum_{n=1}^{N} \mathbf{p}_n^*(I_{C_{n+M_1-1}^{M_1-1}} \otimes \mathbf{e}_{\bar{W}})\frac{1}{n}L_{N-n}(N, \tilde{S}_1)\mathbf{e},$$

$$\varphi_2 = \sum_{r=1}^{R} \mathbf{q}_r^*(I_{C_{r+M_2-1}^{M_2-1}} \otimes \mathbf{e}_{\bar{W}})\frac{1}{r}L_{R-r}(R, \tilde{S}_2)\mathbf{e}.$$

5   Sojourn Time Distribution

Denote by $v_n^{(1)}(\eta^{(1)}, \eta^{(2)}, \ldots, \eta^{(M_1)}, \tilde{\eta}, \nu, s)$ the Laplace-Stieltjes transform (LST) of the virtual sojourn time distribution of a customer of type 1 for which service began with the phase $\tilde{\eta}$, and which found n customers of the first type in the system, the number of customers in phase $m^{(1)}$ equal to $\eta^{(m^{(1)})}$, and the underlying process of the $MMAP$ in the state ν, $n = \overline{0, N-1}$, $\eta^{(m^{(1)})} = \overline{0, n}$, $m^{(1)} = \overline{1, M_1}$, $\nu = \overline{0, W}$.

Similarly, let $v_r^{(2)}(\tau^{(2)}, \ldots, \tau^{(M_2)}, \tilde{\tau}, \nu, s)$ be the Laplace-Stieltjes transform of the virtual sojourn time distribution of a customer of type 2 for which service began with the phase $\tilde{\tau}$, and which found in the system r customers of the second type, the number of customers in the phase $m^{(2)}$ equal to $\tau^{(m^{(2)})}$, and the underlying process of the $MMAP$ in the state ν, $r = \overline{0, R-1}$, $\tau^{(m^{(2)})} = \overline{0, r}$, $m^{(2)} = \overline{1, M_1}$, $\nu = \overline{0, W}$. First we derive formulas of the conditional LSTs $v_n^{(1)}(\eta^{(1)}, \eta^{(2)}, \ldots, \eta^{(M_1)}, \tilde{\eta}, \nu, s)$. Let us arrange these LSTs in the reverse lexicographic order of arguments $\eta^{(1)}, \eta^{(2)}, \ldots, \eta^{(M_1)}$, in the direct lexicographic order of arguments $\tilde{\eta}, \nu$ and form the column vectors

$$\mathbf{v}_n^{(1)}(s), \; n = \overline{0, N-1}, \quad \mathbf{v}^{(1)}(s) = ((\mathbf{v}_0^{(1)}(s))^T, (\mathbf{v}_1^{(1)}(s))^T, \ldots, (\mathbf{v}_{N-1}^{(1)}(s))^T)^T.$$

Similarly, for customers of type 2, we form the column vectors

$$\mathbf{v}_r^{(2)}(s), \; r = \overline{0, R-1}, \quad \mathbf{v}^{(2)}(s) = ((\mathbf{v}_0^{(2)}(s))^T, (\mathbf{v}_1^{(2)}(s))^T, \ldots, (\mathbf{v}_{R-1}^{(2)}(s))^T)^T.$$

Theorem 1. *The Laplace-Stieltjes transform vector* $\mathbf{v}^{(1)}(s)$ *is calculated as follows:*

$$\mathbf{v}^{(1)}(s) = (sI - A^{(1)})^{-1}\mathbf{b}^{(1)}, \tag{1}$$

where

$$A^{(1)} = diag\{[\tfrac{1}{n+1}A_n(N, S_1) + \Delta_n] \oplus S_1 \oplus (D_0 + D_2), n = \overline{0, N-1}\}$$

$$+ diag^-\{\tfrac{1}{n+1}L_{N-n}(N, \tilde{S}_1) \otimes I_{M_1\bar{W}}, n = \overline{1, N-1}\}$$

$$+ diag^+\{P_n(\boldsymbol{\beta}_1) \otimes I_{M_1} \otimes D_1, n = \overline{0, N-2}\}$$

$$+ diag\{O_{\bar{W}\sum_{n=0}^{N-2}C_{n+M_1-1}^{M_1-1}}, I_{C_{N+M_1-2}^{M_1-1}}\} \otimes I_{M_1} \otimes D_1\},$$

$$\mathbf{b}^{(1)} = diag\{I_{C_{n+M_1-1}^{M_1-1}} \otimes \tfrac{1}{n+1}\boldsymbol{S}_0^{(1)} \otimes I_{\bar{W}}, n = \overline{0, N-1}\}\mathbf{e}.$$

Proof. Using the probabilistic interpretation of the Laplace-Stieltjes transform, we obtain the following equations for the vectors $\mathbf{v}_n^{(1)}(s), n = \overline{0, N-1}$:

$$\mathbf{v}_n^{(1)}(s) =$$

$$\int_0^\infty e^{-st} e^{[\frac{1}{n+1}A_n(N,S_1)+\Delta_n]\oplus S_1 t}(I_{C_{n+M_1-1}^{M_1-1}} \otimes \frac{1}{n+1}S_0^{(1)}) \otimes e^{(D_0+D_2)t} dt e$$

$$+ \int_0^\infty e^{-st} e^{[\frac{1}{n+1}A_n(N,S_1)+\Delta_n]\oplus S_1 t}(L_{N-n}(N,\tilde{S}_1) \otimes I_{M_1}) \otimes e^{(D_0+D_2)t} dt e \mathbf{v}_{n-1}^{(1)}(s)$$

$$+ \int_0^\infty (e^{-st} e^{[\frac{1}{n+1}A_n(N,S_1)+\Delta_n]\oplus S_1 t} \otimes e^{(D_0+D_2)t})(P_n(\boldsymbol{\beta}_1) \otimes I_{M_1} \otimes D_1) dt$$

$$\times \mathbf{v}_{\min\{n+1,N-1\}}^{(1)}(s)e. \tag{2}$$

Let us explain the meaning of the terms on the right-hand side of (2):

- the first integral (first term) is the probability that the incoming virtual customer will be serviced before any of the n customers of type 1 that are already on the server at the time of the virtual customer arriving, and during the time of servicing the virtual customer there will be no catastrophe.
- the integral in the second term is the vector of probabilities that after the arrival of the virtual customer one of the n customers of type 1 that are already on the server at the time of the arrival of the virtual customer will be served first, and no catastrophe will occur during the service of this first customer. After the first of the mentioned n customers is served, the server resource is redistributed between the remaining i customers, including the virtual one, and the further scenario of servicing the virtual customer up to the distribution of the $MMAP$ states and service phases will be the same as at the moment of the arrival of a virtual customer that found $n-1$ customers in the system. By definition, the corresponding vector of $LSTs$ is $\mathbf{v}_{n-1}^{(1)}(s)$ The product of the integral and $\mathbf{v}_{n-1}^{(1)}(s)$ will give the required vector of $LSTs$ of the service time distribution of the virtual customer.
- when describing the third term, we will distinguish between the cases $n < N-1$ and $n = N-1$. In both cases, the integral in the third term is a vector of probabilities that after the arrival of the virtual customer, the first event that entails a change in the number of customers on the server will be the arrival of a customer of type 1 and no catastrophe will occur in the time before it arrives. In the case $n < N-1$, after this customer arrives, the server resource will be redistributed between $n+2$ customers, including the virtual one, and the further scenario of servicing the virtual customer up to the distribution of the $MMAP$ states and servicing phases will be the same as at the moment of arrival of the virtual customer that found $n+1$ customers in the system. By definition, the corresponding vector of $LSTs$ is $\mathbf{v}_{n+1}^{(1)}(s)$. The product of

the integral and $\mathbf{v}_{n+1}^{(1)}(s)$ will give the required vector of LSTs of service time of the virtual customer. In the case $n = N - 1$ the received customer will be rejected, since the server already contains N customers, including the virtual one. Then the further scenario of servicing the virtual customer, up to the distribution of the states of the $MMAP$, will be the same as at the moment of the arrival of the virtual customer that found $N - 1$ customers in the system. By definition, the corresponding vector of LSTs is $\mathbf{v}_{N-1}^{(1)}(s)$. The product of the integral and $\mathbf{v}_{N-1}^{(1)}(s)$ will give the required vector of LSTs of service time of the virtual customer.

After calculating the integrals in (2) and a number of algebraic transformations, we obtain the required formula (1).

Corollary 1. *The Laplace-Stieltjes transform vector $\mathbf{v}^{(2)}(s)$ is calculated as follows:*

$$\mathbf{v}^{(2)}(s) = (sI_{\bar{W}} - A^{(2)})^{-1}\mathbf{b}^{(2)},$$

where the matrix $A^{(2)}$ and the vector $\mathbf{b}^{(2)}$ are obtained from the matrix $A^{(1)}$ and the vector $\mathbf{b}^{(1)}$, respectively, by replacing N by R and permutation of indices 1 and 2.

Having known the Laplace-Stieltjes transforms defined in Theorem 1 and Corollary 1, we can find all the moments of the sojourn time, in particular, the mean and the variance of this time.

The corresponding mean (variance) for customers of type 1 we denote as $\bar{v}_n^{(1)}(\eta^{(1)}, \eta^{(2)}, \ldots, \eta^{(M_1)}, \tilde{\eta}, \nu)$ $(d_n^{(1)}(\eta^{(1)}, \eta^{(2)}, \ldots, \eta^{(M_1)}, \tilde{\eta}, \nu))$ and for customers of type 2 as $\bar{v}_r^{(2)}(\tau^{(1)}, \ldots, \tau^{(M_2)}, \tilde{\tau}, \nu)$ $(d_n^{(2)}(\tau^{(1)}, \tau^{(2)}, \ldots, \tau^{(M_2)}, \tilde{\tau}, \nu))$.

We renumber the values $\bar{v}_n^{(1)}(\eta^{(1)}, \eta^{(2)}, \ldots, \eta^{(M_1)}, \tilde{\eta}, \nu)$, $(\bar{v}_n^{(1)}(\eta^{(1)}, \eta^{(2)}, \ldots, \eta^{(M_1)}, \tilde{\eta}, \nu))^2$ and $d_n^{(1)}(\eta^{(1)}, \eta^{(2)}, \ldots, \eta^{(M_1)}, \tilde{\eta}, \nu)$ in the lexicographic order described above and form the corresponding column vectors

$$\bar{\mathbf{v}}_n^{(1)}, \; \bar{\bar{\mathbf{v}}}^{(1)}{}_n, \; \bar{\mathbf{d}}_n^{(1)}, \; n = \overline{0, N-1}.$$

In turn, from these vectors we form the column vectors

$$\bar{\mathbf{v}}^{(1)} = ((\bar{\mathbf{v}}_0^{(1)})^T, (\bar{\mathbf{v}}_1^{(1)})^T, \ldots, (\bar{\mathbf{v}}_{N-1}^{(1)})^T)^T, \; \bar{\bar{\mathbf{v}}}^{(1)} = ((\bar{\bar{\mathbf{v}}}^{(1)}{}_0)^T, (\bar{\bar{\mathbf{v}}}^{(1)}{}_1)^T, \ldots, (\bar{\bar{\mathbf{v}}}^{(1)}{}_{N-1})^T)^T,$$

$$\mathbf{d}^{(1)} = ((\mathbf{d}_0^{(1)})^T, (\mathbf{d}_1^{(1)})^T, \ldots, (\mathbf{d}_{N-1}^{(1)})^T)^T.$$

By analogy we introduce the column vectors $\bar{\mathbf{v}}^{(2)}, \bar{\bar{\mathbf{v}}}^{(2)}, \mathbf{d}^{(2)}$.

Corollary 2. *The vector of conditional means, $\bar{\mathbf{v}}^{(k)}$, and the vector of conditional variances, $\mathbf{d}(k)$, of the sojourn times of a customer of type k are calculated by the following formulas:*

$$\bar{\mathbf{v}}^{(k)} = -(A^{(k)})^{-1}\mathbf{e}, \; \mathbf{d}^{(k)} = 2(A^{(k)})^{-2}\mathbf{e} - \bar{\bar{\mathbf{v}}}^{(k)}, \; k = 1, 2.$$

To calculate the Laplace-Stieltjes transforms of the sojourn time distributions of the customers of type 1 and 2 admitted into the system, we introduce into consideration the vector $\mathbf{p}^+$ ($\mathbf{q}^+$), the components of which define the joint distribution of the number of type 1 (type 2) customers that are in different service phases and the states of the $MMAP$ immediately after the moment the customer of type 1 (type 2) has been admitted into the system. It is easy to see that these vectors are calculated as follows:

$$\mathbf{p}^+ = \lambda_1^{-1}(\mathbf{p}_0^*, \mathbf{p}_1^*, \ldots, \mathbf{p}_{N-1}^*)[diag\{P_n(\boldsymbol{\beta}_1), n = \overline{0, N-1}\} \otimes D_1],$$

$$\mathbf{q}^+ = \lambda_2^{-1}(\mathbf{q}_0^*, \mathbf{q}_1^*, \ldots, \mathbf{q}_{R-1}^*)[diag\{P_r(\boldsymbol{\beta}_2), r = \overline{0, R-1}\} \otimes D_2].$$

Theorem 2. *The Laplace-Stieltjes transformations of the sojourn time distributions of the customers of type 1 and type 2 accepted to the system are calculated as*

$$v^{(1)}(s) = \mathbf{p}^+\mathbf{v}^{(1)}(s), \quad v^{(2)}(s) = \mathbf{q}^+\mathbf{v}^{(2)}(s).$$

Corollary 3. *The means and variances of the sojourn times of customers of type 1 and type 2 accepted to the system are calculated using the following formulas:*

$$\bar{v}^{(1)} = \mathbf{p}^+\bar{\mathbf{v}}^{(1)}, \; d^{(1)} = \mathbf{p}^+\mathbf{d}^{(1)}; \quad \bar{v}^{(2)} = \mathbf{q}^+\bar{\mathbf{v}}^{(2)}, \; d^{(2)} = \mathbf{q}^+\mathbf{d}^{(2)}.$$

6   Numerical Results

In this section we conduct a number of numerical experiments aimed at studying the behavior of the performance characteristics of the system depending on its parameters and at solving optimization problems. To carry out the experiments, a computer program was written in Python using built-in packages for processing matrices, calculating complex mathematical formulas and executed in the PyCharm 2019.3.4 (Professional Edition) program.

In **Experiment 1** we analyse the dependence of the loss probabilities, $P_{loss,k}, k = 1, 2$, and the mean sojourn times, $\bar{v}^{(k)}, k = 1, 2$, on the maximum number of channels allocated for customers of type k. In this experiment we used the following input data.

The $MMAP$ is specified by the matrices D_0, D_1, D_2, where

$$D_0 = \begin{pmatrix} -86 & 0.01 \\ 0.02 & -2.76 \end{pmatrix}, \quad D_1 = \begin{pmatrix} 59.5 & 0.693 \\ 0.14 & 1.778 \end{pmatrix}, \quad D_2 = \begin{pmatrix} 25.5 & 0.297 \\ 0.06 & 0.762 \end{pmatrix}.$$

With such matrices $\lambda = 12.43$, $\lambda_1 = 0.7\lambda$ and $\lambda_2 = 0.3\lambda$, $c_{cor}^{(1)} = 0.39$, $c_{cor}^{(2)} = 0.33$.

The PH distribution of the service time of a single customer of type 1 is given by the vector $\boldsymbol{\beta}^{(1)} = (1, 0)$ and the matrix $S^{(1)} = \begin{pmatrix} -80 & 80 \\ 0 & -80 \end{pmatrix}$. This means that the service time has Erlang distribution of order 2 with parameter 80 and the service rate $\mu_1 = 40$.

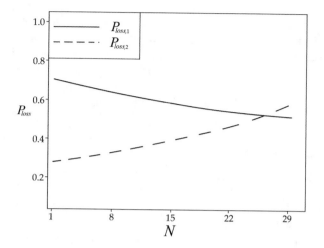

Fig. 1. $P_{loss,1}$ and $P_{loss,2}$ vs N under fixed number of channels $N + R = 30$

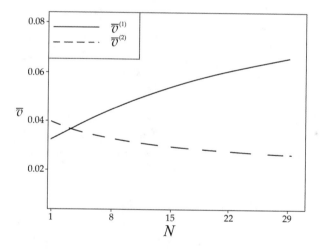

Fig. 2. $\bar{v}^{(1)}$ and $\bar{v}^{(2)}$ vs N under fixed number of channels $N + R = 30$

The PH distribution of the service time of a single customer of type 2 is given by the vector $\boldsymbol{\beta}^{(2)} = (1, 0)$ and the matrix $S^{(2)} = \begin{pmatrix} -20 & 20 \\ 0 & -20 \end{pmatrix}$. This means that the service time has Erlang distribution of order 2 with parameter 20 and the service rate $\mu_2 = 10$.

The total number of channels, into which the throughput of the servers is divided, is $N + R = 30$.

It is seen from Fig. 1 that $P_{loss,1}$ decreases and $P_{loss,2}$ increases. This is due to the fact that with an increase in N the possible number of type 1 customers in the system increases and the smaller part of the customers will be lost. Taking into account the equality $N + R = 30$, with an increase in N the value of R decreases and more and more customers are lost.

Figure 2 shows that $\bar{v}^{(1)}$ is an increasing function of N. This is due to the fact that with an increase in N the throughput allocated for a customer of type 1 decreases and hence the sojourn time increases. Due to the relation $N + R = 30$ when N increases then R decreases. That entails an increase in the throughput available for a customer of type 2 and a decrease in the time for servicing the customer.

Experiment 2. In this experiment, we solve numerically the optimization problem which consists in the optimal sharing of the throughput $\mu = \mu_1 + \mu_2$ of the server between customers of types 1 and 2 and the optimal choice of the maximum numbers of simultaneously served customers of types 1 and 2 under the given restrictions on the minimum throughput allocated for each customer.

As a criterion for the quality of the operation of the system, we use the economic functional, which is the average penalty per unit of time

$$J = a\bar{N} + c_1\lambda_1 P_{loss,1} + c_2\lambda_2 P_{loss,2}, \tag{3}$$

where a is the penalty charged per unit of time spent by one customer of type 1 in the system, c_k is the penalty charged for the lost customer of type k, $k = 1, 2$.

The problem is to choose the parameters μ_1, N and R which provide the minimum to criterion (3) under the following conditions:

$$\mu_1 + \mu_2 = \mu = const, \quad \gamma_1 = \frac{\mu_1}{N} = const, \quad \gamma_2 = \frac{\mu_2}{R} = const.$$

Here γ_k is the minimum throughput of the server that can be used to provide service to a customer of type k.

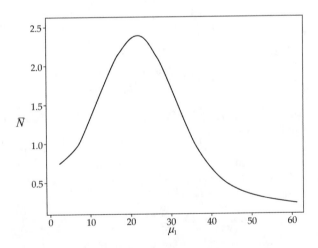

Fig. 3. $\bar{N}$ vs μ_1 under restrictions $\mu = 70, \gamma_1 = 2, \gamma_2 = 7$

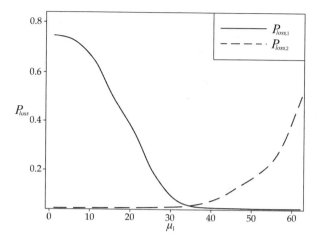

Fig. 4. $P_{loss,1}, P_{loss,2}$ vs μ_1 under restrictions $\mu = 70, \gamma_1 = 2, \gamma_2 = 7$

In the experiment, we will use the $MMAP$ specified in Experiment 1. The shape of service time distributions is the same as in Experiment 1. In the course of the current experiment, we will only change the service rates μ_1 and μ_2 multiplying the matrices S_1, S_2 by the corresponding constants. We fix the values of μ, γ_1, γ_2 as $\mu = 70, \gamma_1 = 2, \gamma_2 = 7$.

For these initial data, let us look at the graphs of the dependence of the mean number of customers of the type 1, $\bar{N}$, and the probabilities of losses of customers of different types, $P_{loss,1}, P_{loss,2}$, which are shown in Fig. 3 and 4.

Having calculated the dependence of $\bar{N}, P_{loss,1}, P_{loss,2}$ on μ_1 we can calculate the dependence of the cost criteria J on μ_1 under different cost coefficients. Let

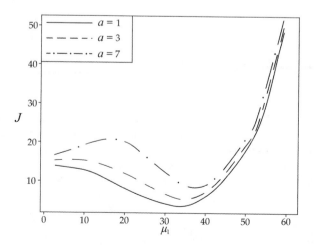

Fig. 5. J vs μ_1 for $c_1 = 1, c_2 = 20, a = 1, 3, 7$ under restrictions $\mu = 70, \gamma_1 = 2, \gamma_2 = 7$

us consider the following cost coefficients: $a = 1, 3, 7, c_1 = 1, c_2 = 20$. The results of calculation are presented in Fig. 5 and in Table 1.

Table 1. Values of N, R, J as functions of μ_1 for $c_1 = 1$, $c_2 = 20$, $a = 1, 3, 7$ under restrictions $\mu = 70, \gamma_1 = 2, \gamma_2 = 7$

μ_1	N	R	$J, a = 1$	$J, a = 3$	$J, a = 7$
2	1	9	13.09	14.60	17.64
5	2	9	12.81	14.76	18.67
10	5	8	11.69	14.91	21.34
15	7	7	9.24	13.58	22.24
20	10	7	6.54	11.40	21.11
25	12	6	3.96	8.17	16.57
30	15	5	2.45	5.58	11.83
35	17	5	**2.15**	**4.04**	7.83
40	20	4	4.13	5.32	**7.72**
45	22	3	8.82	9.65	11.32
50	25	2	16.99	17.62	18.89
55	27	2	22.84	23.35	24.37
60	30	1	44.79	45.22	46.07
63	31	1	50.32	50.70	51.48

It is seen from Fig. 5 and Table 1 that for input data under consideration the server throughput is divided approximately in half between customers of types 1 and 2. In the case a = 1, 3, it is optimal to divide the throughput allocated to customers of types 1 and 2 as 17:5. When a = 7, this proportion changes as 20:4.

7   Conclusion

We analysed a queuing system with the $MMAP$ of customers of two types, processor sharing and a limited number of places for customers of different types. We described the system operation by the multi-dimensional Markov chain, calculated its stationary distribution and the main performance characteristics. The Laplace-Stieltjes transform of the sojourn time distribution is found. Formulas for means and variances of the sojourn time are obtained. We carried out numerical experiments to study the behavior of the system performance characteristics and to find the optimal strategy for sharing the processor throughput between users of different types. The results obtained can be used in the study and planning of telecommunication networks for various purposes, in particular, the Internet of Things.

References

1. Ghosh, A., Banik, A.D.: An algorithmic analysis of the $BMAP/MSP/1$ generalized processor-sharing queue. Comput. Oper. Res. **79**, 1–11 (2017)
2. Telek, M., van Houdt, B.: Response time distribution of a class of limited processor sharing queues. ACM SIGMETRICS Perform. Eval. Rev. **45**, 143–155 (2018). https://doi.org/10.1145/3199524.3199548
3. Yashkov, S., Yashkova, A.: Processor sharing: a survey of the mathematical theory. Autom. Remote. Control. **68**, 662–731 (2007)
4. Zhen, Q., Knessl, C.: On sojourn times in the finite capacity $M/M/1$ queue with processor sharing. Oper. Res. Lett. **37**, 447–450 (2009)
5. Masuyama, H., Takine, T.: Sojourn time distribution in a $MAP/M/1$ processor-sharing queue. Oper. Res. Lett. **31**, 406–412 (2003)
6. Kennedy, E., Bulega, T.: Resource sharing between M2M and H2H traffic under time-controlled scheduling scheme in LTE networks. In: Proceedings of the 8th International Conference on Telecommunication Systems Services and Applications (TSSA) (2014). https://doi.org/10.1109/TSSA.2014.7065909
7. Fawal, A., Najem, M., Mansour, A., Roy, F., Jeune, D.: CTMC modelling for H2H/M2M coexistence in LTE-A/LTE-M networks. J. Eng. **12**, 1954–1962 (2018)
8. Ahmadi, M., Golkarifard, M., Movaghar, A., Yousefi, H.: Processor sharing queues with impatient customers and state-dependent rates. IEEE/ACM Trans. Netw. **29**(6), 2467–2477 (2021). https://doi.org/10.1109/TNET.2021.3091189
9. Dudin, S., Dudin, A., Dudina, O., Samouylov, K.: Analysis of a retrial queue with limited processor sharing operating in the random environment. In: Koucheryavy, Y., Mamatas, L., Matta, I., Ometov, A., Papadimitriou, P. (eds.) WWIC 2017. LNCS, vol. 10372, pp. 38–49. Springer, Cham (2017). https://doi.org/10.1007/978-3-319-61382-6_4
10. Dudin, A., Dudin, S., Dudina, O., Samouylov, K.: Analysis of queuing model with limited processor sharing discipline and customers impatience. Oper. Res. Perspect. **5**, 245–255 (2018)
11. He, Q.M.: Queues with marked customers. Adv. Appl. Probab. **28**, 567–587 (1996)
12. Ramaswami, V.: Independent Markov processes in parallel. Commun. Statist.-Stochastic Models **1**, 419–432 (1985)
13. Ramaswami, V., Lucantoni, D.M.: Algorithms for the multi-server queue with phase-type service. Commun. Statist.-Stochastic Models **1**, 393–417 (1985)
14. Klimenok, V.I., Kim, C.S., Orlovsky, D.S., Dudin, A.N.: Lack of invariant property of Erlang $BMAP/PH/N/0$ model. Queueing Syst. **49**, 187–213 (2005)

Energy Efficiency of a Single-Server with Inactive State by Matrix-Analytic Method

Alexander Golovin[1] and Alexander Rumyantsev[1,2]

[1] Institute of Applied Mathematical Research, Karelian Research Centre of RAS, Petrozavodsk, Russia
`ar0@krc.karelia.ru`
[2] Petrozavodsk State University, Petrozavodsk, Russia

Abstract. We consider a single-server system with energy saving inactive state, non-zero setup, shutoff and hot reserve state. Matrix-analytic method is used to obtain the steady-state performance and average power demand, as well as study the energy-performance tradeoff in explicit way. Numerical results illustrate the model's properties.

Keywords: Matrix-Analytic Method · Explicit Solution · Single Server · Energy-Performance Tradeoff

1 Introduction

Energy efficiency is one of the important subjects in the telecommunication systems development catalyzed by the dramatic increase in the energy consumption of the ICT infrastructure in recent years [2,15,19]. There are many theoretical and engineering approaches to these problems, including the queueing systems analysis, which in some cases allows to derive explicit expressions for the optimal policies.

Matrix-analytic method [6,13,18] is an efficient approach suitable for a detailed modeling of stochastic systems using structured Markov processes. It allows to perform a detailed study of the steady-state characteristics of the model with the help of stochastic and algebraic approaches. While in general the procedure has to be performed numerically and suffers from the curse of dimensionality, in relatively rare cases there are rigorous explicit expressions which allow to avoid numerical computation. In this paper one of such cases is studied.

To address the increasing energy demand, it is typical to introduce various energy saving policies which allow to reduce the average *steady-state energy demand*, however, at the price of the system *performance* degradation. In many cases, there is an optimal trade-off between these two key system characteristics, see e.g. [12]. In the present paper, we address such an issue in a rather simple model of a single-server device capable of a standby regime that allows to save

© Springer Nature Switzerland AG 2022
A. Dudin et al. (Eds.): ITMM 2021, CCIS 1605, pp. 172–184, 2022.
https://doi.org/10.1007/978-3-031-09331-9_14

energy. Inspired by the policy studied in the work [4], we study a slightly different model. As opposed to the one studied in [4], our model is a single-server model with unbounded queue and First-Come-First-Served discipline, setup and shutoff phases, as well as an exponentially distributed idle (e.g. hot reserve) delay before entering the shutoff in an empty system. We treat the rate of such a delay as a management parameter and study its inference on the key system performance/energy efficiency measures.

The paper is indeed an exercise in applying the matrix-analytic method to a rather simplistic model. However, the main contribution of this research is an explicit solution which allows to solve the energy/performance optimization problem in a rigorous way. The second contribution of this research is the rather technical yet interesting result, Lemma 1, for the computation of the steady-state performance in the system in terms of the marginal phase probability.

The structure of the paper is as follows. We perform a very short literature survey in Sect. 1.1 and introduce the notation in Sect. 1.2. We introduce the matrix-analytic method and prove some interesting though technical results that simplify subsequent analysis in Sect. 1.3. The model is stated and analyzed in Sect. 2. The results are numerically illustrated in Sect. 3. We finalize the scope with a conclusion.

1.1   Literature Review

There is a huge body of literature covering various aspects of energy efficiency in the computing systems. Due to a lack of space, below we briefly enumerate some of the papers where the results were obtained by means of applied probability and queueing theory.

There are various energy saving mechanisms that are addressed, e.g. the dynamic voltage and frequency scaling [3,14,21], throttling [9], energy harvesting [5,22], load balancing [10], to name a few. At the same time, in many cases simulation is used since the analytical results are hard to obtain. In particular, a similar model of the server farms with setup costs were studied in [8], where the results were obtained by means of approximation and asymptotic analysis.

To finalize this review, we note that explicit analysis of the energy-performance trade-off is usually complicated due to a sophisticated nature of the models, however, if it is possible, it allows to derive the most general conclusions. At the same time, such an explicit analysis can be augmented by the simulation and technical modeling [20,21] to convert the conclusions obtained to practical recommendations.

1.2   Notation Conventions

Vectors and matrices are highlighted with bold letters, with special notation for a zero matrix, O, zero vector, 0, and vector of ones, 1. Being a column or a row vector should be clear from the context.

We use e_i as the vector having one at ith position and zero elsewhere, i.e. e_i is the ith row of an identity matrix I.

1.3 Matrix-Analytic Method

In this section we briefly describe the necessary results for the matrix-analytic method which is used to study the discrete or continuous-time discrete state space Markov chains having a specific structure. In particular, it is useful for the celebrated level-independent Quasi-Birth-Death (QBD) process which is a continuous-time Markov chain living in a discrete state space $E = \{0, \mathcal{Y}_0\} \times \bigcup_{i \geq 1} \{i, \mathcal{Y}\}$ with the infinitesimal generator matrix Q having block-tridiagonal structure [6, 13]

$$Q = \begin{bmatrix} A^{0,0} & A^{0,1} & O & O & \cdots \\ A^{1,0} & A^{1,1} & A^{(1)} & O & \cdots \\ O & A^{(-1)} & A^{(0)} & A^{(1)} & \cdots \\ O & O & A^{(-1)} & A^{(0)} & \cdots \\ \vdots & \vdots & \vdots & \vdots & \ddots \end{bmatrix}. \tag{1}$$

The blocks $A^{0,1}, A^{1,0}$ are the (non-sqare in general) matrices of the transition rates to/from the level zero, whereas $A^{(i)}$ are the square matrices defining the transition rates from the level $k \geq 1$ to $k + i$, $i = -1, 0, 1$. Finally, $A^{i,i}$ are the square matrices describing the transition rates for the phases within the level $i = 0, 1$. Due to the properties of a generator matrix,

$$Q\mathbf{1} = \mathbf{0}. \tag{2}$$

As such, the diagonal elements of the blocks $A^{i,i}, i = 0, 1$ and $A^{(0)}$ are non-positive, while all the remaining elements are of Q are non-negative. The stability criterion of such a process is given by the celebrated Neuts ergodicity condition,

$$\alpha A^{(1)} \mathbf{1} < \alpha A^{(-1)} \mathbf{1}, \tag{3}$$

where α is the stochastic vector solving the system

$$\alpha (A^{(-1)} + A^{(0)} + A^{(1)}) = \mathbf{0}. \tag{4}$$

Provided the stability condition (3) holds, the steady-state probability vector $\pi = (\pi_0, \pi_1, \dots)$, where $\pi_0 = \|\pi_{0,y}\|_{y \in \mathcal{Y}_0}$ and $\pi_i = \|\pi_{i,y}\|_{y \in \mathcal{Y}}, i \geq 1$, can be obtained using the so-called matrix-geometric recursive solution

$$\pi_{i+1} = \pi_i R. \tag{5}$$

The so-called *rate matrix* R is the minimal (in terms of the spectre of its eigenvalues) non-negative solution of a matrix quadratic equation with matrix unknown,

$$R^2 A^{(-1)} + R A^{(0)} + A^{(1)} = O. \tag{6}$$

The vectors π_0 and π_1 are to be found with a boundary condition system

$$(\pi_0, \pi_1) \begin{bmatrix} A^{0,0} & A^{0,1} \\ A^{1,0} & A^{1,1} + R A^{(-1)} \end{bmatrix} = \mathbf{0}, \tag{7}$$

$$\pi_0 \mathbf{1} + \pi_1 (I - R)^{-1} \mathbf{1} = 1. \tag{8}$$

Finally, the performance measures of the system can be obtained using the steady-state vector $\boldsymbol{\pi}$, in particular,

$$\mathrm{E}X_e = \sum_{n \geq 1} n\boldsymbol{\pi}_n \mathbf{1} = \boldsymbol{\pi}_1(\boldsymbol{I} - \boldsymbol{R})^{-2}\mathbf{1}, \tag{9}$$

where X_e is the number of customers in the system in steady state.

In general, the matrix $\boldsymbol{R}$ is obtained by one of the numerical methods [13]. However, there are nice special cases when the matrix $\boldsymbol{R}$ can be found in explicit form, in particular, when matrix $\boldsymbol{A}^{(1)}$ or $\boldsymbol{A}^{(-1)}$ is of rank 1, the latter case is considered in the following theorem.

Theorem 1. *[16, Theorem 5] Let $\boldsymbol{A}^{(-1)} = \boldsymbol{cr}$, where $\boldsymbol{c}$ is the column vector and $\boldsymbol{r}$ is the row vector such that $\boldsymbol{r}\mathbf{1} = 1$. Let the matrix $\boldsymbol{G}$ of order $|\mathcal{Y}|$ be the minimal non-negative solution of the matrix quadratic equation*

$$\boldsymbol{A}^{(-1)} + \boldsymbol{A}^{(0)}\boldsymbol{G} + \boldsymbol{A}^{(1)}\boldsymbol{G}^2 = \boldsymbol{O}. \tag{10}$$

Then

$$\boldsymbol{G} = \mathbf{1}\boldsymbol{r}. \tag{11}$$

The matrix $\boldsymbol{G} = \|G_{i,j}\|_{i,j \in \mathcal{Y}}$ is a stochastic matrix that consists of the conditional probabilities $G_{i,j}$ of entering the level $k-1$ by entering the state $(k-1, j)$ starting from the state $(k, i), k \geq 1$ [13,16]. The matrix $\boldsymbol{R}$ can be obtained from the matrix $\boldsymbol{G}$ using the following relation [11]

$$\boldsymbol{R} = -\boldsymbol{A}^{(1)}(\boldsymbol{A}^{(0)} + \boldsymbol{A}^{(1)}\boldsymbol{G})^{-1}. \tag{12}$$

In particular, if (11) holds, the matrix $\boldsymbol{R}$ can be obtained from (12) in an explicit form.

The system (7)–(8) can be simplified if the matrix (1) has a more simple structure, namely, if

$$\boldsymbol{A}^{0,1} = \boldsymbol{A}^{(1)}, \quad \boldsymbol{A}^{1,0} = \boldsymbol{A}^{(-1)}, \quad \boldsymbol{A}^{1,1} = \boldsymbol{A}^{(0)}. \tag{13}$$

In such a case, $\mathcal{Y}_0 = \mathcal{Y}$ and the following simplified system can be used:

$$\begin{cases} \boldsymbol{\pi}_0(\boldsymbol{A}^{0,0} + \boldsymbol{R}\boldsymbol{A}^{(-1)}) = 0 \\ \boldsymbol{\pi}_0(\boldsymbol{I} - \boldsymbol{R})^{-1}\mathbf{1} = 1. \end{cases} \tag{14}$$

It is interesting to suggest an alternative approach to $\boldsymbol{\pi}_0$ calculation which avoids the matrix inversion. Note that $\pi_{0,y} = \mathrm{P}\{X_e = 0, Y_e = y\}$, $y \in \mathcal{Y}$, where (X_e, Y_e) are the corresponding steady-state variables. Consider the marginal probability $\pi_y^{(Y)} = \mathrm{P}\{Y_e = y\}$, $y \in \mathcal{Y}$ and note that

$$\boldsymbol{\pi}^{(Y)} = \|\pi_y^{(Y)}\|_{y \in \mathcal{Y}} = \sum_{i=0}^{\infty} \boldsymbol{\pi}_i = \boldsymbol{\pi}_0(\boldsymbol{I} - \boldsymbol{R})^{-1}. \tag{15}$$

Then the system (14) is equivalent to

$$\begin{cases} \pi^{(Y)}(I - R)(A^{0,0} + RA^{(-1)}) = 0 \\ \pi^{(Y)}1 = 1. \end{cases} \tag{16}$$

Using (6), we obtain from (16) an equivalent system

$$\begin{cases} \pi^{(Y)}B = 0, \\ \pi^{(Y)}1 = 1, \end{cases} \tag{17}$$

where
$$B = A^{0,0} + A^{(1)} + R(A^{(-1)} + A^{(0)} - A^{0,0}).$$

Note that the solution of (17) can be computed without the matrix inversion. Moreover, in some cases it is possible to use regenerative approach to obtain $\pi^{(Y)}$ explicitly without (17), see [17]. Finally, the vector π_0 can be computed using the expression

$$\pi_0 = \pi^{(Y)}(I - R). \tag{18}$$

Using $\pi^{(Y)}$, it is possible to rewrite (9), taking into account (18), into the following expression:

$$EX_e = \pi^{(Y)}(I - R)^{-1}1 - 1. \tag{19}$$

The expression (19) requires a matrix inversion, whereas the explicit expression for R given in Theorem 1 requires one more inversion. Thus, we find it useful to derive the following technical result.

Lemma 1. *The expression (19) has the following form:*

$$EX_e = \pi^{(Y)}\left[I - A^{(1)}\left(A^{(0)} + A^{(1)}(G + I)\right)^{-1}\right]1 - 1.$$

Proof. Using (12) and denoting $T = A^{(0)} + A^{(1)}G$, transform (19):

$$EX_e = \pi^{(Y)}\left[(T + A^{(1)})T^{-1}\right]^{-1}1 - 1.$$

Equivalently,
$$EX_e = \pi^{(Y)}T(T + A^{(1)})^{-1}1 - 1.$$

Finally, adding and substracting $A^{(1)}$ to the multiplier T, obtain the desired statement.   □

2   Model

We study a system with an energy-saving state referred below as the *inactive state* which is common both in the IoT systems and in conventional battery-powered devices such as the laptops/smartphones (some examples are technically known as hibernate, sleep, suspend states etc.). In such a state, the system cannot

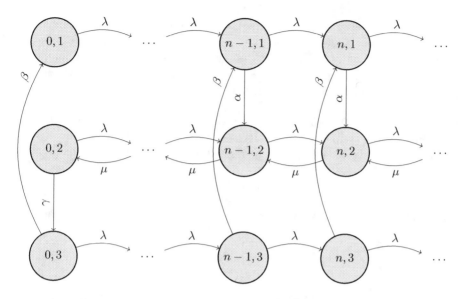

Fig. 1. State transition diagram of the Markov chain $\{X(t), Y(t)\}_{t \geqslant 0}$.

serve the customers. Moreover, entering and exiting from the inactive state takes some time during which the service is also not possible. As such, there is a decline in the system performance which, however, may result in energy saving. Below we study this tradeoff.

In what follows, to save the space we simultaneously describe the model and give the necessary parameters of the distributions of random variables involved which are in most cases *exponential* if not given otherwise explicitly. Let $X(t)$ be the number of customers in the system referred below as the *level* and $Y(t) \in \{1, 2, 3\} =: \mathcal{Y}$ be the *phase* of the system encoded as follows:

1 – the setup phase (exiting from the inactive phase) if $X(t) > 0$ and inactive phase if $X(t) = 0$;
2 – the working phase;
3 – the shutoff phase (entering the inactive phase).

The transitions (Fig. 1) are possible from a fixed phase $(x, y) \in E := \mathbb{Z}_0 \times \mathcal{Y}$ to the states

- $(x + 1, y)$ with rate λ, due to a customer arrival;
- $(x - 1, y)$ with rate μ, if $x > 0$ and $y = 2$, due to a departure of customer;
- $(x, 2)$ with rate α, if $x > 0$ and $y = 1$, due to a completion of the setup period;
- $(x, 1)$ with rate β, if $y = 3$, due to a completion of the shutoff phase and entering inactive phase if $x = 0$ or entering the setup phase if $x > 0$, respectively;
- $(0, 3)$ with rate γ, if $y = 2$ and $x = 0$.

It should be noted that the phase 3 can be entered only when there are no customers left in the system, and thus γ is the parameter of an exponentially distributed *idle* period before the beginning of a shutoff. Thus, hereinafter we consider γ as the *management parameter* of the system and study the dependence of system performance on γ in a rigorous way.

Since the transitions of the system are only possible within the two adjacent *levels*, the process $\{X(t), Y(t)\}_{t \geq 0}$ is indeed a QBD with a generator matrix of the form (1). Below we define the blocks of the generator $\boldsymbol{Q}$ explicitly. The following matrices define the transitions related to arrival/departure of a customer

$$\boldsymbol{A}^{(1)} = \lambda \boldsymbol{I}, \quad \boldsymbol{A}^{(-1)} = \operatorname{diag}(\mu \boldsymbol{e}_2), \tag{20}$$

whereas the matrix $\boldsymbol{A}^{(0)}$ gives the transition rates related to a phase change,

$$\boldsymbol{A}^{(0)} = -\boldsymbol{A}^{(1)} - \boldsymbol{A}^{(-1)} + \boldsymbol{A}, \tag{21}$$

where

$$\boldsymbol{A} = \begin{bmatrix} -\alpha & \alpha & 0 \\ 0 & 0 & 0 \\ \beta & 0 & -\beta \end{bmatrix}. \tag{22}$$

It is easy to see that $\boldsymbol{A} = \boldsymbol{A}^{(-1)} + \boldsymbol{A}^{(0)} + \boldsymbol{A}^{(1)}$ and, moreover, the vector $\boldsymbol{\alpha}$ solving (4) equals $\boldsymbol{\alpha} = (0, 1, 0)$ which gives a rather simple and expected stability condition

$$\rho := \frac{\lambda}{\mu} < 1. \tag{23}$$

Since $A^{(-1)} = \boldsymbol{cr}'$, where the column vector $\boldsymbol{c} = (0, \mu, 0)$ and the row vector $\boldsymbol{r}' = (0, 1, 0)$ follow immediately, the matrix $\boldsymbol{G} = \boldsymbol{1r}'$ as given by (11) in Theorem 1. Now using (12), the matrix $\boldsymbol{R}$ has the following explicit form:

$$\boldsymbol{R} = \begin{bmatrix} \frac{\lambda}{\alpha+\lambda} & \rho & 0 \\ 0 & \rho & 0 \\ \frac{\beta\lambda}{(\alpha+\lambda)(\beta+\lambda)} & \rho & \frac{\lambda}{\beta+\lambda} \end{bmatrix}. \tag{24}$$

To obtain the steady-state probability vector, we need to define the matrices $\boldsymbol{A}^{i,j}, i, j \in \{0, 1\}$. The matrix $\boldsymbol{A}^{0,0}$ defining the transitions of an idle system is given as follows

$$\boldsymbol{A}^{0,0} = -\boldsymbol{A}^{(1)} + \begin{bmatrix} 0 & 0 & 0 \\ 0 & -\gamma & \gamma \\ \beta & 0 & -\beta \end{bmatrix}. \tag{25}$$

It remains to note that the condition (13) holds good, and thus the steady-state vector $\boldsymbol{\pi}$ can be found in explicit form using one of the approaches introduced in Sect. 1.3. At the same time, the steady-state performance (number of customers in the system) is given by Lemma 1.

Now we define the steady-state energy demand to state the optimization problem. Assume that $\boldsymbol{d}^{(0)} = (d_1^{(0)}, d_2^{(0)}, d_3^{(0)})$ is the (column vector of) energy demand at the standby/idle/"idle shutoff" state, while $\boldsymbol{d} = (d_1, d_2, d_3)$ is the energy demand in the setup/active/shutoff states, respectively. Note that the "idle shutoff" state is the state of an empty system experiencing the shutoff period preceding the standby state, whereas the "shutoff" corresponds to the same state of a system with a non-empty queue. It is possible that some of these quantities are identical, say, $d_3^{(0)}$ might be equal to d_3, but we do not impose such restrictions to stay general. Then the average steady-state energy demand in the system, $E\mathcal{E}_e$ equals

$$E\mathcal{E}_e(\gamma) = \boldsymbol{\pi}_0 \boldsymbol{d}^{(0)} + (\boldsymbol{\pi}^{(Y)} - \boldsymbol{\pi}_0)\boldsymbol{d}, \tag{26}$$

where we stress the dependence on γ in the notation. At that, the steady-state performance is obtained by Lemma 1 and can also be denoted $EX_e(\gamma)$.

Following the procedure, the expression for $\boldsymbol{\pi}^{(Y)}$ can be obtained in an explicit form, denoting $b = \beta + \lambda$,

$$\pi_1^{(Y)} = C^{-1}\beta\gamma(\alpha\beta + \lambda b)(1 - \rho), \tag{27}$$

$$\pi_2^{(Y)} = \rho + C^{-1}\alpha\beta\lambda b(1 - \rho), \tag{28}$$

$$\pi_3^{(Y)} = C^{-1}\alpha\gamma\lambda b(1 - \rho), \tag{29}$$

where

$$C = \beta\gamma\lambda b + \alpha(\gamma\lambda^2 + \beta^2(\gamma + \lambda) + \beta\lambda b).$$

Moreover, the vector $\boldsymbol{\pi}_0 = (\pi_{0,1}, \pi_{0,2}, \pi_{0,3})$ can also be obtained explicitly,

$$\pi_{0,1} = C^{-1}\alpha\beta^2\gamma(1 - \rho), \tag{30}$$

$$\pi_{0,2} = C^{-1}\alpha\beta\lambda b(1 - \rho), \tag{31}$$

$$\pi_{0,3} = C^{-1}\alpha\beta\gamma\lambda(1 - \rho), \tag{32}$$

and it is clear that $\rho = \pi_2^{(Y)} - \pi_{0,2}$ is the busy probability. We note that working with symbolic expressions can be performed using Wolfram Cloud engine. Finally, using (27)–(32), the expression (26) also becomes explicit.

Consider now the partial derivative which, after some algebra, can be obtained as follows,

$$\frac{\partial EX_e(\gamma)}{\partial\gamma} = C^{-2}\lambda^2 b(\lambda b(\alpha^2 + \beta^2) + \alpha\beta(\beta b + \lambda^2)) > 0.$$

Similarly a second derivative can be shown to be negative, which gives the optimal point at $\gamma = 0$ (that is, the server is never using the inactive phase), as expected. Following [7], we name the case $\gamma = 0$ as NEVEROFF, while $\gamma = \infty$ can be named INSTANTOFF.

Consider now the partial derivative

$$\frac{\partial E\mathcal{E}_e(\gamma)}{\partial\gamma} = C^{-2}\alpha\beta\gamma b(\beta(d_2^{(0)} - d_1)\lambda b + \alpha(\beta^2(d_2^{(0)} - d_1^{(0)})$$

$$+ \beta\lambda(d_2^{(0)} - d_3^{(0)}) + \lambda^2(d_2^{(0)} - d_3)))(\rho - 1). \tag{33}$$

Interestingly, the derivative does not depend on the value d_2. However, the sign of the derivative depends on the exact values of the per-state energy demands. Specifically, if

$$d_2^{(0)} > \max(d_1, d_3, d_1^{(0)}, d_3^{(0)}),$$

then this partial derivative is always negative, which gives the monotone decrease of the energy w.r.t. γ. Indeed, in such a case due to a high energy demand of the idle regime, it is preferable to use the standby regime at earliest (that is, INSTANTOFF would be efficient if the performance is not taken into account). Then the optimal γ is the one satisfying the steady-state performance constraints expressed, say, in multiplicative form

$$EX_e(\gamma) \le (1 + \varepsilon)EX_e(0), \tag{34}$$

for some small $\varepsilon > 0$. If, on the contrast,

$$d_2^{(0)} < \min(d_1, d_3, d_1^{(0)}, d_3^{(0)}),$$

then this partial derivative is always positive, which gives the optimum at $\gamma = 0$ (in such a case, the EX_e is also minimal), that is, NEVEROFF is the optimal policy. Indeed, in such a configuration, the energy demand in idle regime is less than the demand in standby regime, and thus there is no reason for switching to standby. This case is, however, not realistic. The most realistic assumption would be the following,

$$d_1^{(0)} < d_2^{(0)} < \min(d_1, d_3, d_3^{(0)}). \tag{35}$$

It is clear from (33) that, since $C > 0$ and $\rho < 1$, $\alpha\beta\gamma b \ge 0$, the sign of the partial derivative depends on the expression

$$\phi(\lambda) := \beta(d_2^{(0)} - d_1)\lambda b + \alpha(\beta^2(d_2^{(0)} - d_1^{(0)}) + \beta\lambda(d_2^{(0)} - d_3^{(0)}) + \lambda^2(d_2^{(0)} - d_3)), \tag{36}$$

which is a quadratic polynomial of λ. Let us analyze this expression. Denote $\phi(\lambda) = c_2\lambda^2 + c_1\lambda + c_0$, where

$$c_2 = \alpha(d_2^{(0)} - d_3) + \beta(d_2^{(0)} - d_1),$$
$$c_1 = \beta^2(d_2^{(0)} - d_1) + \alpha\beta(d_2^{(0)} - d_3^{(0)}),$$
$$c_0 = \alpha\beta^2(d_2^{(0)} - d_1^{(0)}).$$

It follows from (35) that $c_2 < 0$ and $c_1 < 0$, while $c_0 > 0$. Hence $D := c_1^2 - 4c_2c_0 > 0$ and the polynomial $\phi(\lambda)$ has two roots. Then $\phi(0) = c_0 > 0$. Since $-c_1/(2c_2) < 0$, one of the two roots of $\phi(\lambda)$ is positive, that is, $\phi(\lambda^*) = 0$, where

$$\lambda^* = \frac{-c_1 - \sqrt{D}}{2c_2}. \tag{37}$$

As such, $\phi(\lambda) > 0$ for $\lambda < \lambda^*$, hence it follows from (33) that the partial derivative is negative. Thus, the mean steady-state energy demand decreases with γ

if the input rate $\lambda < \lambda^*$, while it increases with γ for $\lambda > \lambda^*$. This allows to formulate a threshold policy which selects the INSTANTOFF or NEVEROFF policy according to the input rate, or selects the γ using the restriction (34) accordingly. We also note that a decision should also take into account the stability condition (23). In the next section we illustrate this most interesting case numerically.

3   Numerical Illustration

To extend the understanding of the findings of Sect. 2, we numerically illustrate the model in the most interesting case (35), where the balance between INSTANTOFF and NEVEROFF policies is attained depending on the input rate λ.

In the following experiment we configure the constant parameters α, β, μ in such a way that $\lambda^* < \mu$ and plot the graphs of $E\mathcal{E}_e(\gamma)$ and $EX_e(\gamma)$ vs. γ for several values of λ s.t. $\lambda < \mu$. We (arbitrarily) fix the values $\alpha = 1, \beta = 2, \mu = 5$ and select the demands of the system states according to a specification of a HP ProBook 450 G8 Notebook PC [1] as follows (the numbers are in Watts):

$$d_1^{(0)} = 0.384, d_2^{(0)} = 2.184, d_1 = d_2 = d_3 = d_3^{(0)} = 4.164.$$

As such, (35) is satisfied and it follows from (37) that $\lambda^* \approx 0.48732$. Now we vary $\lambda \in \{0.3, 0.4, \lambda^*, 0.6\}$, *ceteris paribus*, and depict the corresponding dependency of $EX_e(\gamma)$ on γ in Fig. 2; $E\mathcal{E}_e(\gamma)$ on γ in Fig. 3, for $\gamma \in [0, 5]$.

A brief look at the Figs. 2–3 confirms the monotone dependency of $EX_e(\gamma)$ on γ in steady state, and the change of convexity/concavity for λ below and above

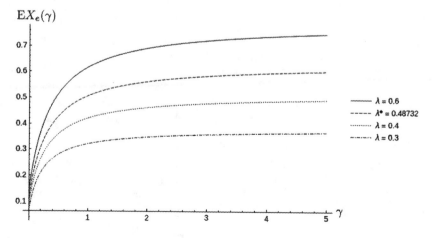

Fig. 2. Dependency of the mean number of customers in the system, $EX_e(\gamma)$, given in (19), on $\gamma \in [0, 5]$ for $d_1^{(0)} = 0.384, d_2^{(0)} = 2.184, d_1 = d_2 = d_3 = d_3^{(0)} = 4.164, \alpha = 1, \beta = 2, \mu = 5$ and $\lambda \in \{0.3, 0.4, 0.48732, 0.6\}$.

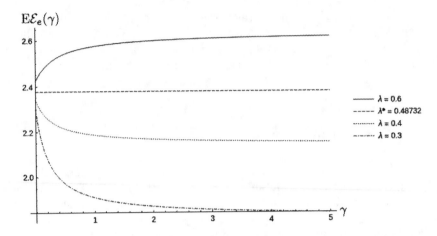

Fig. 3. Dependency of the mean energy demand of the system in steady state, $E\mathcal{E}_e(\gamma)$, given in (26), on $\gamma \in [0, 5]$ for $d_1^{(0)} = 0.384, d_2^{(0)} = 2.184, d_1 = d_2 = d_3 = d_3^{(0)} = 4.164$, $\alpha = 1$, $\beta = 2$, $\mu = 5$ and $\lambda \in \{0.3, 0.4, 0.48732, 0.6\}$.

the value λ^*. As such, for relatively large input rate $\lambda > \lambda^*$ the most energy efficient state that simultaneously offers the best performance is $\gamma = 0$, i.e. the NEVEROFF policy. For input rates smaller than λ^* the energy savings (i.e. the decrease of mean energy demand compared to a system with NEVEROFF policy) for $\gamma > 0$ increase with decreasing input rate. The non-linear dependence is also visible. It can be seen from (27)–(32) and (19), (26) that the dependency of mean steady-state demand, as well as performance, is indeed hyperbolic on γ.

4   Conclusion

In this paper, a single-server system with energy saving inactive state, non-zero setup, shutoff and hot reserve state was studied explicitly. The properties of the model allowed to use the matrix-analytic method to obtain the steady-state performance and average power demand, as well as study the energy-performance tradeoff. It might be interesting to continue this research towards multiserver systems, as well as study the system in more general case with non-exponential distributions of the random sequences involved. However, we leave this opportunity for future research.

Acknowledgements. The publication has been prepared with the support of Russian Science Foundation according to the research project No.21-71-10135 https://rscf. ru/en/project/21-71-10135/. The authors thank the referees for carefully reading the paper and for suggestions which helped to improve the paper.

References

1. HP ProBook 450 G8 Notebook PC, c06907888 - DA16756 - Worldwide - Version 18 (2021). https://h20195.www2.hp.com/v2/GetPDF.aspx/c06907888.pdf

2. Andrae, A.S.G., Edler, T.: On global electricity usage of communication technology: trends to 2030. Challenges **6**(1), 1–41 (2015). https://doi.org/10.3390/challe6010117

3. Basmadjian, R., Niedermeier, F., de Meer, H.: Modelling performance and power consumption of utilisation-based DVFS using M/M/1 queues. In: Proceedings of the Seventh International Conference on Future Energy Systems. e-Energy 2016. Association for Computing Machinery, New York (2016). https://doi.org/10.1145/2934328.2934342

4. Daraseliya, A.V., Sopin, E.S., Samuylov, A.K., Shorgin, S.Y.: Comparative analysis of the mechanisms for energy efficiency improving in cloud computing systems. In: Galinina, O., Andreev, S., Balandin, S., Koucheryavy, Y. (eds.) NEW2AN/ruSMART -2018. LNCS, vol. 11118, pp. 268–276. Springer, Cham (2018). https://doi.org/10.1007/978-3-030-01168-0_25

5. Fourneau, J.M.: Modeling green data-centers and jobs balancing with energy packet networks and interrupted poisson energy arrivals. SN Comput. Sci. **1**(1), 28 (2020)

6. Latouche, G., Ramaswami, V.: Introduction to Matrix Analytic Methods in Stochastic Modeling. ASA-SIAM, Philadelphia (1999)

7. Gandhi, A., Gupta, V., Harchol-Balter, M., Kozuch, M.A.: Optimality analysis of energy-performance trade-off for server farm management. Perform. Eval. **67**(11), 1155–1171 (2010)

8. Gandhi, A., Harchol-Balter, M., Adan, I.: Server farms with setup costs. Perform. Eval. **67**(11), 1123–1138 (2010)

9. Gandhi, A., Harchol-Balter, M., Das, R., Kephart, J.O., Lefurgy, C.: Power capping via forced idleness. In: Proceedings of Workshop on Energy Efficient Design, pp. 1–6 (2009). http://repository.cmu.edu/compsci/868/

10. Gebrehiwot, M.E., Aalto, S., Lassila, P.: Energy efficient load balancing in web server clusters. In: 2017 29th International Teletraffic Congress (ITC 29), vol. 3, pp. 13–18 (2017). https://doi.org/10.23919/ITC.2017.8065804

11. Gillent, F., Latouche, G.: Semi-explicit solutions for M/PH/1-like queuing systems. Eur. J. Oper. Res. **13**(2), 151–160 (1983)

12. Harrison, P.G., Patel, N.M., Knottenbelt, W.J.: Energy-Performance trade-offs via the EP queue. ACM Trans. Model. Perform. Eval. Comput. Syst. **1**(2), 1–31 (2016)

13. He, Q.M.: Fundamentals of Matrix-Analytic Methods. Springer, New York (2014). https://doi.org/10.1007/978-1-4614-7330-5

14. Kuehn, P.J., Mashaly, M.: DVFS-power management and performance engineering of data center server clusters. In: 2019 15th Annual Conference on Wireless Ondemand Network Systems and Services (WONS), pp. 91–98 (2019). https://doi.org/10.23919/WONS.2019.8795470

15. Lima, J.: Data centres of the world will consume 1/5 of earth's power by 2025. Technical report, Data Economy (2017). https://www.broad-group.com/data/news/documents/b1m2y6qlx5dv5t

16. Liu, D., Zhao, Y.Q.: Determination of Explicit Solutions for a General Class of Markov Processes. In: Chakravarthy, S., Alfa, A.S. (eds.) Matrix-Analytic Methods in Stochastic Models, 1 edn, pp. 363–378. CRC Press (1996). https://doi.org/10.1201/b17050-21. https://www.taylorfrancis.com/books/9781482292176/chapters/10.1201/b17050-21

17. Morozov, E., Rumyantsev, A., Dey, S., Deepak, T.: Performance analysis and stability of multiclass orbit queue with constant retrial rates and balking. Perform. Eval. **134**, 102005 (2019)

18. Neuts, M.F.: Matrix-Geometric Solutions in Stochastic Models: An Algorithmic Approach. Johns Hopkins University Press, Baltimore (1981)

19. Nguyen, M.H., Gruber, J., Fuchs, J., Marler, W., Hunsaker, A., Hargittai, E.: Changes in digital communication during the COVID-19 global pandemic: implications for digital inequality and future research. Soc. Media Soc. **6**(3), 2056305120948255 (2020). https://doi.org/10.1177/2056305120948255
20. Rumyantsev, A., Basmadjian, R., Golovin, A., Astafiev, S.: A three-level modelling approach for asynchronous speed scaling in high-performance data centres. In: Proceedings of the Twelfth ACM International Conference on Future Energy Systems, e-Energy 2021, pp. 417–423. Association for Computing Machinery, New York (2021). https://doi.org/10.1145/3447555.3466580
21. Rumyantsev, A., Zueva, P., Kalinina, K., Golovin, A.: Evaluating a single-server queue with asynchronous speed scaling. In: German, R., Hielscher, K.-S., Krieger, U.R. (eds.) MMB 2018. LNCS, vol. 10740, pp. 157–172. Springer, Cham (2018). https://doi.org/10.1007/978-3-319-74947-1_11
22. Zheng, X., Zhou, S., Jiang, Z., Niu, Z.: Closed-form analysis of non-linear age of information in status updates with an energy harvesting transmitter. IEEE Trans. Wirel. Commun. **18**(8), 4129–4142 (2019). https://doi.org/10.1109/TWC.2019.2921372. Place: Piscataway Publisher: Ieee-Inst Electrical Electronics Engineers Inc WOS:000480661000026

Optimization of the Adaptive Control System for Conflict Cox–Lewis Flows

Evgeniy Kudryavtsev[ID] and Mikhail Fedotkin[(✉)][ID]

Lobachevsky State University of Nizhny Novgorod, Gagarin Avenue 23,
Nizhny Novgorod 603950, Russia
`fma5@rambler.ru`

Abstract. In the article the queuing system for non-ordinary Poisson flows is described using the Lyapunov-Yablonsky's cybernetic approach. The system model is a multidimensional Markov chain. The analytical properties of this Markov chain have been studied. In this work, a simulation model of the system has been studied. The transient process is investigated and an algorithm for searching the moment of the end of the transient process is proposed. By the numerical optimization, the quasi-optimal parameters of the system were found according to the condition of the minimum average waiting time for service.

Keywords: conflicting flows · cybernetic system · simulation · quasi-optimal parameters

1 Introduction

This work is related to the important problem of creating algorithms in intelligent transport systems that control conflicting flows [1] at the intersections of highways in large cities. An adaptive algorithm for controlling this kind of flows is proposed. The control algorithm takes into account not only the lengths of the queues, but also the order in which requests arrive in the system. A mathematical model of such a flow control system from non-homogeneous requests has been built and studied.

Input flows are two independent conflicting non-ordinary Poisson flows Π_1 and Π_2. For $t_0 \geq 0$, $t > 0$ and $j = 1, 2$ the probability $P_j(t, k)$ of k requests for the time interval $[t_0, t_0 + t)$ along the flow Π_j is obtained of the following form

$$P_j(t, k) = e^{-\lambda_j t} \sum_{n=0}^{[k/2]} \alpha_j^n \frac{(\lambda_j t p_j)^{k-n}}{n!(k-2n)!}$$

$$+ e^{-\lambda_j t} \sum_{n=0}^{[k/2]} \alpha_j^n \sum_{m=1}^{\min\{k-2n,n\}} \beta_j^m \sum_{l=0}^{k-2n-m} \gamma_j^l \frac{(\lambda_j t p_j)^{k-n-m-l} C_{m+l-1}^l}{(n-m)!m!(k-2n-m-l)!}, \quad (1)$$

where α_j, β_j, γ_j, λ_j and $p_j = (1 + \alpha_j + \alpha_j \beta_j/(1 - \gamma_j))^{-1}$ are distribution parameters. The properties of such flows with non-homogeneous requests are

A. Dudin et al. (Eds.): ITMM 2021, CCIS 1605, pp. 185–196, 2022.
https://doi.org/10.1007/978-3-031-09331-9_15

studied in [2,3]. For input flow Π_j, the mathematical expectation of the number of requests at the moment of arrival is

$$M_j = (1 + 2\alpha_j + \alpha_j\beta_j(\frac{2}{1-\gamma_j} + \frac{1}{(1-\gamma_j)^2}))p_j.$$

The adequacy of the representation of real flows using the distribution (1) is shown on examples of tables from [4].

2   Mathematical Model of the Control System

2.1   Cybernetic Approach of Lyapunov–Yablonsky

The cybernetic approach of Lyapunov–Yablonsky is used to construct and describe a mathematical model of a discrete system for adaptive control of conflict flows and service of non-homogeneous requests [5]. The application of this approach to the system under study is described in detail in [1].

According to the cybernetic approach, we will consider the system at random discrete times τ_i or at intervals $[\tau_i, \tau_{i+1})$ for $i = 0, 1, \dots$ Here the value τ_0 is the initial moment of time, and τ_i, $i > 0$ are the moments of changing the states of the server. Let $y_0 = (0,0)$, $y_1 = (1,0)$, $y_2 = (0,1)$ and X is an integer one-dimensional non-negative lattice. Now let's define the following random variables and elements:

1. $\Gamma_i \in \Gamma = \{\Gamma^{(1)}, \Gamma^{(2)}, \dots, \Gamma^{(8)}\}$—the state of the service device in the interval $[\tau_i, \tau_{i+1})$;
2. $\eta_{j,i} \in X$—the number of requests from the flow Π_j that entered the system during the interval $[\tau_i, \tau_{i+1})$, and $\eta_i = (\eta_{1,i}, \eta_{2,i})$;
3. η_i'—a random vector taking the value y_0, if no orders have been received in the system at the i-th time step $[\tau_i, \tau_{i+1})$, and the value y_j, if on the i-th step the first request came (or requests) of the flow Π_j;
4. $\kappa_{j,i} \in X$—the number of requests of the flow Π_j in the system at the moment τ_i, and $\kappa_i = (\kappa_{1,i}, \kappa_{2,i})$;
5. $\xi_{j,i}$—the maximum possible number of requests from the flow Π_j that the system can serve on the interval $[\tau_i, \tau_{i+1})$, and $\xi_i = (\xi_{1,j}, \xi_{2,j})$.

Let us now define the sequence $\{\tau_i; i > 0\}$ moments of state change of the server. For this purpose, we present the meaningful meaning of each state from the set Γ. The $\Gamma^{(3j-2)}$ state corresponds to the first stage of the thread's service period Π_j. The duration of servicing one request arriving from the queue is equal to a constant value $\mu_{j,1}^{-1}$. Duration of stay in $\Gamma^{(3j-2)}$ is equal to T_{3j-2}. The $\Gamma^{(3j-1)}$ state corresponds to the second stage of the thread's service period Π_j. The duration of servicing one request is equal to $\mu_{j,2}^{-1} < \mu_{j,1}^{-1}$. The duration of stay in this state is a random variable taking values in the set $\{kT_{3j-1}; k = \overline{1, n_j}\}$, where n_j is the maximum number of renewals and T_{3j-1} is the duration of one renewal. Renewal occurs in 2 cases: 1) the length of the queue along the flow Π_j is not less than the constant integer parameter $K_j > 0$, 2) at the previous stage of renewals, there were requests that need to be serviced. The $\Gamma^{(3j)}$ state corresponds to the changeover mode for the Π_j flow, during which only additional servicing of the requests of the Π_j flow is possible. The duration of stay in this state is T_{3j}. The $\Gamma^{(6+j)}$ corresponds to the first stage of the service period of the flow Π_j, in the case when an instant transition to the state $\Gamma^{(3j)}$ is possible. The duration of stay in $\Gamma^{(6+j)}$ is a random variable. The maximum time spent in this state is T_{3j-2}. In this state, the queue for the serviced flow is empty, and the server monitors the order of arrival of requests. If during the time T_{3j-2} the first request of the flow Π_j arrived, then in T_{3j-2} from the time τ_i there will be a transition to the state $\Gamma^{(3j-1)}$. If a request from another thread arrived first, then an instant transition to the state $\Gamma^{(3j)}$. And, finally, if during this time not a single request arrives for both flows, then the server will also switch to the state in $\Gamma^{(3j)}$. Constants T_k, $k = \overline{1,6}$, it is advisable to choose in the following form

$$T_{3j-2} = \mu_{j,1}^{-1} + l_{3j-2}\theta_j\mu_{j,1}^{-1}, \quad T_{3j-1} = l_{3j-1}\theta_j\mu_{j,2}^{-1}, \quad T_{3j} = l_{3j}\theta_j\mu_{j,2}^{-1}, \quad (2)$$

where $l_{3j-2} \in X$, l_{3j-1}, $l_{3j} \in \{1, 2, \ldots\}$ are parameters. The value $0 < \theta_j \le 1$ denotes the part of service that a request needs to go through in order to start serving the next request. In the case $\theta_j < 1$ several requests can be served simultaneously. The relations (2) means that a change in the state of the server occurs at the moment when one of the requests is finished servicing. We get that the maximum possible number of serviced requests is equal $1 + l_{3j-2}$ for the state $\Gamma^{(3j-2)}$, is equal to kl_{3j-1} for the state $\Gamma^{(3j-1)}$ and is equal to the integer part $1/\theta_j$ for the state $\Gamma^{(3j)}$.

2.2   Recurrent Relations

The cybernetic approach allows us to obtain [1] the following theorem.

Theorem 1. *For each* $i = 1, 2, \ldots$ *and* $j, s = 1, 2$, $j \ne s$,

$$\Gamma_{i+1} = u(\Gamma_i, \kappa_i, \eta_i')$$

$$= \begin{cases} \Gamma^{(3j-2)}, & \left\{[\Gamma_i = \Gamma^{(3s)}] \, \& \, [(\kappa_{j,i} > 0) \vee (\kappa_{s,i} \ge K_s) \vee (\eta_i' = y_j)]\right\} \vee \\ & \vee \left\{[\Gamma_i = \Gamma^{(3j)}] \, \& [\kappa_{s,i} = 0] \& [\kappa_{j,i} \le K_j] \& [\eta_i' = y_j]\right\}, \\ \Gamma^{(3j-1)}, & \{\Gamma_i = \Gamma^{(3j-2)}\} \vee \left\{[\Gamma_i = \Gamma^{(6+j)}] \, \& [\eta_i' = y_j]\right\}, \\ \Gamma^{(3j)}, & \{\Gamma_i = \Gamma^{(3j-1)}\} \vee \left\{[\Gamma_i = \Gamma^{(6+j)}] \, \& [\eta_i' \ne y_j]\right\}, \\ \Gamma^{(6+j)}, & [\Gamma_i = \Gamma^{(3s)}] \, \& [\kappa_{j,i} = 0] \& [\kappa_{s,i} < K_s] \& [\eta_i' = y_0]; \end{cases} \quad (3)$$

$$\kappa_{j,i+1} = v_j(\Gamma_i, \kappa_i, \eta_i, \xi_i)$$

$$= \begin{cases} \max\{0, \kappa_{j,i} + \eta_{j,i} - \xi_{j,i}\} & \text{if } \Gamma_i \in \Gamma \backslash \{\Gamma^{(3)}, \Gamma^{(6)}\}, \\ \eta_{j,i} + \max\{0, \kappa_{j,i} - \xi_{j,i}\} & \text{if } \Gamma_i \in \{\Gamma^{(3)}, \Gamma^{(6)}\}. \end{cases} \quad (4)$$

Hereinafter in the article $j, s = 1, 2$, $j \ne s$. Using the functional recurrent in i relation (3), the adaptive algorithm for changing the states of the server can

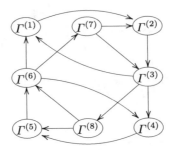

Fig. 1. The graph of the server state change algorithm.

be displayed by a graph in Fig. 1. Note that the state of the server at the next step depends on the state at the previous step, the lengths of the queues, and the order of arrival of requests.

The relations (3) and (4) allow us to consider and study the limiting properties of the vector sequence $\{(\Gamma_i, \kappa_i); i = 0, 1, \ldots\}$, which is a probabilistic model of the considered system of adaptive control of conflicting flows and service of heterogeneous requests [1]. Here are some of the theorems on the this Markov sequence $\{(\Gamma_i, \kappa_i); i = 0, 1, \ldots\}$.

Theorem 2. *If there is a limiting distribution of the vector Markov sequence $\{(\Gamma_i, \kappa_i); i \geq 0\}$, then*

$$\frac{\theta_1 \lambda_1 M_1}{\mu_{1,2}} + \frac{\theta_2 \lambda_2 M_2}{\mu_{2,2}} < 1.$$

Corollary from the Theorem 2 is

$$\theta_j \lambda_j M_j < \mu_{j,2}, \ j = 1, 2.$$

We introduce the values

$$T = T_1 + n_1 T_2 + T_3 + T_4 + n_2 T_5 + T_6,$$
$$L_j = l_{3j-2} + n_j l_{3j-1} + l_{3j}.$$

Theorem 3. *For the existence of the limit distribution of the vector Markov sequence $\{(\Gamma_i, \kappa_i); i \geq 0\}$ it suffices to satisfy inequalities*

$$\lambda_j M_j T - L_j < 0, \ j = 1, 2.$$

Theorem 4. *If there is a limiting distribution of the vector Markov sequence $\{(\Gamma_i, \kappa_i); i \geq 0\}$, then for some $j = 1, 2$*

$$\lambda_j M_j T - L_j < 0.$$

3 Numerical System Investigation

3.1 System Parameters

Unfortunately, it is not possible to analytically find such important characteristics as the average sojourn time of an arbitrary request in the system and the average length of queues across flows.

To solve the questions posed, a simulation model of the adaptive control system for conflicting flows of non-homogeneous requests [7] has been implemented in the C++ program. The simulation model allows not only to study the control process of servicing of non-homogeneous requests, but also to obtain realizations of the vector sequence $\{(\Gamma_i, \kappa_i); i \geq 0\}$. Each implementation is specified using the following inputs:

1. Input flows parameters are α_j, β_j, γ_j, λ_j;
2. System parameters are T_1, T_2, ..., T_6, $\mu_{j,1}$, $\mu_{j,2}$, θ_j, K_j, n_j;
3. Initial values $\Gamma^{(r)}$, x_1, x_2 of random elements Γ_0, $\kappa_{1,0}$, $\kappa_{2,0}$.

In contrast to the cybernetic approach, simulation allows observing the processes in the system at each moment of time, and not only at specially selected moments of the discrete time scale. Therefore, with the help of simulation modeling, it is possible to build a model closer to the real system. The disadvantages of simulation is the impossibility of obtaining new analytical results and the results obtained are approximate.

In the cybernetic approach, the input pole is given by the input and saturation flows. In the program, the input flows are generated before the start of the simulation of the system operation. In this case, the intervals between the groups of requests are modeled using the inverse function method, and the number of requests in the group is modeled using the method of modeling discrete distributions. Saturation flows are not explicitly entered in the program. The description of the congestion flows is specified using the parameters $\mu_{j,1}$, $\mu_{j,2}$ and θ_j, through which the duration of servicing requests and the moments of release of the server are determined.

External memory describes the state of the queues in the system. At the same time, the moments of arrival and the moments of the beginning of servicing of requests are stored in the implementation of queues. These points are necessary to calculate the numerical characteristics of the functioning of the system. The internal memory is determined by the state of the server The program for determining the current state of the server also observes the current simulation time and the point in time when the server becomes free to service the next request. Modeling the operation of the studied adaptive algorithm is performed step by step between the moments of changing the states of the server. During the simulation of one step, the moments of arrival of requests in the system and the moments of release of the server are processed. If service is not possible during the arrival of a request, then the request is added to the queue; otherwise, it is sent for service. When the server is released, the first request from the queue is sent for servicing if the queue is not empty.

3.2   Transient Process and Stationary Mode

First of all, it is necessary to study the characteristics of the functioning of the system in a stationary mode. Therefore, an important task is to determine the time to reach a stationary regime and to study the transient process. In simulation, we will consider two implementations of the Cox–Lewis adaptive flow control process with two types of initial conditions:

1. Zero $\Gamma^{(r)} = \Gamma^{(1)}$, $x_1 = 0$, $x_2 = 0$;
2. Shifted $\Gamma^{(r)} = \Gamma^{(1)}$, $x_1 = K_1' > K_1$, $x_2 = K_2' > K_2$.

The same implementation of input flows are used. Let's define the completion moment of the transient process as follows. Let us denote by $\gamma_j^0(l)$ and $\gamma_j^+(l)$ the time spent in the system by the request with the number $l = 1, 2, \ldots$ of the flow Π_j, that entered the system after the beginning of the simulation, with zero initial conditions and with shifted initial conditions of the second type. The values

$$\overline{\gamma}_j^0(n) = \frac{1}{n} \sum_{l=1}^{n} \gamma_j^0(l), \quad \overline{\gamma}_j^+(n) = \frac{1}{n} \sum_{l=1}^{n} \gamma_j^+(l) \tag{5}$$

determine the sample mean times of sojourn in the system of the first n requests of the flow Π_j under the initial conditions of the first and, accordingly, of the second type. If the condition

$$\left| \overline{\gamma}_j^+(n) - \overline{\gamma}_j^0(n) \right| \leq \delta \overline{\gamma}_j^0(n), \tag{6}$$

is satisfied for the proximity parameter $\delta > 0$, then at the meaningful level we can assume that the initial conditions have ceased to affect the sample mean residence time for the requests of the flow Π_j. If a stationary regime exists in the system, the value $n_j(d)$ determines the number of the request under which the condition (6) is first fulfilled d times in a row, where d—is a constant natural number. Let t_j—be the moment of completion of servicing the request with the number $n_j(d)$ of the flow Π_j and $t^* = \max(t_1, t_2)$. We will assume that t^* determines the moment of the end of the transient process in the system for a given implementation of input flows. Note that the simulation model makes it possible to find the dependence of the duration of the transient process on the parameters δ and d. Different realizations of input flows correspond to different values of estimates for the duration of the system's transient process. Let's consider l independent simulations. The estimate $\overline{\gamma}_j^*$ of the average sojourn time for the requests of the flow Π_j is calculated by the formula

$$\overline{\gamma}_j^* = \frac{1}{l} \sum_{i=1}^{l} \overline{\gamma}_{j,i}^0(N_i).$$

Value $\overline{\gamma}_{j,i}^0(N_i)$ is mean times from (5) for i-th simulation. request number N_i arrives after time t_i^* for i-th simulation. The estimate $\overline{\gamma}^*$ of the average sojourn time of an arbitrary request will be calculated using the formula for the weighted average

$$\overline{\gamma}^* = \frac{\lambda_1 M_1 \overline{\gamma}_1^* + \lambda_2 M_2 \overline{\gamma}_2^*}{\lambda_1 M_1 + \lambda_2 M_2}.$$

Next, we determine the sample average queue length for the flow Π_j. Let the imitation of the system operation lasted for the time t. At the same time, along the flow Π_j, a queue of length $k = 0, 1, \ldots$ was observed in $t_j^{(k)}$ units of time. Then the sample average length of the queue $\overline{\kappa}_j^*$ along the flow Π_j will be calculated by the following formula

$$\overline{\kappa}_j^* = \frac{1}{t} \sum_{k=0}^{\infty} kt_j^{(k)}.$$

The sample average length of the queue $\overline{\kappa}^*$ for the entire system is defined as the arithmetic mean of the sample lengths of all queues

$$\overline{\kappa}^* = \frac{1}{2}(\overline{\kappa}_1^* + \overline{\kappa}_2^*).$$

Sample average residence times and sample average queue lengths can serve as estimates for the respective characteristics.

3.3   Simulation Example

Let's give an example of the results of simulation modeling with the following set of parameters:

1. Input flow parameters are $\alpha_1 = 0.8$, $\beta_1 = 0.7$, $\gamma_1 = 0.5$, $\lambda_1 = 0.6$, $\alpha_2 = 0.6$, $\beta_2 = 0.5$, $\gamma_2 = 0.2$, $\lambda_2 = 0.3$;
2. System parameters are $T_1 = 1$, $T_2 = 2$, $T_3 = 1$, $T_4 = 1$, $T_5 = 3$, $T_6 = 1$, $\mu_{1,1} = 0.5$, $\mu_{2,1} = 1$, $\mu_{1,2} = 0.3$, $\mu_{2,2} = 0.6$, $\theta_1 = 1$, $\theta_2 = 0.5$, $K_1 = 10$, $K_2 = 10$, $n_1 = 10$, $n_2 = 10$;
3. Initial values of random elements Γ_0, $\kappa_{1,0}$, $\kappa_{2,0}$ are $\Gamma^{(r)} = \Gamma^{(1)}$, $x_1 = 0$, $x_2 = 0$.

Figure 2 and 3 show the dynamics of the queue length for the flow Π_1 and the dynamics of the average waiting time for servicing the requests of this flow, provided that the stationary mode exists. The abscissa shows the number of the algorithm step, and the ordinate shows the tracked characteristic. In this case, the queues have steady oscillations that do not depend on the initial conditions. The average waiting time for servicing differs significantly for different initial conditions at the start of the simulation; later on, the characteristics converge.

Let's increase the intensity of the input flows $\lambda_1 = 0,8$ and $\lambda_2 = 0,7$. In this case, there is no stationary mode in the system. Figure 4 and 5 show the dynamics of the queue length for the flow Π_1 and the dynamics of the average waiting time for servicing the requests of this flow in the absence of a stationary mode in the system. Similarly, the abscissa shows the number of the algorithm

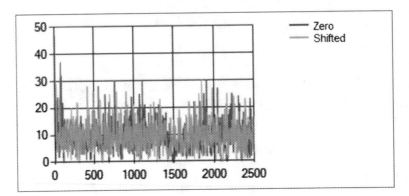

Fig. 2. Dynamics of the queue length along the flow Π_1 under the condition of the existence of a stationary mode.

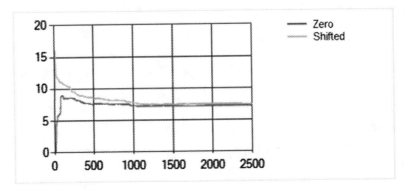

Fig. 3. Dynamics of the average waiting time for servicing requests of the flow Π_1 under the condition of the existence of a stationary regime.

step, and the ordinate shows the tracked characteristic. In the absence of a stationary mode, the queues tend to grow indefinitely with an increase in the duration of the simulation. The average service wait time also increases with long simulations.

Let us now investigate the dependence of the duration of the transient process on the parameters d and δ in the case of a stationary regime in the system. With a fixed value of $\delta = 0.05$, we obtain the dependence on d of the duration of the transient process, shown in Fig. 6. The abscissa is the parameter d, and the ordinate is the number of the request on which the condition for reaching the stationary was fulfilled. The blue graph corresponds to the flow Π_1, the orange one—to the flow Π_2. With a fixed value of $d = 10$, we obtain the dependence of the duration of the transient process on δ, shown in Fig. 7. From the graphs in Fig. 6 and 7 we obtain that with an increase in the d parameter, the duration of the transient process will be longer. Also, with an increase in the δ parameter, an inverse relationship is observed.

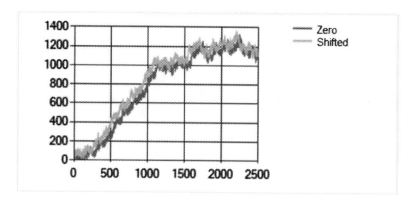

Fig. 4. Dynamics of the queue length along the flow Π_1 in the absence of a stationary mode.

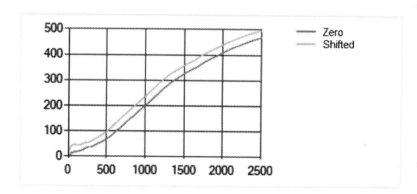

Fig. 5. Dynamics of the average waiting time for servicing requests of the flow Π_1 in the absence of a stationary mode.

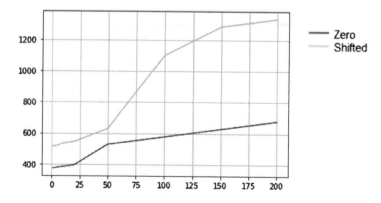

Fig. 6. Dependence of the duration of the transient process on the parameter d.

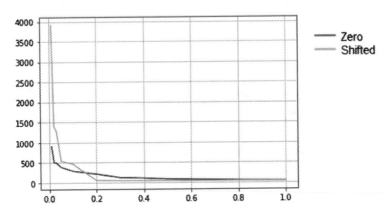

Fig. 7. Dependence of the duration of the transient process on the parameter δ.

With parameters 1–3 of the control system, we obtain the following estimates of the average waiting time for servicing

$$\overline{\gamma}_1^* = 7.1684, \quad \overline{\gamma}_2^* = 12.8481, \quad \overline{\gamma}^* = 8.6625,$$

and estimates for the average queue lengths

$$\overline{\kappa}_1^* = 10.4691, \quad \overline{\kappa}_2^* = 6.7333, \quad \overline{\kappa}^* = 8.6012.$$

3.4   Search for Quasi-Optimal System Parameters

Below is an example of searching for quasi-optimal parameter values. For the system under study, the duration of the first stage of servicing cannot be arbitrarily changed, since it is determined by the properties of the input flow. Also, the duration of the changeover cannot be arbitrarily changed due to the conflicting flows. Insufficient changeover times can lead to accidents. Therefore, we fix the values $T_1 = 1, T_3 = 1, T_4 = 1$ and $T_6 = 1$. Thus, the following parameters are available for optimization $T_2, T_5, n_1, n_2, K_1, K_2$. Optimization is performed step by step by parameter pairs (T_2, T_5), (n_1, n_2) and (K_1, K_2) by means of reduced search. After searching for the optimal parameters for one of the pairs with fixed values of the remaining parameters, the search for the optimal parameters for the other pair is performed. The algorithm for searching for quasi-optimal parameters ends when the next optimization fails to improve the characteristics of the system. Using the reduced search algorithm, the following quasi-optimal parameters were obtained

$$T_2^* = 5, T_5^* = 1, n_1^* = 3, n_2^* = 10, K_1^* = 1, K_2^* = 4.$$

The following estimates of average waiting time for servicing by flows Π_1, Π_2 and, accordingly, an estimate of the average waiting time for servicing an arbitrary request correspond to these quasi-optimal parameters:

$$\overline{\gamma}_1^* = 2.47241, \quad \overline{\gamma}_2^* = 5.13818, \quad \overline{\gamma}^* = 3.17367.$$

The Table 1 shows the steps of the optimization algorithm.

Table 1. Search for quasi-optimal parameters.

T_2	T_5	n_1	n_2	K_1	K_2	$\overline{\gamma}_1^*$	$\overline{\gamma}_2^*$	$\overline{\gamma}^*$
2	3	10	10	10	10	7.91365	12.9163	9.22965
9	**6**	10	10	10	10	5.10318	11.3382	6.74334
9	6	**6**	**9**	10	10	5.06648	11.4252	6.73919
9	6	6	9	**3**	**6**	3.79753	6.81921	4.59241
4	**2**	6	9	3	6	3.09255	5.72206	3.78427
4	2	**8**	**9**	3	6	3.01794	5.58343	3.69281
4	2	8	9	**1**	**4**	2.81271	4.77639	3.32928
5	**1**	8	9	1	4	2.48334	5.29321	3.2225
5	1	**3**	**10**	1	4	2.47241	5.13818	**3.17367**
5	1	3	10	**1**	**8**	2.48873	5.21597	3.20615

The values $T_2 = 2$, $T_5 = 3$, $n_1 = 10$, $n_2 = 10$, $K_1 = 10$, $K_2 = 10$ in the second row of the table correspond to the initial parameters of the algorithm. At the first step of the algorithm, the parameters $n_1 = 10$, $n_2 = 10$, $K_1 = 10$, $K_2 = 10$ were fixed. The pair of parameters (T_2, T_5) was optimized. The characteristics of the system were calculated for $T_2 = 1, 2, \ldots, 10$ and $T_5 = 1, 2, \ldots, 10$. The smallest value of the estimate $\overline{\gamma}^*$ corresponds to $T_2 = 9$ and $T_5 = 6$. The found parameters, which were optimized in the first step, are highlighted in bold in the third line. At the second step, the parameters $T_2 = 9$, $T_5 = 6$, $K_1 = 10$, $K_2 = 10$ were fixed and a pair of parameters (n_1, n_2) were optimized. Values $n_1 = 5, 6$, $\ldots, 15$ and $n_2 = 5, 6, \ldots, 15$ have been tested. The values $n_1 = 6$ and $n_2 = 9$ turned out to be optimal of the average waiting time for servicing an arbitrary request. These parameters are marked in bold in the fourth line of the table. The next steps are performed in a similar way for parameter pairs. At the last two steps, it was not possible to reduce the average waiting time for servicing an arbitrary request. Therefore, the quasi-optimal parameters are given in the penultimate line of the table.

4   Conclusion

The results of an analytical and numerical study of the Cox-Lewis flow control system from non-homogeneous requests were presents. The mathematical model of the control system was built using the cybernetic approach and the apparatus of the queuing theory. The algorithm for determining the completion moment of the transient process was described. Numerical optimization was carried out on the basis of a software-implemented simulation model.

References

1. Kudryavtsev, E.V., Fedotkin, M.A.: Analysis of a discrete model of an adaptive control system for conflicting nonhomogeneous flows. Mosc. Univ. Comput. Math. Cybern. **43**(1), 17–24 (2019). https://doi.org/10.3103/S0278641919010047
2. Fedotkin, M.A., Fedotkin, A.M., Kudryavtsev, E.V.: Construction and analysis of a mathematical model of spatial and temporal characteristics of traffic flows. Autom. Control. Comput. Sci. **48**(6), 358–367 (2014). https://doi.org/10.3103/S0146411614060030
3. Fedotkin, M.A., Fedotkin, A.M., Kudryavtsev, E.V.: Nonlocal description of the time characteristic for input flows by means of observations. Autom. Control. Comput. Sci. **49**(1), 29–36 (2015). https://doi.org/10.3103/S0146411615010034
4. Cox, D., Lewis, P.: The Statistical Analysis of Series of Events. Springer, Netherlands (1966)
5. Lyapunov, A.A., Yablonsky, S.V.: Theoretical problems of cybernetics. Problems of Cybernetics, Physmatgiz, Moscow, pp. 5–22 (1963). (in Russian)
6. Fedotkin, M.A., Kudryavtsev, E.V.: Investigation of the transient process of adaptive flow control of non-homogeneous requests by simulation. In: Proceedings of the XVIII International Conference "Information Technologies and Mathematical Modeling (ITMM-2019)", Part 2, pp. 207–212. NTL, Tomsk (2019). (in Russian)
7. Fedotkin, M.A., Kudryavtsev, E.V.: Simulation of adaptive control system with conflict flows of non-homogeneous requests. In: Proceedings of the Twelfth International Conference "Computer Data Analysis and Modeling: Stochastics and Data Science", pp. 163–166. Minsk, BSU (2019)

Synergistic Effects in Queuing Systems and Related Statistical Problems

Gurami Tsitsiashvili$^{(\boxtimes)}$ [ID]

Institute for Applied Mathematics, Far Eastern Branch of Russian Academy Sciences,
Radio Street 7, Vladivostok, Russia
guram@iam.dvo.ru

Abstract. In this paper, there is a approach to detect the results of elements interaction in multi-channel queuing system with large load and small queue. This method is extended to statistical estimates of characteristics of non-uniform Poisson flow, describing distribution of animals in some areas, a resolution of the most powerful decision rule for constructing of technical systems "friend – foe". Such approach gives possibility to expand applications area and to simplify using methods of research. These methods consists of structural analysis and construction of upper bounds of objective functions. It permits to shorten numerical calculations and to obtain explicit results.

Keywords: Multi-server queuing system · Almost deterministic one-server queuing system · Most powerful decision rule

1 RQ-Queuing Systems with a Large Number of Servers

Consider an RQ-system, i.e., a queuing system with orbit in which customer, which has not possibility to be served is directed to the orbit. When some server is released, the customer may be directed to the server in accordance with some protocol [1–3]. RQ-systems attract attention of specialists in queuing theory last years (see, for example, materials of Conference ITMM 2018 in Tomsk and 12th International Workshop on Retrial Queues and Related Topics (WRQ 2018). But calculations of RQ-systems with large number of servers are sufficiently complicated. To decrease a complexity of these calculations we use the theorem on the asymptotic behaviour of an n-server queuing system for $n \to \infty$. In this theorem, it is proved that at $T > 0$ for $n \to \infty$, the probability $P_n(T)$ of customers direction to the orbit during time interval $[0, T]$ tends to zero. So used theorem gives possibility to change objective functions of multi-channel RQ-system from its limit distribution to probability of customers direction to the orbit during time interval T.

1.1 Preliminaries

Consider n - server queuing systems with the parameter $n \to \infty$. Assume that an intensity of input flow is proportional to n and $e_n(t)$ is a number of input

© Springer Nature Switzerland AG 2022
A. Dudin et al. (Eds.): ITMM 2021, CCIS 1605, pp. 197–207, 2022.
https://doi.org/10.1007/978-3-031-09331-9_16

flow customers arriving until the moment t, $e_n(0) = 0$. Suppose that $q_n(t)$ is a number of working servers at the moment t, $q_n(0) = 0$, τ_j is the service time of j-th arriving customer and τ_j, $j \geq 1$, is a sequence of independent and identically distributed random variables (s.i.i.d.r.v.'s) with the distribution function (d.f.) $F(t)$ ($\overline{F} = 1 - F$). Here $F(t)$ has continuous density $f(t) \leq \bar{f}$, where $0 \leq \bar{f} < \infty$. This section is based on [4, Chapter II, § 1, Theorem 1]

Theorem 1. *Assume that the following conditions are true.*

(1) For some $a > 0$ we have $Ee_n(t) = nat$, $t \geq 0$.
(2) There is $B(n)$ such that $A(n) = \max(n^{1/2}, B(n))$ satisfies the relation for $n \to \infty$

$$\frac{B(n)}{A(n)} \to B \geq 0, \quad \frac{\sqrt{n}}{A(n)} \to K \geq 0, \quad \frac{n}{A(n)} \to \infty.$$

and $\max(B, K) = 1$).

(3) Random processes $x_n(t) = \dfrac{e_n(t) - Ee_n(t)}{B(n)}$ C-converges to the centred Gaussian process $z(t)$, when $n \to \infty$.

(4) Random process $\zeta(t) = \displaystyle\int_0^t \overline{F}(t - u)dz(u) + K\Theta(t)$, $0 \leq t \leq T$, where $\Theta(t)$ is centred Gaussian process independent with $z(t)$, and its covariance function $R(t, t + u) = \displaystyle\int_0^t \overline{F}(v + u)F(v)adv$ and satisfies the formula $P(\sup_{0 \leq t \leq T} \zeta(t) > L) \to 0$, $L \to \infty$.

(5) If $\rho = aE\tau_j < 1$, then for any $T > 0$ we have $P\left(\sup_{0 \leq t \leq T} q_n(t) \geq n\right) \to 0$, $n \to \infty$.

Designate $\mathcal{F}_1$ the space of deterministic functions on the segment $[0, T]$ with uniform metric ρ and denote $\mathcal{F}$ the set of bounded functional's f defined on $\mathcal{F}_1$ and continuous in the metric ρ : if $z = z(t), z_1 = z_1(t), z_2 = z_2(t), \ldots \in \mathcal{F}_1$ and $\rho(z, z_n) \to 0$, $n \to \infty$, then $f(z_n) \to f(z)$, $n \to \infty$. Say that the sequence of random processes $z_n = z_n(t)$, $n \geq 1$, C - converges to the random process $z = z(t)$ if for any functional $f \in \mathcal{F}$ we have that $Ef(z_n) \to Ef(z)$, $n \to \infty$.

1.2   Main Results

In this subsection we used the following obvious inequality for RQ-systems

$$P_n(T) \leq P\left(\sup_{0 \leq t \leq T} q_n(t) \geq n\right), \quad n \geq 1.$$

Then from Theorem 1 it is possible to prove the relation

$$P\left(\sup_{0 \leq t \leq T} q_n(t) \geq n\right) \to 0, \quad n \to \infty \tag{1}$$

for n-channel RQ-systems with different input flows and so $P_n(T) \to 0$, $n \to \infty$..

Deterministic Input Flow of Customer Groups. Suppose that at the moments $1, 2, \ldots$, groups of customers of the size $\eta_1 \geq 0, \eta_2 \geq 0, \ldots$ arrive in the n-channel RQ system. Here $\eta_1, \eta_2, \ldots$ are i.i.d.r.v.'s with integer values, $E\eta_1 = a$, $Var\,\beta_1 < \infty$. Define deterministic input flow as follows by the equality

$$e_n(t) = \sum_{k=1}^{[nt+\psi]} \eta_k, \ t \geq 0, \text{ where } \psi \text{ is independent of } \eta_k, \ k \geq 1, \ \tau_j, \ j \geq 1, \text{ r.v.}$$

with uniform distribution on $[0, 1]$ and $[g]$ is the integer part of the real number g. For the n-channel RQ system with arbitrary protocol of customers direction to servers after their being in orbit the relation (1) is proved in [5].

Alternating Input Flow. This flow is defined by ON and OFF periods alternating with lengths $X_0 \geq 0$, $X_1 \geq 0, X_2 \geq 0, \ldots$, and $Y_0 \geq 0$, $Y_1 \geq 0$, $Y_2 \geq 0, \ldots$ respectively. In [6,7] a continuous random flow with ON and OFF period is defined. Denote $F_1(t) = P(X_1 < t)$, $F_2(t) = P(Y_1 < t)$, $t \geq 0$, and suppose that

$$\overline{F}_1(t) = t^{-\alpha_1} L_1(t), \ \overline{F}_2(t) = t^{-\alpha_2} L_2(t), \ 1 < \alpha_1 < \alpha_2 < 2,$$

with $L_1(t) \to l_1 > 0$, $t \to \infty$, and $L_2(t)$ - slowly varying function and $b(t)$ is the inverse $1/\overline{F}_1(t)$: $b(1/\overline{F}_1(t)) = t$.

Introduce i.r.v.'s B, X, Y, and r.v. Y_0 independent of X_n, Y_n, $n \geq 1$, so that $P(B{=}1) = \dfrac{\mu_1}{\mu}$, $P(B{=}0) = \dfrac{\mu_2}{\mu}$, $\mu{=}\mu_1 + \mu_2$, $\mu_1 = EX_1$, $\mu_2 = EY_1$,

$$P(X{\leq}x) = \frac{1}{\mu_1} \int_0^x \overline{F}_1(s)ds, \ P(Y{\leq}x) = \frac{1}{\mu_2} \int_0^x \overline{F}_2(s)ds.$$

Then random sequence (X_k, Y_k), $k \geq 0$ generates the ON–OFF process $W(t)$ as follows

$$W(t) = BI_{[0,X)}(t) + \sum_{n=0}^{\infty} I_{[T_n, T_n+X_{n+1})}(t), \ t \geq 0 \text{ where } T_0 = B(X + Y_0) + (1 -$$

$B)Y$, $T_n = T_0 + \displaystyle\sum_{i=1}^{n}(X_i + Y_i)$, $n \geq 1$ and $I_A(t) = 1$ if $t \in A$ and $I_A(t) = 0$ else. The process $W(t)$ satisfies equalities $W(t) = 1$ if t is in ON-period, $W(t) = 0$ if t is in off-period, and stationary and $EW(t) = \mu_1/\mu = \alpha$.

Denote $A(t) = \displaystyle\int_0^t W(s)ds$, then $EA(t) = \alpha t$, $t \geq 0$. Let $n = n(N) = NM(N)$, $M = M(N) = [N^\gamma]$, $\gamma > 0$, and assume that random functions $A_m(t)$, $m = 1, ..., M$, are independent copies of $A(t)$, $e_n(t) = \left[\displaystyle\sum_{m=1}^{M} A_m(Nt) + \psi \right]$ For so defined alternating input fLow the formula (1) is proved in [5].

Erlang Input Flow. Assume that $E_n(t)$ is Poisson flow intensity $n\alpha$ and $e_n(t) = \left[\dfrac{E_n(t)}{r} + \psi \right]$, $t \geq 0$, with random variable ψ independent of η_k, $k \geq 1$, τ_j, $j \geq 1$. and integer r. In [8] it is proved the formula (1) in condition $\alpha E\tau_j < 1$.

Consequently if the objective function of multi-channel RQ-system is $P_n(T)$, then it is possible to replace complicated calculations by known asymptotic Theorem 1.

2 Alternative Designs of High Load Queuing Systems with Small Queue

It is well known that queuing systems in high-load mode have long queues. A large number of publications are devoted to the study of asymptotic regimes in such systems (see, for example, [9]. Therefore, such modes of operation of these systems, that do not have large queues, are of great interest. These modes are convenient from an economic point of view, since the service device is almost fully loaded. On the other hand, this mode is also convenient for users which waiting times become small.

Multi-channel Queuing System $M|M|n|\infty$. Consider n – channel system with a Poisson input flow of intensity $n\lambda$ and the service time has an exponential distribution $1-\exp(-\mu_t)$. Such a system can be considered as an aggregation (Fig. 1, right) of n single-channel systems $M|M|1|\infty$ (Fig. 1, left) with Poisson input flows of λ intensity and a similar distribution of service times. Here, aggregation of n single-channel systems is understood as combining their input flows and combining service channels into a multi-channel system. Denote $\rho = \lambda/\mu$ load factor of the system $M|M|n|\infty$ and put A_n the stationary average waiting time, B_n the stationary average queue length.

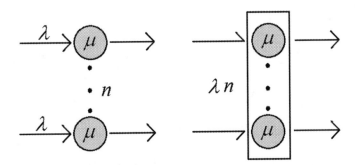

Fig. 1. Transformation of n single-channel systems $M|M|1|\infty$ into aggregated n - channel system $M|M|n|\infty$.

The following are obtained in [10].

Theorem 2. *1) If $\rho < 1$, then for some $c < \infty$, $q < 1$ the relation holds $A_n \leq c\,q^n$, $n \geq 1$, 2) If $\rho = 1 - n^{-\alpha}$, $0 < \alpha < \infty$, then for $n \to \infty$*

$$A_n \to \begin{cases} 0, & \alpha < 1, \\ 1/\mu, & \alpha = 1, \\ \infty, & \alpha > 1. \end{cases} \quad B_n \to \begin{cases} 0, & \alpha < 1/2, \\ \infty, & \alpha \geq 1/2. \end{cases}$$

This theorem develops and specifies the results of [11,12] in the direction of determining the changed structure of the queuing system.

It is clear that an alternative to the described mode of operation of a queuing system with a large load and a small queue can serve as an almost deterministic queuing system. Such a system operates on a specific schedule and its maintenance processes are almost cyclical [13]. The question arises as how to randomly perturb cyclic processes in order to keep a small queue in them along with a large load. Obviously, such perturbations will strongly depend on the distributions of random fluctuations.

Almost Deterministic Single-Channel Queuing System. Despite the importance of Theorem 2, such a queuing system design assumes its large size, which is not always convenient from an application point of view. It is clear that an alternative to the described mode of operation of a queuing system with a large load and a small queue can serve as an almost deterministic one channel queuing system (see, for example [13]).

Let's describe the single-channel queuing system $G|G|1|\infty$ by Lindley chain of waiting times for the service: $w_{i+1} = \max(0, w_i + \eta_i - \tau_i)$. Here τ_i is the interval between the arrival of i - th and $(i+1)$ - th customers, $M\tau_i = a$, and η_i – service time of i - th customer, $M\eta_i = b$, $0 < a - b = \varepsilon$. Assume that random deviations from the distributions means are reduced as follows:

$$\eta_i^\varepsilon = b + \varepsilon^\alpha(\eta_i - b), \ \tau_i^\varepsilon = a + \varepsilon^\alpha(\tau_i - a)$$

and introduce Markov chain w_i^ε, $i \geq 0$, $w_0^\varepsilon = 0$, describing almost deterministic single-channel queuing system

$$w_{i+1}^\varepsilon = \max(0, w_i^\varepsilon + \eta_i^\varepsilon - \tau_i^\varepsilon) = \max(0, w_i^\varepsilon + \varepsilon^\alpha\delta_i).$$

Here $\delta_0, \delta_1, \ldots$, is a sequence of independent and identically distributed random variables, $\delta_i = \eta_i - \tau_i + \varepsilon$, $M\delta_i = 0$. In high load mode, when the load factor $\rho = \dfrac{b}{a}$ is close to one, the positive parameter $\varepsilon = (1 - \rho)a$ is small: $\varepsilon \ll 1$. Value $\alpha > 0$ characterizes the rate of decreasing random perturbations with increasing loading.

Due to known results for a single-channel queuing system $G|G|1|\infty$ Markov chain w_i^ε, $i \geq 0$ has given for any $\varepsilon, \alpha : 0 < \varepsilon, 0 < \alpha$ the stationary distribution $\lim_{i\to\infty} \mathcal{P}\{w_i^\varepsilon > y\} = \mathcal{P}\{W_\alpha(\varepsilon) > y\}$, $y \geq 0$. Using [15–25] it is possible to formulate following statements.

Statement 1. Let for some positive constants β, $c < \infty$ the inequality $M|\delta_1|^{2+\beta} \leq c$ takes place. Then for any $y \geq 0$ we have $\mathcal{P}\{\varepsilon W_0(\varepsilon) > y\} \to e^{-2y/d}$, $\varepsilon \to 0$.

Statement 2. If for some fixed ν, $1 < \nu < 2$; $h_\nu > 0$, the following relations are true when $y \to \infty$ $P(\eta_1 > y) \sim h_\nu y^{-\nu}$; $P(\tau_1 > y) = o(P(\eta_1 > y))$, or $P(\tau_1 > y) \sim h_\nu y^{-\nu}$; $P(\eta_1 > y) = o(P(\tau_1 > y))$. Then there is a tail $R(y)$ of non - degenerate distribution function and $\Delta_\nu(\varepsilon) \sim c\varepsilon^{1/(\nu-1)}$, $\varepsilon \to 0$, such that for any $y \geq 0$ we have $\mathcal{P}\{\Delta_\nu(\varepsilon)W_0(\varepsilon)/b > y\} \to R(y)$, $\varepsilon \to 0$ or $\mathcal{P}\{\Delta_\nu(\varepsilon)W_0(\varepsilon)/a > y\} \to R(y)$, $\varepsilon \to 0$.

Using Statement 1 it is possible to prove Theorem 3.

Theorem 3. *Assume that in a single-channel queuing system* $G|G|1|\infty$ *conditions of Statement 1 are true. Then the following limit relations are valid:* $W_\alpha(\varepsilon) \Rightarrow$ *(convergence in distribution)* $+\infty$, $0 \leq \alpha < 1/2$; $W_\alpha(\varepsilon) \Rightarrow 0$, $1/2 < \alpha$; $W_\alpha(\varepsilon) \Rightarrow \eta$, $\mathcal{P}\{\eta > y\} = e^{-2y/d}$, $\alpha = 1/2$.

Using Statement 2 it is possible to prove Theorem 4.

Theorem 4. *Assume that in a single-channel queuing system* $G|G|1|\infty$ *conditions of Statement 2 are true. Then the following limit relations are valid:* $W_\alpha \Rightarrow +\infty$, $0 \leq \alpha < 1/\nu$; $W_\alpha \Rightarrow 0$, $1/\nu < \alpha$; $\varepsilon \to 0$.

The most simple variant of these theorems proves are based on following well known and elementary statement [14, Exercises 15-19 on pages 184-185].

Statement 3. Suppose that X_n, $n \geq 1$, is a sequence of positive real-valued random variables that converges in distribution to a non degenerate limit random variable X as $n \to \infty$. Then if a_n are positive real numbers with $a_n \to \infty$, then it follows that $a_n X_n \Rightarrow \infty$ and $X_n/a_n \Rightarrow 0$ as $n \to \infty$.

Thus, a parameter α, characterizing either the rate of convergence of the load factor to one in the system $M|M|n|\infty$, or a random fluctuation in the system $G|G|1|\infty$, allows to detect the convergence of the stationary waiting time to either zero or infinity.

3   Related Statistical Problems

In this section statistical estimates of characteristics of non-uniform Poisson flow, describing distribution of animals in some areas and resolution of the most powerful decision rule for constructing of technical systems discriminating "friend – foe". Main idea of this consideration is in a choice of convenient objective functions for next estimates. Such objective functions may be as relative errors of mean number of points of Poisson flow in some area so a calculation of the most powerful decision rule in a construction of technical system for discriminating "friend – foe". This results are based on the classification of statistical problems proposed in the monographs [32,33] and on the ideas of testing statistical hypotheses in the processing of physical and physico-technical observations [34,35].

Estimates of the Mean Number of Poisson Flow Points in Some Area.
In geographical and geological investigations (see, for example [27]) there is a problem to estimate mean number of points in some area and to evaluate its quality. Let the study area is divided into m cells, and the number of points in the cell k is n_k, $k = 1, \ldots, m$. As we deal with Poisson flow then the random variables $n_1, \ldots, n_m$ are independent with Poisson distributions which have the parameters $\lambda_1, \ldots, \lambda_m$. Consequently the random variable $N = \sum_{k=1}^{m} n_k$ has a Poisson distribution with the parameter $\Lambda = \sum_{k=1}^{m} \lambda_k$ and so $EN = \Lambda$, $VarN = \Lambda$.

Consider now random variable $\dfrac{N}{EN} = \dfrac{N}{\Lambda}$ and calculate its variance $Var\dfrac{N}{\Lambda} = \dfrac{1}{\Lambda}$. Consequently the following relation is true $\sqrt{Var\dfrac{N}{\Lambda}} = \dfrac{1}{\sqrt{\Lambda}}$.

From Chebyshev-Bienome inequality we have $P\left(\left|\dfrac{N}{\Lambda} - 1\right| > \Lambda^{-1/3}\right) \leq \Lambda^{-1/3} \to$ 0, $\Lambda \to \infty$. Therefore, the relative error of this estimate, constructed for non uniform Poisson flow decreases with the growth of total Λ.

Resolution of the Most Powerful Decision Rule. In the papers [28–31], a neural network converter 'Biometrics access code'is built on the basis of an electroencephalogram. The main indicator of the effectiveness of this converter is the probability of errors of the first α_1 kind when the probability of errors of the second kind α_2 is chosen by experts to distinguish between simple hypotheses "friend - foe". This distinction of hypotheses is made using the most powerful decision rule. A special role here is played by a set of sample characteristics, with the help of which these hypotheses are distinguished.

In this paper, we introduce a characteristic A of the resolution of the most powerful decision rule. The value of A is determined by the probability α_2, by the sample size n from independent and equally normally distributed random variables with variance σ^2 and the difference of the average $a_1 - a_2$ of these random variables when performing alternative hypotheses. It is established that the probability of errors of the first kind strongly (approximately as $\dfrac{\exp(-A^2)}{A\sqrt{2\pi}}$) depends on the resolution A of the most powerful solving rule.

This work is based on the classification of statistical problems proposed in the monographs [32,33], the Neumann-Pearson lemma and the well-known rule for finding the most powerful solving rule by the Bayesian solving rule. An important role here is played by the idea of testing statistical hypotheses when processing physical and physico-technical observations [34,35]. The main characteristic that determines the distinguishing ability of A in this statistical problem is the difference of the averages $a_1 - a_2$. This difference of parameters corresponding to the hypotheses "friend - foe" plays an important role in the design of the technical system that specifies the access code, thus $A = A(a_1 - a_2, \alpha_2, n, \sigma)$.

Consider a sample $x_1, \ldots, x_n$, consisting of independent random variables having a normal distribution with an average a and a known variance σ^2. From two hypotheses $H_1 = (a = a_1)$, $H_2 = (a = a_2)$, $a_1 > a_2$, the most likely hypothesis is selected. This choice is made under the assumption that the probability of an error of the second kind is $P(H_1/H_2) = \alpha_2$, where the value of α_2 is determined by experts (and in accordance with the requirements of GOST). In this assumption, we are looking for a decisive rule that minimizes the probability of a first-kind error $P(H_2/H_1)$. The search for the most powerful solving rule is based on the Neumann-Pearson lemma [32, chapter 3, § 1, 2] and is searched in the form

$$\frac{1}{n}\sum_{i=1}^{n} x_i > C \Rightarrow H_1, \quad \frac{1}{n}\sum_{i=1}^{n} x_i \leq C \Rightarrow H_2. \tag{2}$$

The constant C is determined by the probability of an error of the second kind α_2 from the relations

$$\alpha_2 = P\left(\frac{1}{n}\sum_{i=1}^{n}x_i > C/H_2\right) = P\left(\frac{1}{\sqrt{n}}\sum_{i=1}^{n}\frac{\sqrt{(x_i - a_2)}}{\sigma} > \frac{\sqrt{n}(C - a_2)}{\sigma}/H_2\right)$$

Let's denote X a random variable with a normal distribution having zero mean and unit variance. Then from the above equalities we get

$$\alpha_2 = P\left(X > \frac{\sqrt{n}(C - a_2)}{\sigma}\right) = \int_{t(\alpha_2)}^{\infty}\frac{\exp(-u^2/2)}{\sqrt{2\pi}}du, \; t(\alpha_2) = \frac{\sqrt{n}(C - a_2)}{\sigma}.$$
$$(3)$$

It follows from the formula (2) that the constant C, defining the decisive rule (2), satisfies the equality

$$C = a_2 + \frac{t(\alpha_2)\sigma}{\sqrt{n}}.$$
$$(4)$$

Consequently we have

$$\alpha_1 = P\left(\frac{1}{n}\sum_{i=1}^{n}x_i \leq C/H_1\right) = P\left(\frac{1}{\sqrt{n}}\sum_{i=1}^{n}\frac{\sqrt{(x_i - a_1)}}{\sigma} \leq \frac{\sqrt{n}(C - a_1)}{\sigma}/H_1\right).$$

Hence the equality follows

$$\alpha_1 = P\left(X \leq \frac{\sqrt{n}(C - a_1)}{\sigma}\right) = P\left(X \geq \frac{\sqrt{n}(a_1 - C)}{\sigma}\right) =$$

$$= \int_{t(\alpha_1)}^{\infty}\frac{\exp(-u^2/2)}{\sqrt{2\pi}}du, \; t(\alpha_1) = \frac{\sqrt{n}(a_1 - C)}{\sigma}.$$
$$(5)$$

Substituting the formula (4) into the formula (5), we get

$$t(\alpha_1) = A(a_1, a_2, \alpha_2, n, \sigma),$$
$$(6)$$

where the value

$$A(a_1, a_2, \alpha_2, n, \sigma) = \frac{\sqrt{n}}{\sigma}(a_1 - a_2) - t(\alpha_2)$$
$$(7)$$

defines the resolution of the most powerful decision rule (2).

Let us now consider how strong is the dependence of the probability of errors of the first kind on this value. To do this, we calculate for $t > 0$

$$J(t) = \int_{t}^{\infty}\frac{\exp(-u^2/2)}{\sqrt{2\pi}}du = \int_{t}^{\infty}\frac{\exp(-u^2/2)}{\sqrt{2\pi}u}d\frac{u^2}{2} \leq \frac{\exp(-t^2/2)}{t\sqrt{2\pi}},$$

from here we get

$$J(t) = \int_{t}^{\infty}\frac{\exp(-u^2/2)}{\sqrt{2\pi}u}d\frac{u^2}{2} \geq \frac{\exp(-t^2/2)}{t\sqrt{2\pi}}\left(1 - \frac{1}{t^2}\right).$$

Combining the obtained inequalities, we find

$$\left(1 - \frac{1}{t^2}\right)\frac{\exp(-t^2/2)}{t\sqrt{2\pi}} \leq J(t) \leq \frac{\exp(-t^2/2)}{t\sqrt{2\pi}}, \; t > 0. \qquad (8)$$

It follows that the function $J(t)$, determining the probabilities of errors of the first and second kind decreases very quickly with the growth of t.

Let's take $\alpha_2 = 10^{-9}$, $\alpha_1 = 7 \cdot 10^{-4}$ as a numerical example (these values are taken from [31]), then we can build an approximation $J(t) \approx \dfrac{\exp(-t^2/2)}{t\sqrt{2\pi}}$ and with an accuracy of 10^{-2}, get the values $t(\alpha_1) = 5.99781$, $t(\alpha_2) = 3.19465$. As a result, we come to equality

$$A(a_1, a_2, \alpha_2, n) = \frac{\sqrt{n}}{\sigma}(a_1 - a_2) - t(\alpha_2) = 9.19246.$$

Since $\alpha_1 = 7 \cdot 10^{-4}$, then combined with the formula (8) from the inequality

$$\alpha_1 \leq \frac{\exp(-A^2(a_1, a_2, \alpha_2, n)/2)}{\sqrt{A(a_1, a_2, \alpha_2, n)}}$$

it can be seen how much the resolution of $A(a_1, a_2, \alpha_2, n)$ affects the probability of an error of the first kind α_1, which is the main indicator in this statistical problem.

The formula (7), which determines the resolution of $A(a_1, a_2, \alpha_2, n)$, specifying the probability of an error of the first kind, despite its simplicity, contains a whole series of characteristics: the difference of the averages $a_1 - a_2$, variance σ^2, sample size n (and the probability of an error of the second kind α_2). Therefore, the choice of the characteristics of $a_1 - a_2$, σ^2, n becomes a rather difficult task of designing the technical system described in [30]. Moreover, a special role here is played by the difference $a_1 - a_2 > 0$ of the average a_1, a_2, characterizing the distributions of samples describing the "friend - foe" states shared by the technical system.

4   Conclusion

The results presented in this paper go beyond the theory of probability and queuing. In these results, the main focus is not on proving probabilistic theorems of the greatest generality, but on obtaining explicit estimates of the comparison of queuing systems, statistical algorithms and programs before and after the transformation of their structure. The peculiarity of such results, and it is convenient to call them synergetic effects, is the strong dependence of the compared performance indicators when a certain parameter tends to zero or to infinity. However, this circumstance in no way reduces the requirements for the accuracy of the estimates obtained. According to the author, such estimates are most convenient to carry out during computational experiments. Another thing is that it is convenient to conduct such computational experiments working with complex systems if there are some analytical estimates of the marginal behavior of performance indicators.

References

1. Dudin, A., Nazarov, A.: On a tandem queue with retrials and losses and state dependent arrival, service and retrial rates. Int. J. Oper. Res. **29**, 170–182 (2017)
2. Artalejo, J., Gomez-Corral, A.: Retrial Queueing Systems. A Computational Approach. Springer, Heidelberg (2008). https://doi.org/10.1007/978-3-540-78725-9
3. Nazarov, A., Moiseeva, E.: Investigation of the RQ-system $MMPP|M|1$ by the method of asymptotic analysis in the condition of a large load. Izv. Tomsk. Polytech. Univ. **322**, 19–23 (2013). (in Russian)
4. Borovkov, A.: Asymptotic Methods in Queueing Theory. Wiley, New York (1984)
5. Tsitsiashvili, G., Osipova, M.: Synergetic effects for number of busy servers in multiserver queuing systems. Commun. Comput. Inf. Sci. Ser. **564**, 404–414 (2015)
6. Heath, D., Resnick, S., Samorodnitsky, G.: Heavy tails and long range dependence in on/off processes and associated fluid models. Math. Oper. Res. **23**, 145–165 (1998)
7. Mikosch, T., Resnick, S., Rootzen, H., Stegeman, A.: Is network traffic approximated by stable Levy motion or fractional Brownian motion? Ann. Appl. Probab. **12**, 23–68 (2002)
8. Tsitsiashvili, G., Markova, N.: Synergistic effects in a multi-channel queuing system with an Erlangian input flow. Bull. Pac. State Univ. **4**, 17–22 (2015). (in Russian)
9. Borovkov, A.A.: Asymptotic methods in queueing theory. Science, Moscow (1980). (in Russian)
10. Tsitsiashvili, G., Osipova, M.: Phase transitions in multiserver queuing systems. In: Dudin, A., Gortsev, A., Nazarov, A., Yakupov, R. (eds.) ITMM 2016. CCIS, vol. 638, pp. 341–353. Springer, Cham (2016). https://doi.org/10.1007/978-3-319-44615-8_30
11. Halfin, S., Whitt, W.: Heavy-traffic limits for queues with many exponential servers. Oper. Res. **29**(3), 567–588 (1981)
12. Jagerman, D.L.: Some properties of the erlang loss function. Bell Syst. Tech. J. **53**, 525–551 (1974)
13. Afanasyeva, L.G.: Queuing Systems with cyclic control processes. Cybern. Syst. Anal. **41**(1), 54–68 (2005). (in Russian)
14. Chung, K.L.: A Course in Probability Theory, 3rd edn. Stanford University, Academic Press (2001)
15. Borovkov, A.A.: Some limit theorems of the queueing theory. Probab. Theory Appl. **IX**(4) (1964), X (3) (1965). (in Russian)
16. Prokhorov, Y.V.: Transient phenomena in queuing processes. Litov. Math. Collection III **1**, 199–205 (1963). (in Russian)
17. Harrison, J.M.: The heavy traffic approximation for single server queues in series. J. Appl. Probab. **10**(3), 613–629 (1973)
18. Abate, J., et al.: Calculation of the Gl/G/1 waiting time distribution and its cumulants from Pollaczek's formulas. Arch. Elektr. Uebertragung **47**(Pollaczek memorial volume), 311–321 (1993)
19. Beran, J., et al.: Long-range dependence in variable-bit-rate video. IEEE Trans. Commun. **43**, 1566–1579 (1995)
20. Bingham, N.H., et al.: Regular Variation. Cambridge University Press, Cambridge (1987)
21. Boxma, O.J.: Fluid queues and regular variation. Perform. Eval. **27**(28), 699–712 (1996)

22. Boxma, O.J., Cohen, J.W.: The M/G/1 queue with heavy-tailed service time distribution. IEEE J. Sel. Areas Commun. **16**, 749–763 (1998)
23. Cohen, J.W.: Some results on regular variation for distributions in queueing and fluctuation theory. J. Appl. Probab. **10**, 343–353 (1973)
24. Cohen, J.W.: Superimposed renewal processes and storage with gradual input. Stochastic Process. Appl. **2**, 31–58 (1974)
25. Boxma, O.J., Cohen, J.W.: Heavy-traffic analysis for the GI/G/1 queue with heavy-tailed distributions. Queueing Syst. **33**, 177–204 (1999)
26. Gnedenko, B.V., Korolev, V.Y.: Random Summation. CRC Press, Boca Raton (1996)
27. Stoyan, D., Kendall, W., Mecke, J.: Stochastic Geometry and Its Applications. Wiley, New York (1987)
28. Goncharov, S.M., Borshevnikov, A.E.: Construction of a neural network converter "Biometrics access code" based on the parameters of the visual evoked potential of an electroencephalogram. Reports of the Tomsk State University of Control Systems and Radioelectronics, vol. 2, no. 32, pp. 51–55 (2014). (in Russian)
29. Goncharov, S.M., Borshevnikov, A.E.: Prediction of the output parameters of the neural network converter "Biometrics-access code" based on an electroencephalogram. Inf. Secur. **21**(3), 302–307 (2018). (in Russian)
30. Borshevnikov, A.E., Goncharov, S.M., Dobrzhinsky, Yu.V.: On the requirements for the formation of synthetic images of an electroencephalogram for the tasks of highly reliable biometric authentication. Modern science: actual problems of theory and practice. Series: Nat. Tech. Sci. **2**, 27–29. (in Russian)
31. Borshevnikov, A.E.: Neural network model of a highly reliable biometric user authentication system based on an encephalogram. Oral report at the general Institute seminar of the IAM FEB RAS (2021)
32. Borovkov, A.A.: Mathematical statistics. Evaluation of parameters. Hypothesis testing. Moscow: Nauka. The main editorial office of the physical and mathematical literature (1984). (in Russian)
33. Borovkov, A.A.: Mathematical statistics. Additional chapters. Nauka, Moscow (1984). (in Russian)
34. Gortsev, A.M., Nezhelskaya, L.A., Shevchenko, T.I.: Estimation of the states of the MC-flow of events in the presence of measurement errors. Izv. Vuzov. Phys. **36**(12), 67–85 (1993). (in Russian)
35. Gortsev, A.M., Nezhelskaya, L.A.: Optimization of adapter parameters during observations of the MS flow. In Sat. Stochastic and deterministic models of complex systems, pp. 20–32. Publishing house of the Computing Center of the Siberian Branch of the USSR Academy of Sciences, Novosibirsk (1988). (in Russian)

An Explicit Solution for an Inventory Model with Positive Lead Time and Server Interruptions

E. Sandhya[1]([⊠])(iD), C. Sreenivasan[2](iD), Sajeev S. Nair[3](iD), and M. P. Rajan[3]

[1] Department of Mathematics, Government College, Chittur, University of Calicut, Chittur, India
esandhya79@gmail.com
[2] Department of Mathematics, Government Victoria College, Palakkad, India
[3] Department of Mathematics, Government Engineering College, Thrissur, India

Abstract. A single server queuing system with inventory is considered. Customers arrive according to a Poisson process and service times follow exponential distribution. Inventory is replenished according to (s, S) policy with positive lead time which follows exponential distribution. Interruption to service process and repair of interrupted service are considered, times between two interruptions and repairs both follow exponential distributions. We assume that during interruption, the customer being served waits there until his service is completed, no inventory is lost due to interruption, no arrivals are allowed and order placed if any is cancelled. We also assume that no arrival is entertained when inventory level is zero. Stability of the above system is analyzed and the steady state vector is calculated explicitly. Expressions for several system performance measures such as expected number of customers in the system, expected inventory level, expected interruption rate etc. are obtained. Even though explicit expressions are obtained several other performance measures are calculated numerically as well.

Keywords: (s,S) inventory model · Server interruptions · Positive lead time · Explicit solution

1 Introduction

The pioneers in the study of queueing inventory models are Melikov and Molchanov [12] and Sigman and Simchi- Levi [16]. In Sigman and Simchi- Levi customers are allowed to join even when there is no inventory in the system. They also discuss the case of non exponential lead time distribution. Later Berman and et al. [2] considered an inventory system where a processing time is required for serving the inventory. Here they considered deterministic service time and the model was discussed as a dynamic programming model. Berman and Kim [3] and Berman and Sapna [4] later discussed inventory queueing systems with exponential service time distribution and with arbitrary distribution.

© Springer Nature Switzerland AG 2022
A. Dudin et al. (Eds.): ITMM 2021, CCIS 1605, pp. 208–220, 2022.
https://doi.org/10.1007/978-3-031-09331-9_17

There are several papers on inventory queueing models by Krishnamoorthy and his co-authors [1,5–11,13]. They mainly used Matrix Analytic Methods to study these models. In most of the models service time for providing the invento-ried item is assumed. Schwarz et al. [15] considered a queueing inventory model with Poisson arrivals and exponentially distributed service and lead times. They could obtain a product form solution for the system steady state. But they assumed that no customers join the system when the inventory level is zero.

2   Mathematical Model

The system under consideration is described as below. There is a single server counter where inventory is served to which customers arrive for service. The number of arrivals by time t follows a Poisson process with parameter λt. The service times are independently and identically distributed exponential random variables with parameter μ. Inventory is replenished according to (s, S) policy, in the sense that whenever inventory level drops to s an order is placed, order quantity being fixed as $Q = S - s$. The replenishment times follow exponential distribution with parameter η. While a customer is being served by the server, the service may be interrupted, the interruption rate being exponential with rate δ_1. Following a service interruption the service restarts at an exponential rate δ_2.

We make the following assumptions for the model under consideration.

i) There is no loss of inventory due to a service interruption.
ii) The customer being served when interruption occurs waits there until his service is completed.
iii) No arrival is entertained when the inventory level is zero.
iv) An order placed if any is cancelled while the server is on interruption.
v) We also assume that there are no arrivals while the server is on interruption.

We denote by $N(t)$ the number of the customers in the system including the one being served (if any), $L(t)$ the inventory level and $S(t)$ the server status at time t.

$$\text{Let } S(t) = \begin{cases} 0 & \text{if the server is idle} \\ 1 & \text{if the server is busy} \\ 2 & \text{if the server is on interruption} \end{cases}$$

Then $\Omega = X(t) = ((N(t), S(t), L(t))$ will be a Markov chain. The state space of this Markov chain can be described as $E = \{(0, 0, k) : 0 \le k \le S\} \cup \{(i, 0, 0) : i \ge 1\} \cup \{(i, j, k) : i \ge 1, j = 1, 2; 1 \le k \le S\}$. The above state space can be partitioned into levels $L(i)$ where $L(0) = ((0, 0, 0), (0, 0, 1), \dots, (0, 0, S))$ and $L(i) = ((i, 0, 0), (i, 1, 1), (i, 1, 2), \dots, (i, 1, S), (i, 2, 1), (i, 2, 2), \dots, (i, 2, S)) ; i \ge 1$. The Markov chain Ω described above is a level independent quasi birth death process whose infinitesimal generator matrix is given by

$$T = \begin{bmatrix} B_0 & B_1 & 0 & 0 & . & . & . \\ B_2 & A_1 & A_0 & 0 & 0 & . & . \\ 0 & A_2 & A_1 & A_0 & 0 & . & 0 \\ 0 & 0 & A_2 & A_1 & A_0 & 0 & . \\ & & & . & . & . & \\ & & & & . & . & . \end{bmatrix}$$

Here B_0, B_1, B_2 are matrices of orders $(S+1) \times (S+1), (S+1) \times (2S+1)$ and $(2S+1) \times (S+1)$ respectively. All other matrices are square matrices of order $2S + 1$. The different transitions in the Markov chain $\Omega = X(t) = ((N(t), S(t), L(t))$ are given below.

i) Transitions due to arrival of customers

$$(i, j, k) \xrightarrow{\lambda} (i+1, j, k); i \geq 0, 0 < k \leq S, j = 0, 1$$

ii) Transitions due to service completion of customers

$$(i, j, k) \xrightarrow{\mu} (i-1, j, k-1); i > 0, 0 < k \leq S, j = 1$$

iii) Transitions due to replenishment of inventory

$$(i, j, k) \xrightarrow{\eta} (i, j, k+Q); i \geq 0, 0 \leq k \leq S, j = 0, 1$$

iv) Transitions due to server interruption

$$(i, 1, k) \xrightarrow{\delta_1} (i, 2, k); i \geq 1, 0 < k \leq S$$

v) Transitions due to restart of service after a service interruption

$$(i, 2, k) \xrightarrow{\delta_2} (i, 1, k); i \geq 1, 0 < k \leq S$$

The matrix B_0 contains the transition rates within level $L(0)$, B_1 records the transition rates from $L(0)$ level to $L(1)$ and B_2 that from $L(1)$ to $L(0)$. Similarly the matrices $A_0.A_1, A_2$ contains the transitions from levels $L(i)$ to $L(i+1)$, $L(i)$ to itself and $L(i+1)$ to $L(i)$ for $i \geq 1$.

3   Analysis of the Model

Stability condition
Define $A = A_0 + A_1 + A_2$ and

$$\pi = (\pi(0,0), \pi(1,1), \pi(1,2), \ldots, \pi(1,S), \pi(2,1), \pi(2,2), \ldots, \pi(2,S))$$

be the steady state vector of A. We know the QBD process with generator matrix T is stable if and only if $\pi A_0 e < \pi A_2 e$ [14]. That is if and only if $\lambda [\pi(1,1) + \pi(1,2) + \ldots + \pi(1,S)] < \mu [\pi(1,1) + \pi(1,2) + \ldots + \pi(1,S)]$, that is if and only if $\lambda < \mu$.

Thus we have the following theorem for the stability of the system under study.

Theorem 1. *The Markov chain is stable if and only if $\lambda < \mu$.*

4   Computation of Steady State Vector

We first consider a system identical to the above system except for service time is negligible. For this system $\tilde{\Omega} = \tilde{X}(t) = (S(t), L(t))$ will be a Markov chain where $S(t)$ and $L(t)$ are as defined for the original system. The state space of this Markov chain can be described as

$$\tilde{E} = \{(0,0), (1,1), (1,2) \ldots, (1,S), (2,1), (2,2), \ldots, (2,S)\}.$$

The infinitesimal generator matrix of the process is given by $\tilde{T} = \begin{bmatrix} \tilde{B}_0 & \tilde{B}_1 \\ \tilde{B}_2 & \tilde{B}_3 \end{bmatrix}$, where

$\tilde{B}_1 = \begin{bmatrix} 0 \\ \delta_1 I_s \end{bmatrix}_{(S+1) \times S}$, $\tilde{B}_2 = \begin{bmatrix} 0 & \delta_2 I_s \end{bmatrix}_{S \times (S+1)}$, $\tilde{B}_3 = -\delta_2 I_s$, $\tilde{B}_0 = \begin{bmatrix} C_1 & C_2 \\ C_3 & C_4 \end{bmatrix}$ Here

$C_1 = \begin{bmatrix} -\eta & 0 \\ 0 & -(\lambda + \eta + \delta_1)I_{s-1} \end{bmatrix}_{(s+1) \times (s+1)} + \begin{bmatrix} 0 & 0 \\ \lambda I_{s-1} & 0 \end{bmatrix}_{(s+1) \times (s+1)}$, $C_4 = -(\lambda +$

$\delta_1) I_Q + \begin{bmatrix} 0 & 0 \\ \lambda I_{Q-1} & 0 \end{bmatrix}_{Q \times Q}$, $C_3 = \begin{bmatrix} 0 & \lambda \\ 0 & 0 \end{bmatrix}_{Q \times (s+1)}$, $C_2 = \begin{bmatrix} 0 & \eta I_{s+1} \end{bmatrix}_{(s+1) \times Q}$ Let $x =$
$(x(0,0), x(1,1), \ldots, x(1,S), x(2,1), \ldots, x(2,S))$ be the steady state probability vector of the process $\tilde{\Omega}$. Then $x\tilde{T} = 0$ and $xe = 1$ gives

$$x(1,i) = \frac{\eta}{\lambda} \left(\frac{\eta + \lambda}{\lambda} \right)^{i-1} x(0,0) ;\ 1 \le i \le s+1$$

$$x(1, s+1) = x(1, s+2) = \ldots = x(1, Q)$$
$$x(1, Q+i) = x(1, Q) - x(1, i) ;\ 1 \le i \le s$$
$$x(2,i) = \frac{\delta_1}{\delta_2} x(1,i) ;\ 1 \le i \le S$$

where $x(0,0) = \left[1 + Q\frac{\eta}{\lambda} \left(\frac{\eta + \lambda}{\lambda} \right)^s \left(\frac{\delta_1 + \delta_2}{\delta_2} \right) \right]^{-1}$.

Let $\pi = (\pi_0, \pi_1, \pi_2, \ldots)$ be the steady state probability vector of the process Ω, where $\pi_0 = (\pi(0,0,0), \pi(0,0,1), \ldots, \pi(0,0,S))$ and $\pi_i = (\pi(i,0,0), \pi(i,1,1), \pi(i,1,2), \ldots, \pi(i,1,S), \pi(i,2,1), \pi(i,2,2), \ldots, \pi(i,2,S)); i \ge 1$. Then π satisfies $\pi T = 0$ and $\pi e = 1$. We have the equations

$$\pi_0 B_0 + \pi_1 B_2 = 0$$
$$\pi_0 B_1 + \pi_1 A_1 + \pi_2 A_2 = 0$$
$$\pi_i A_0 + \pi_{i+1} A_1 + \pi_{i+2} A_2 = 0;\ i \ge 1$$

All the above equations are satisfied by taking

$$\pi_0 = \zeta(x(0,0), x(1,1), x(1,2), \ldots, x(1,S))$$
$$\pi_i = \zeta \left(\frac{\lambda}{\mu} \right)^i \Big(x(0,0), x(1,1), x(1,2), \ldots,$$
$$x(1,S), \frac{\delta_1}{\delta_2} x(1,1), \frac{\delta_1}{\delta_2} x(1,2), \ldots, \frac{\delta_1}{\delta_2} x(1,S) \Big);\ i \ge 1$$

The value of ζ is obtained from $\pi e = 1$ as $\zeta = \dfrac{(\mu - \lambda)\delta_2}{\delta_2\mu + \delta_1\lambda[1 - x(0,0)]}$

5   System Performance Measures

5.1   Expected Waiting Time of a Customer in the Queue

First we compute the expected waiting time of a customer who joins the queue as the rth person. For that consider a Markov process $\psi = (\hat{N}(t), S(t), L(t))$, where $\hat{N}(t)$ represent the rank of the customer in the queue, $S(t)$ the server status and $L(t)$ the inventory level. The state space of the above Markov chain is $\hat{E} = \{(i, 0, 0), 1 \le i \le r - 1\} \cup \{(i, j, k), 1 \le i \le r; j = 1, 2; 1 \le k \le S\} \cup \Delta$, where Δ correspond to the state, the rth customer is taken for service. The generator matrix of the Markov chain is given by $\hat{Q} = \begin{bmatrix} T & T^0 \\ 0 & 0 \end{bmatrix}$, where T^0 is an $(r(2S+1) - 1) \times 1$ matrix with $T^0(i, 1) = \mu; 2 \le i \le S + 1$ and

$$T = \begin{bmatrix} B & 0 & 0 & - & - & - & 0 \\ A_2 & B & 0 & - & - & - & 0 \\ 0 & A_2 & B & 0 & - & - & 0 \\ 0 & 0 & A_2 & B & & & \\ & & & - & - & - & - \\ & & & & - & - & - \\ & & & & & - & - \\ 0 & 0 & - & - & - & \hat{A}_2 & \hat{B} \end{bmatrix}$$

The different transitions in T are as follows.

i) $(i, 0, k) \xrightarrow{\eta} (i, 1, k + Q); 1 \le i \le r; 0 \le k \le s$

ii) $(i, j, k) \xrightarrow{\eta} (i, j, k + Q); 1 \le i \le r; j = 1; 0 \le k \le s$

iii) $(i, 1, k) \xrightarrow{\delta_1} (i, 2, k); 1 \le i \le r; 1 \le k \le S$

iv) $(i, 2, k) \xrightarrow{\delta_2} (i, 1, k); 1 \le i \le r; 1 \le k \le S$

v) $\hat{B}(i, j) = B(i + 1, j + 1); \ \hat{A}(i, j) = A_2(i + 1, j)$

Now the waiting time of the customer who joins as the rth customer is given by $W^r = \hat{I}_{2S}(-T^{-1}e)$, where $\hat{I}_{2S} = \begin{bmatrix} 0 & I_{2S} \end{bmatrix}_{(2S) \times (r(2S+1) - 1)}$.

So the expected waiting time of a general customer is given by $E(W_L) = \sum_{r=1}^{\infty} \hat{\pi}_r W^r$, where $\hat{\pi}_r(i) = \pi_r(i + 1)$. Similarly the variance of waiting time of a general customer is also calculated numerically.

5.2   Other Performance Measures

1. The expected number of customers in the system,

$$L_s = \sum_{i=1}^{\infty} \sum_{j=1}^{S} i \left\{ \pi(i,1,j) + \pi(i,2,j) \right\} + \sum_{i=1}^{\infty} i\pi(i,0,0)$$

$$= \zeta \frac{\lambda}{\mu} \left(\frac{\mu}{\mu - \lambda} \right)^2 \left[1 + \frac{\delta_1}{\delta_2} (1 - x(0,0)) \right].$$

2. The expected inventory level in the system,

$$INV_{mean} = \sum_{i=1}^{\infty} \sum_{j=1}^{S} j\{\pi(i,1,j) + \pi(i,2,j)\} + \sum_{j=1}^{S} j\pi(0,0,j)$$

$$= \zeta \frac{\lambda}{\mu} Q \left\{ 1 + \frac{\delta_1}{\delta_2} \frac{\lambda}{\mu} \right\} \left(\frac{(S+s+1)}{2} \frac{\eta}{\lambda} \left(\frac{\eta + \lambda}{\lambda} \right)^s + \right.$$

$$\left. \left[1 - \left(\frac{\eta + \lambda}{\lambda} \right)^s \right] \right) x(0,0).$$

3. The expected rate of ordering, $E_{or} = \sum\limits_{i=1}^{\infty} \mu\pi(i,1,s+1)$.

4. The expected replenishment rate,

$$REP_{mean} = \sum_{i=0}^{\infty} \sum_{j=0}^{s} \eta \left\{ \pi(i,0,j) + \pi(i,1,j) \right\}.$$

5. The expected interruption rate, $INT_{mean} = \sum\limits_{i=1}^{\infty} \sum\limits_{j=1}^{S} \delta_1\pi(i,1,j) = \delta_1 P(busy)$.

6. The loss rate of customers,

$$LOSS_{mean} = \sum_{i=0}^{\infty} \lambda\pi(i,0,0) + \sum_{i=1}^{\infty} \sum_{j=1}^{S} \lambda\pi(i,2,j) = \lambda\xi \frac{\mu}{\mu - \lambda} x(0,0) + \lambda P(int).$$

7. The probability that the server is busy,

$$P_{busy} = \sum_{i=1}^{\infty} \sum_{j=1}^{S} \pi(i,1,j) = \frac{\delta_2}{\delta_2\mu + \delta_1(1 - x(0,0))} \frac{\lambda}{\mu - \lambda} Q \frac{\eta}{\lambda} \left(\frac{\eta + \lambda}{\lambda} \right)^s.$$

8. The probability that the server is on interruption,

$$P_{int} = \sum_{i=1}^{\infty} \sum_{j=1}^{S} \pi(i,2,j) = \frac{\delta_1}{\delta_2} P(busy).$$

5.3   Cost Analysis

We considered the following
Cost function $Cost = CI \times INV_{mean} + CN \times L_s + CR \times E_{INTR} + (K + (S - s)K_1) \times E_{OR} + CL \times Loss_{mean}$, where

CI : Cost of holding Inventory
CN : Cost of holding customers
CR : Cost incurred due to interruption of service
K : Fixed cost of ordering
K_1 : Cost of a single inventory
CL : Cost incurred due to loss of customers when inventory level drops to zero.
The effect of various parameters on the cost were studied.

6   Numerical Illustration

Eventhough we have explicit expressions for most of the system performance measures we provide numerical illustration of the effect of different parameters on the system performance measures in this section.

6.1   Effect of Arrival Rate λ

In Table 1 we see that as arrival rate increases, there is an increase in both $P(busy)$, $P(int)$ and L_s. The increase in server busy probability is as expected since when arrival rate increases the mean number of customers in the system obviously increases and so the probability that server is busy increases. $P(int)$ is also seen to increase which may be due to the fact that an interruption to service occurs only when the server is busy. Also the decrease in INV_{mean} is due to the fact that the more customers get service when $P(busy)$ increases. Also notice the increase in mean waiting of a customer in the system due to increase in mean number of customers in the system.

6.2   Effect of Service Rate μ

In Table 2 we see that as service rate increases, $P(busy)$, $P(int)$, L_s and $WAIT_{mean}$ all decrease. As the service rate increases, customers leave the system after getting service at a faster rate. Hence the mean waiting time in the system clearly decreases. Also the probability that the server is idle increases with increase in service rate and so $P(busy)$, $P(int)$ and L_s all decrease. It is seen from the tables that μ has no effect on INV_{mean}.

6.3   Effect of Interruption Rate δ_1

In Table 3 we see that as interruption rate increases, $P(busy)$ increases whereas $P(int)$, $WAIT_{mean}$ and L_s decrease. The reason for decrease in the mean number of customers in the system is due to our assumption that when the server is on interruption no arrivals are entertained. The decrease in mean waiting time of a customer in the system is due to the increase in $P(busy)$. Also as mean number of customers in the system decreases, probability that server is idle increases and so $P(int)$ decreases. The interruption rate seems to have no effect on average inventory level in the system.

6.4 Effect of Reorder Level S

In Table 4 we see that s has no considerable effect on the system performance measures $P(busy)$, $P(int)$ and L_S. The expected inventory level in the system increases with increase in re order level is as expected since orders are placed early with increase in s (Table 5, 6, 7, 8, 9 and Figs. 1, 2, 3, 4, 5).

Table 1. Effect of arrival rate on various performance measures $\mu = 10 \; \eta = 2 \; \delta_1 = 6 \; \delta_2 = 7 \; s = 5 \; S = 12$

λ	$P(busy)$	$P(int)$	INV_{mean}	REP_{mean}	L_s	E_{OR}	$WAIT_{mean}$	$WAIT_{var}$
2	0.18749	0.06249	8.7142	0.2678	0.3125	0.2678	0.1597	0.0675
2.2	0.20496	0.06832	8.6856	0.2928	0.3504	0.2928	0.1643	0.0715
2.4	0.2222	0.074	8.6571	0.3174	0.3899	0.3174	0.1693	0.0761
2.6	0.2392	0.0797	8.6284	0.3417	0.4311	0.3417	0.1747	0.0812
2.8	0.2561	0.0853	8.5999	0.3658	0.4742	0.3658	0.1806	0.0868
3	0.2726	0.0909	8.5713	0.3895	0.5194	0.3895	0.187	0.0931

Table 2. Effect of service rate on various performance measures $\lambda = 3 \; \eta = 2 \; \delta_1 = 6 \; \delta_2 = 7 \; s = 5 \; S = 12$

μ	$P(busy)$	$P(int)$	INV_{mean}	REP_{mean}	L_s	E_{OR}	$WAIT_{mean}$	$WAIT_{var}$
9	0.2999	0.0999	8.5713	0.3856	0.5999	0.3856	0.2191	0.1256
9.2	0.2941	0.098	8.5713	0.3865	0.5819	0.3865	0.2118	0.1178
9.4	0.2884	0.0961	8.5713	0.3873	0.5649	0.3873	0.205	0.1107
9.6	0.2829	0.0943	8.5713	0.388	0.5489	0.3881	0.1986	0.1043
9.8	0.2777	0.0926	8.5713	0.3888	0.5337	0.3888	0.1927	0.0984
10	0.2727	0.0909	8.5713	0.3896	0.5195	0.3895	0.187	0.0931

Table 3. Effect of interruption rate on various performance measures $\lambda = 3 \; \mu = 9 \; \eta = 2 \; \delta_2 = 7 \; s = 5 \; S = 12$

δ_1	$P(busy)$	$P(int)$	INV_{mean}	REP_{mean}	L_s	E_{OR}	$WAIT_{mean}$	$WAIT_{var}$
6	0.2726	0.1818	8.5714	0.3506	0.6818	0.3506	0.2918	0.2219
6.2	0.2743	0.1769	8.5714	0.3526	0.6769	0.3526	0.2862	0.2128
6.4	0.2758	0.1723	8.5714	0.3546	0.6723	0.3546	0.281	0.2044
6.6	0.2772	0.168	8.5714	0.3564	0.668	0.3564	0.2761	0.1968
6.8	0.2786	0.1639	8.5714	0.3582	0.6639	0.3582	0.2716	0.1898
7	0.2799	0.1599	8.5714	0.3599	0.6599	0.3599	0.2673	0.1833

Table 4. Effect of reorder level on various performance measures $\lambda = 3$ $\mu = 9$ $\eta = 2$ $\delta_1 = 6$ $\delta_2 = 7$ $S = 21$

s	$P(busy)$	$P(int)$	INV_{mean}	REP_{mean}	L_s	E_{OR}	$WAIT_{mean}$	$WAIT_{var}$
5	0.2999	0.0999	13.071	0.1687	0.5999	0.1687	0.2197	0.1265
6	0.2999	0.0999	13.571	0.1799	0.5999	0.1799	0.2197	0.1265
7	0.2999	0.0999	14.071	0.1928	0.5999	0.1928	0.2196	0.1264
8	0.2999	0.0999	14.571	0.2076	0.5999	0.2076	0.2196	0.1263
9	0.2999	0.0999	15.071	0.2249	0.5999	0.2250	0.2195	0.1262
10	0.2999	0.0999	15.571	0.2454	0.5999	0.2454	0.2195	0.1261

Table 5. Effect of arrival rate on Cost

$CI = 40$	$CN = 30$	$CR = 75$	$K = 500$	$K_1 = 35$	$CZ = 750$
$\mu = 10$	$\eta = 2$	$\delta_1 = 6$	$\delta_2 = 7$	$s = 5$	$S = 12$

λ	2	2.2	2.4	2.6	2.8	3
Cost	566	586	605	624	643	662

Table 6. Effect of service rate on Cost

$CI = 40$	$CN = 30$	$CR = 75$	$K = 500$	$K_1 = 35$	$CZ = 750$
$\lambda = 3$	$\eta = 2$	$\delta_1 = 6$	$\delta_2 = 7$	$s = 5$	$S = 12$

μ	9	9.2	9.4	9.6	9.8	10
Cost	663.47	663.26	663.06	662.89	662.73	662.59

Table 7. Effect of interruption rate on Cost

$CI = 40$	$CN = 30$	$CR = 75$	$K = 500$	$K_1 = 35$	$CZ = 750$
$\lambda = 3$	$\eta = 2$	$\mu = 9$	$\delta_2 = 7$	$s = 5$	$S = 12$

δ_1	6	6.2	6.4	6.6	6.8	7
Cost	663.47	663.87	664.26	664.62	664.96	665.29

Table 8. Effect of repair rate on Cost

$CI = 40$	$CN = 30$	$CR = 75$	$K = 500$	$K_1 = 35$	$CZ = 750$
$\lambda = 3$	$\eta = 2$	$\mu = 9$	$\delta_1 = 6$	$s = 5$	$S = 12$

δ_2	6	6.2	6.4	6.6	6.8	7
Cost	663.47	663.91	664.34	664.74	665.12	665.49

Table 9. Effect of reorder level on Cost

$CI = 40$	$CN = 30$	$CR = 75$	$K = 500$	$K_1 = 35$	$CZ = 750$
$\lambda = 3$	$\eta = 2$	$\mu = 9$	$\delta_1 = 6$	$\delta_2 = 7$	$S = 21$

s	5	6	7	8	9	10
Cost	734	760	786	814	842	873

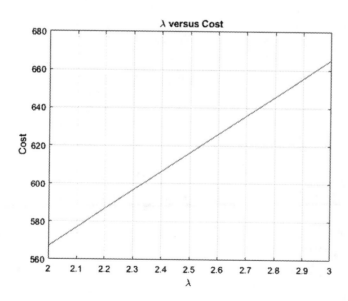

Fig. 1. Arrival rate versus Cost

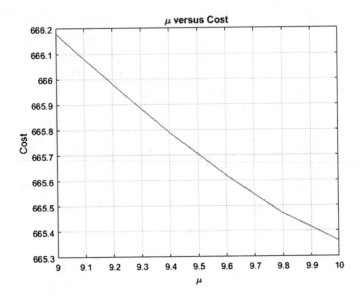

Fig. 2. Service rate versus Cost

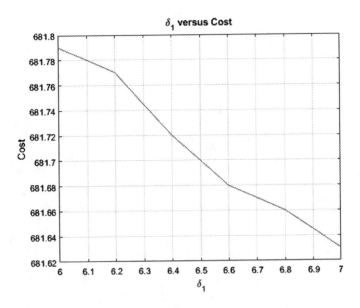

Fig. 3. Interruption rate versus Cost

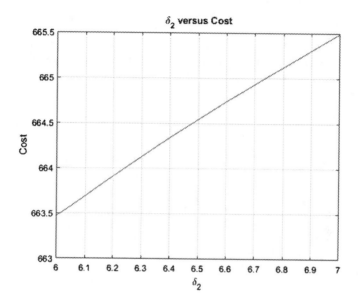

Fig. 4. Repair rate versus Cost

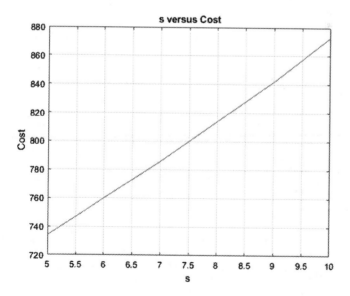

Fig. 5. Reorder level versus Cost

7 Conclusion

We studied a single server queueing model with positive service time, positive lead time and service interruptions. We could arrive at an explicit expression for the steady state probability vector. We wish to extend this model by considering retrials as well.

References

1. Krishnamoorthy, A., Viswanath, C.N., Islam, M.: Retrial production inventory with map and service times; queues, flows, systems, networks. In: Proceedings of the International Conference "Modern Mathematical Methods of Analysis and Optimization of Telecommunication Networks", pp. 148–156 (2003)
2. Berman, O., Kaplan, E.H., Shevishak, D.G.: Deterministic approximations for inventory management at service facilities. IIE Trans. 25(5), 98–104 (1993)
3. Berman, O., Kim, E.: Stochastic models for inventory management at service facilities. Stoch. Model. 15(4), 695–718 (1999)
4. Berman, O., Sapna, K.: Inventory management at service facilities for systems with arbitrarily distributed service times. Stoch. Model. 16(3–4), 343–360 (2000)
5. Deepak, T., Krishnamoorthy, A., Narayanan, V.C., Vineetha, K.: Inventory with service time and transfer of customers and/inventory. Ann. Oper. Res. 160(1), 191–213 (2008)
6. Krishnamoorthy, A., Deepak, T., Narayanan, V.C., Vineetha, K.: Effective utilization of idle time in an (s, s) inventory with positive service tim. J. Appl. Math. Stochast. Anal. 2006 (2006)
7. Krishnamoorthy, A., Islam, M., Narayanan, V.C.: Retrial inventory with batch Markovian arrival and positive service time. Stoch. Model. Appl. 9(2), 38–53 (2006)
8. Krishnamoorthy, A., Nair, S.S., Narayanan, V.C.: An inventory model with retrial and orbital search. Bulletin of Kerala Mathematics association, Special issue, pp. 47–65 (2009)
9. Krishnamoorthy, A., Narayanan, V.C., Deepak, T., Vineetha, P.: Control policies for inventory with service time. Stoch. Anal. Appl. 24(4), 889–899 (2006)
10. Krishnamoorthy, A., Narayanan, V.C., Islam, M.: On production inventory with service time and retrial of customers. In: Proceedings of the 11th International Conference on Analytical and Stochastic Modelling Techniques and Analysis, pp. 238–247. SCS-Publishing House (2004)
11. Krishnamoorthy, A., Jose, K., Narayanan, V.C.: Numerical investigation of a PH/PH/1 inventory system with positive service time and shortage. Neural Parallel Sci. Comput. 16(4), 579 (2008)
12. Melikov, A., Molchanov, A.: Stock optimization in transportation/storage systems. Cybern. Syst. Anal. 28(3), 484–487 (1992)
13. Narayanan, V.C., Deepak, T., Krishnamoorthy, A., Krishnakumar, B.: On an (s, S) inventory policy with service time, vacation to server and correlated lead time. Qual. Technol. Quant. Manag. 5(2), 129–143 (2008)
14. Neuts, M.F.: Matrix-geometric solutions in stochastic models: an algorithmic approach. Bull. Amer. Math. Soc. 8, 97–99 (1983)
15. Schwarz, M., Sauer, C., Daduna, H., Kulik, R., Szekli, R.: M/M/1 queueing systems with inventory. Queueing Syst. 54(1), 55–78 (2006)
16. Sigman, K., Simchi-Levi, D.: Light traffic heuristic for an M/G/1 queue with limited inventory. Ann. Oper. Res. 40(1), 371–380 (1992)

Optimization of the Transmission of Messages Divided into Different Shares, Transmitted on Two Different Channels

Vladimir Zadorozhnyi and Dmitriy Sagaydak[✉]

Omsk State Technical University, 11 Mira Avenue, Omsk 644050, Russia
sagaydak.dmitriy@gmail.com

Abstract. The paper considers the transmission of messages with demultiplexing over two communication channels with different throughput capacities. The channel with the highest throughput receives the largest chunks of messages resulting from the demultiplexing, and the channel with the smallest throughput receives the smallest chunks. The problem of calculating the optimal channels throughputs is solved by taking into account the characteristics of the transmitted traffic.

Keywords: data flow distribution · communication channel throughput · demultiplexing · multiplexing · secret sharing scheme

1 Introduction

Due to the self-isolation and quarantine regimes implemented during the current COVID-19 pandemic, there is an increased demand for Internet connection services, data transfer speeds augmentation, throughput expansion, and additional communication channels purchase [1–3]. The most popular transmitted content is video data, for example, online broadcasts of cinemas, educational webinars. Broadcasting is carried out using client-server applications, in which the content can be pre-transformed using any algorithms, and only then transmitted to the user. The preliminary content transformation can be carried out in order to compress it, in other words, to reduce the transmitted traffic, as well as to ensure confidentiality, i.e., to perform cryptographic transformations. In such situations, even the choice of optimal cryptographic algorithms can lead to significant delays in the playback of the video data stream due to the fact that the reverse cryptographic conversion must be performed on the client side. The use of an additional communication channel makes it possible to organize distributed data transmission, which allows to solve the problem of ensuring confidentiality, but there arise some questions related to the efficiency of the use of computing resources, optimization of channel throughput, synchronization of transmitted streams, etc.

© Springer Nature Switzerland AG 2022
A. Dudin et al. (Eds.): ITMM 2021, CCIS 1605, pp. 221–237, 2022.
https://doi.org/10.1007/978-3-031-09331-9_18

2   Problem Statement

Let the sender (a person or an automatic device) transmit to the receiver a high-quality uncompressed media stream, which is a sequence of images (frames). The transmission is carried out over the Internet, and it is required that no one except the recipient can access the contents of the transmitted data. To meet this requirement, it is possible to organize secure media data streaming using cryptographic methods. However, when using such methods, there may arise problems related to ensuring the stability of the selected algorithms and, accordingly, with the availability of sufficiently powerful computing resources on the receiving side to guarantee timely data decryption. In such a situation, it is advisable to consider the possibility of using other methods of information protection that are not related to classical cryptography, e.g., secret sharing schemes (SSS) [4], demultiplexing.

Algorithms for dividing video data into unequal shares are proposed in [5–7] which will allow the sender and receiver to carry out the separation of the transmitted TCP / IP traffic over these channels, using two communication channels with different throughput, as, for example, it is described in [8–11]. Further, we will assume that SSS for unequal shares can be used not only for transmitting video frames, but also for transmitting streams of any messages, and all the transformations described in [5–7] are performed directly on the bit representation of these messages. When messages are divided into unequal shares, a smaller share of each message is transmitted over a lower throughput channel, while a larger share is transmitted over a higher throughput channel. Such message transmission from the sender to the recipient is carried out at the transport level of the seven-level OSI network model [12], where the TCP protocol provides guaranteed data delivery. When using two communication channels at the same time, there arise questions related to the efficiency of computing resources, optimization of channel throughput, synchronization of transmitted streams, buffering. These issues can be solved by implementing appropriate client-server applications and optimizing the throughput of communication channels. It is advisable to optimize the throughput capacities according to cost minimization criteria, one part of which is associated with message delays in the network (the growth of which leads to a delay in the recipient's response to messages and corresponding losses), the other part is related to the payment for channel throughput, which increases with throughput growth.

As a mathematical model for optimizing a two-channel SSS, a network with splitting requests (S-network) with two single-channel queuing systems (QS) is proposed (see Fig. 1). In terms of queuing theory (QT), we will call messages and their parts *requests*, demultiplexing messages - *splitting requests*, multiplexing messages *assembling requests*. Two requests corresponding to two parts of the same divided message will be referred to as *conjugate requests*. We define the discipline of servicing queues in front of the channels as FIFO (first in - first out) discipline. Requests are transmitted over two channels with different throughputs C_1, C_2 measured, for example, in Kbit/s. Unlike traditional QS networks, at

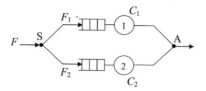

Fig. 1. Network with splitting requests. S - split point, A - assemble point

Point S, the request does not go to one of the branches, but is split into two requests, one of which arrives at QS1 and the other goes to QS2. At Point A, two conjugate requests "merge" and turn into one request. Accordingly, the incoming Traffic F (Kbit/s) is divided into two parts F_1 and F_2, where $F_1 + F_2 = F$. The moment of entry of the request into the network is simultaneously the moment of its splitting and the moment of entry of the resulting conjugate requests into each of the two branches of the network (into each of the two QSs). Consequently, the Intensities λ, λ_1 and λ_2 of the request flows entering the network are the same in QS1 and in the QS2, respectively, and all three request flows are described by the same probabilistic law. Another feature of the considered S-network, not shown in Fig. 1, is that before Point A, two more queues are formed (one on each branch) - synchronization queues. At the moment of exit from the first (second) channel, the request enters the first (second) synchronization queue before Point A, where it remains until its conjugate "half" is found in another synchronization queue. In other words, one of the two conjugate requests that arrived first at Point A waits for the second conjugate request to arrive. At the moment of its arrival, both conjugate requests are merged into one request leaving the network, and the transfer of the request is completed.

Note that all requests arrive in each synchronization queue in the same sequence that they enter the network. Therefore, if at least one request is pending in one synchronization queue, the other synchronization queue is empty. At any finite time interval, either the first synchronization queue is empty, or the second, or both queues are empty. Both of these queues can be non-empty at the same time only at one point in time: when the condition "at the selection point there is a pair of requests conjugated with each other" is fulfilled. It follows from this that of the two conjugate requests, the one that arrives later is not delayed in the synchronization queue. Consequently, the Time u of the message transmission (in terms of QT, the time the request is in the network, i.e., the time elapsed from the moment the request arrives in Point S until the moment it leaves Point A) is determined by the formula:

$$u = \max(u_1, u_2), \tag{1}$$

where u_1 is the time the request was in QS1:

$$u_1 = w_1 + x_1, \tag{2}$$

u_2 is the sojourn time of the conjugate request in QS2:

$$u_2 = w_2 + x_2, \tag{3}$$

w_1 is the request waiting time in Queue 1, x_1 is the request service time in Channel 1, w_2 is the waiting time of the conjugate request in queue 2, x_2 is the service time of the conjugate request in Channel 2.

The average Time U of staying in the S-network, according to (1), is expressed by the formula:

$$U = M\left[\max(u_1, u_2)\right] = M\left[\max(w_1 + x_1, w_2 + x_2)\right]. \tag{4}$$

Time U depends on the Throughputs C_1, C_2:

$$U = U(C_1, C_2).$$

Let the price of the throughput of any channel, calculated for the network operation time, be equal to m c.u./(Kbit/s). Then the problem of optimizing Throughputs C_1, C_2 of the S-network channels (or, in other words, the problem of optimizing the S-network) can be formulated as follows:

$$f = lU\left(C_1, C_2\right) + mC_1 + mC_2 \rightarrow \min_{C_1, C_2}, \tag{5}$$

$$\begin{cases} C_1 \geq F_1, \\ C_2 \geq F_2, \end{cases} \tag{6}$$

where $U(C_1, C_2) = M\left[max(u_1, u_2)\right]$, l (c.u./s) is the cost of the average network delay per second. Cost Coefficient l is equal to losses (arising from waiting for applications) calculated for the period of network operation.

Thus, the problem (5), (6) is posed as the problem of minimizing the average costs over the network operation time. A network with optimal channel capacity will be called optimal.

The non-triviality of the problem posed is due to the absence in the QT of explicit formulas that allow, directly or by means of appropriate transformations, to accurately calculate the average Time U of requests in the S-network under some general and natural assumptions about the incoming flow of requests and methods for their splitting. To solve this problem, it is necessary to develop appropriate exact or approximate methods. Further development and research of such methods is ongoing.

3   Exponential Network with Independent Branches

3.1   Problem Statement

Consider a network with independent branches (Fig. 2), which makes sense to study as a simplified first approximation of the S-network shown in Fig. 1.

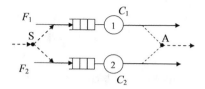

Fig. 2. S-network with independent branches

In this network with independent branches, QS1 and QS2 operate independently, each serving its own stream of requests. Each of these QSs is individually equivalent to the corresponding QS in an S-network, but a pair of QSs in a network with independent branches (Fig. 2) is not equivalent to a pair of QSs in an S-network (Fig. 1). In a network with independent branches, the QSs function independently; in an S-network the processes in one and the other QS are statistically dependent. Statistical independence of the network branches in Fig. 2 simplifies its analysis.

The dashed lines in Fig. 2 show a single passage through the network of a single request, divided into parts, in a stationary mode of network operation at a random time. At the moment of its arrival, the conjugate parts of the split request arrive at the corresponding QSs, then pass through the queues and service, and are assembled into one request, as described above. The travel time of the split request through the network is expressed by formula (1), the optimization problem for such a network is posed in the form (5), (6). In order to solve this problem by analytical methods, let us express $M\left[max(u_1, u_2)\right]$, in terms of Channel Throughputs C_1, C_2.

3.2 Network with Independent Branches Optimization

The calculation of the system in Fig. 2 contains the following steps. First, we can find the distribution functions of the sojourn time of entire requests in QS1, QS2. Since these QSs are exponential, the required distribution functions are known. Since Quantities u_1 and u_2 are independent, we find the distribution of the maximum of these quantities, and through it the desired $M\left[max(u_1, u_2)\right]$.

The distribution function of the sojourn time in QS1 has the form [13]:

$$P(u_1 \leq t) = 1 - e^{-\mu_1(1-\rho_1)t}, \tag{7}$$

similarly in QS2 it is described as:

$$P(u_2 \leq t) = 1 - e^{-\mu_2(1-\rho_2)t}, \tag{8}$$

where μ_1, μ_2 are the service intensity in the first and second channels, ρ_1, ρ_2 are the load factors of the first and second channels.

We find the Distribution Function $max(u_1, u_2)$ as the probability of simultaneous occurrence of two independent events: as the probability that the first value does not exceed t and that the second value does not exceed t:

$$P\left[max(u_1, u_2) \leq t\right] = \left[1 - e^{-\mu_1(1-\rho_1)t}\right]\left[1 - e^{-\mu_2(1-\rho_2)t}\right]. \tag{9}$$

Using the well-known formula for calculating the mathematical expectation of a positive random variable (without calculating the probability density), we find

$$M\left[max(u_1, u_2)\right] = \int_0^\infty \left[1 - \left(1 - e^{-\mu_1(1-\rho_1)t}\right)\left(1 - e^{-\mu_2(1-\rho_2)t}\right)\right] dt =$$

$$= \frac{1}{\mu_1(1-\rho_1)} + \frac{1}{\mu_2(1-\rho_2)} - \frac{1}{\mu_1(1-\rho_1) + \mu_2(1-\rho_2)}. \tag{10}$$

The resulting expression can be substituted into the problem (5), (6) to solve it for a network with independent branches by analytical methods. Before doing this, let us move on to the expression (10) and the parameters used in the problem (5), (6): $\rho_1 = F_1/C_1$, $\rho_2 = F_2/C_2$, $\mu_1 = C_1/H_1$, $\mu_2 = C_2/H_2$. Moving on to these designations in the expression (10) and substituting it into the problem (5), (6), we obtain:

$$f = \frac{lH_1}{(C_1 - F_1)} + \frac{lH_2}{(C_2 - F_2)} - \frac{l}{H_1^{-1}(C_1 - F_1) + H_2^{-1}(C_2 - F_2)}$$

$$+ mC_1 + mC_2 \to \min_{C_1, C_2}, \tag{11}$$

$$\begin{cases} C_1 \geq F_1, \\ C_2 \geq F_2. \end{cases} \tag{12}$$

The point (C_1, C_2) of the local minimum of this positive function can be found from the system of equations

$$\frac{\partial f}{\partial C_1} = 0,$$

$$\frac{\partial f}{\partial C_2} = 0,$$

that is, from the equations

$$-\frac{lH_1}{(C_1 - F_1)^2} + \frac{lH_1^{-1}}{\left(H_1^{-1}(C_1 - F_1) + H_2^{-1}(C_2 - F_2)\right)^2} + m = 0, \tag{13}$$

$$-\frac{lH_2}{(C_2 - F_2)^2} + \frac{lH_2^{-1}}{\left(H_1^{-1}(C_1 - F_1) + H_2^{-1}(C_2 - F_2)\right)^2} + m = 0. \tag{14}$$

Let us denote $C_1 - F_1$ by x and $C_2 - F_2$ by y. As a result, the system (13), (14) takes the following form, which is well solved by numerical methods:

$$-\frac{lH_1}{x^2} + \frac{lH_1^{-1}}{\left(H_1^{-1}x + H_2^{-1}y\right)^2} + m = 0, \tag{15}$$

$$-\frac{lH_2}{y^2} + \frac{lH_2^{-1}}{\left(H_1^{-1}x + H_2^{-1}y\right)^2} + m = 0. \tag{16}$$

One of the main methods for reducing the delay time of conjugate requests in synchronization queues is to introduce the maximum positive correlation between the processes of moving conjugate requests along the network branches. This method is explored in the following sections of the article.

4 S-Networks with Synchronous Branches

4.1 Fundamentals

Definition. Consider an S-network in which each incoming request has a random size h (Kbit) and is split into two conjugate requests so that the same proportion is always maintained between their sizes h_1 and h_2 (where $h_1 + h_2 = h$):

$$h_2/h_1 = \gamma = const. \tag{17}$$

If, in this case, the throughput of the channels is connected by the condition

$$C_2 = \gamma C_1, \tag{18}$$

then the service time x_1 of the request in Channel 1 and the Service Time x_2 of the conjugate request in Channel 2 coincide:

$$x_1 = h_1/C_1,$$
$$x_2 = h_2/C_2 = (\gamma h_1)/(\gamma C_1) = h_1/C_1 = x_1. \tag{19}$$

Since the equality (19) is satisfied for each pair of conjugate requests, then in each pair both conjugate requests enter the queues to the channels, into the channels, and to the assembly point simultaneously. We call such a network an S-network with synchronous branches or an Ss-network.

It is easy to see that in the Ss network, $C_2 = \gamma C_1$ implies $C = (1 + \gamma)C_1$, where $C = C_1 + C_2$ is the total throughput of the channels. Similarly, from $h_2/h_1 = \gamma$ it follows that Traffic F_2 entering the second branch and all Traffic F entering the network are expressed in terms of F_1 by the relations $F_2 = \gamma F_1$, $F = (1 + \gamma)F_1$.

Ss-Network Optimization Problem . The problem (5), (6) of optimizing an S-network with synchronous branches can be solved exactly for any flow of requests for which both QSs in the branches can be calculated using exact QT methods.

Indeed, since Sojourn Time u_1 of any request in QS1 and Sojourn Time u_2 of the corresponding conjugate request in QS2 in the Ss-network coincide, then in (1) we have $u = \max(u_1, u_2) = u_1$ and, therefore, in (4) $U = M[\max(u_1, u_2)] = M[u_1] = U_1$, where U_1 is the average sojourn time of the request in QS1. Performing the substitutions $U(C_1, C_2) = U_1(C_1)$, $C_2 = \gamma C_1$ and $F_2 = \gamma F_1$ in the problem (5), (6), we obtain its equivalent formulation using only one variable parameter C_1:

$$f = lU_1(C_1) + m(1 + \gamma)C_1 \to \min_{C_1}, \tag{20}$$

$$C_1 \geq F_1. \tag{21}$$

Note that after these substitutions are performed, the second constraint in (6) becomes equivalent to the first and, therefore, is absent in the constraints (21).

The solution to the problem (20), (21) of optimization of the Ss-network determines the optimal throughput C_1 of the first channel and, at the same time, the corresponding throughput $C_2 = \gamma C_1$ and $C = (1 + \gamma)C_1$.

4.2   Ss-Network with Regular Incoming Flow and Fixed Order Size

Optimization. With a regular incoming flow, Time τ between arrivals of requests to the network (and, therefore, to each of the two branches) is constant: $\tau = const$. The size of requests arriving in QS1 is also fixed ($h_1 = const$), so the service time $x_1 = h_1/C_1$ is also fixed in QS1. It follows from the restriction $C_1 \geq F_1$ that $x_1 \leq \tau_1$, i.e., each request arriving in QS1 is serviced before the next one arrives. Therefore, the queue in front of Channel 1 is not formed, and the average sojourn time of the request in $U_1 = x_1 = h_1/C_1$. Substituting this expression for U_1 in (20), (21) instead of $U_1(C_1)$, we obtain the problem

$$f = l\frac{h_1}{C_1} + m(1 + \gamma)C_1 \to \min_{C_1}, \tag{22}$$

$$C_1 \geq F_1, \tag{23}$$

whose solution is reduced to solving the algebraic equation

$$\frac{\partial f(C_1)}{\partial C_1} = 0 \ or \ -l\frac{h_1}{C_1^2} + m(1 + \gamma) = 0,$$

determining the point of the local minimum

$$C_1 = \sqrt{\frac{lh_1}{m(1 + \gamma)}}. \tag{24}$$

If the obtained value C_1 satisfies the constraint $C_1 \geq F_1$, then (24) is the solution to problem (22), (23). Otherwise, the solution to this problem is the smallest value C_1 closest to the point (24) that satisfies the constraint $C_1 \geq F_1$, i.e., value $C_1 = F_1$.

Comparison with the Single-Channel Version. The two-channel implementation of the SSS, compared to the single-channel implementation, significantly increases the security of the transmitted data from unauthorized use. To estimate the losses due to which this is achieved, let us compare the costs obtained in the optimal Ss-network with the costs characterizing the corresponding basic single-channel optimal system.

In the considered case of a regular incoming flow and a fixed size of requests, the average Time U of message transmission over one channel for $C \geq F$ is equal to the average request service time (since there is no queue in front of the channel). I.e. $U = h/C$, where $h = h_1 + h_2 = const$ is the size of requests. Therefore, the optimization problem for a basic single-channel system takes the form

$$f = lh/C + mC \to \min_C, \tag{25}$$

$$C \geq F. \tag{26}$$

The local minimum of the objective function (25) is attained at the point

$$C = \sqrt{\frac{lh}{m}}. \tag{27}$$

Theorem 1. Transmission of a regular flow of fixed-size requests through the optimal Ss-network leads to the same costs as transmission through the optimal single-channel system. In this case, the total throughput of the optimal Ss-network is equal to the throughput of the optimal single-channel system.

Proof of the Theorem. To prove the theorem, it suffices to note that the statement of the problem (22), (23) for optimizing the Ss-network differs from the statement of the problem (25), (26) for optimizing a single-channel system only due to the formulation of the problem (22), (23) in terms of the optimal choice of throughputs abilities C_1. But since any of the parameters C_1, C_2, and C uniquely determine the other two parameters in the Ss-network, the problem of its optimization can be formulated in terms of the optimal choice of any of these three parameters. When choosing the variable C as a variable parameter - the total throughput of the channels - the formulation of the Ss-network optimization problem becomes equivalent to the formulation of the optimization problem for a single-channel system.

Indeed, in the problem (22), (23) $C_1 = C/(1 + \gamma)$, $F_1 = F/(1 + \gamma)$ and $h_1 = h/(1 + \gamma)$. Carrying out the corresponding changes in the problem (22), (23), we obtain its formulation

$$f = l\frac{h/(1 + \gamma)}{C/(1 + \gamma)} + m(1 + \gamma)C/(1 + \gamma) \to \min_C,$$

$$C/(1 + \gamma) \geq F/(1 + \gamma).$$

equivalent to the formulation of the problem (25), (26). From the equivalence of the formulations of the two fundamentally different problems under consideration, the numerical coincidence of their solutions follows. The theorem is proved.

The two problems under consideration are optimization problems for two different systems, the Ss-network and a single-channel system. The formal coincidence of their solutions means that the transmission of a regular incoming flow with a fixed request size through the optimal Ss-network leads to exactly the same costs as its transmission through the optimal single-channel system.

4.3   Ss-Network with an Arbitrary Incoming Flow of Requests

Theorem 1 is generalized by the following theorem.

Theorem 2. The transmission of any request flow through the optimal Ss-network leads to the same costs as its transmission through the optimal single-channel system. In this case, the total throughput of the optimal Ss-network is equal to the throughput of the optimal single-channel system.

The proof of the theorem is based on a comparison of the processes of passing through the Ss-network and through a single-channel system of the same implementation of the incoming request flow. Then, provided that the throughput of the single-channel system is equal to the throughput of the Ss-network, the advancement of requests in each branch of the Ss-network occurs synchronously with the advancement of requests in the single-channel system. Therefore:

- the equivalence of the network optimization problem and the optimization problem for a single-channel system (including when they are considered independently, i.e. when independent implementations of the same request flow are fed to the network input and to the single-channel system input);
- the coincidence of the total throughput of the optimal network channels with the throughput of the optimal single-channel system;
- the coincidence of the costs calculated for the period of operation of the optimal Ss-network and the costs of the optimal single-channel system for the same period.

A detailed presentation of the proof is beyond the scope of this article.

4.4   Example of Exponential Ss Network Optimization

Definition. An Ss network is said to be exponential if it includes a Poisson request flow and the request sizes are distributed exponentially. Accordingly, both QSs in such a network are M/M/1 systems. Their calculation is carried out according to the well-known formulas [13].

Ss-Network Optimization. The average Sojourn Time U_1 in QS1 of the exponential Ss-network is [13]:

$$U_1 = \frac{1/\mu_1}{1 - \rho_1} = \frac{1}{\mu_1 - \lambda_1} = \frac{H_1}{H_1\mu_1 - H_1\lambda_1} = \frac{H_1}{C_1 - F_1}, \tag{28}$$

where μ_1 is the intensity of servicing requests in the QS1, $\rho_1 = \lambda_1/\mu_1 = F_1/C_1-$ is the load factor of QS1, $H_1 = M(h_1)$ is the average size of requests arriving in QS1.

Therefore, the problem (20), (21) as applied to the exponential Ss-network is specified as follows:

$$f = l\frac{H_1}{C_1 - F_1} + m(1 + \gamma)C_1 \to \min_{C_1}, \tag{29}$$

$$C_1 \geq F_1. \tag{30}$$

The only minimum of objective function (29), determined from the equation

$$\frac{\partial f}{\partial C_1} = -\frac{lH_1}{(C_1 - F_1)^2} + m(1 + \gamma) = 0 \tag{31}$$

is reached at the point

$$C_1 = F_1 + \sqrt{\frac{lH_1}{m(1 + \gamma)}} \tag{32}$$

and is the solution to the problem (29), (30),since it satisfies the constraint (30).

Substituting the throughput (32) into the objective function expression (29), we find the costs of using the optimal exponential Ss-network:

$$f = l\frac{H_1}{\sqrt{\frac{lH_1}{m(1+\gamma)}}} + m(1 + \gamma)\sqrt{\frac{lH_1}{m(1 + \gamma)}} = 2\sqrt{(1 + \gamma)mlH_1}. \tag{33}$$

Optimization of a Single-Channel Exponential System. The flow of requests included in the considered exponential Ss-network has intensity $\lambda = F/H = F_1/H_1$ and average request size $H = (1 + \gamma)H_1$. When this stream is transmitted over a single-channel system, the average request transmission time is $U = \frac{1/\mu}{1-\rho} = \frac{H}{C-F}$. The optimization problem for such a single-channel QS has the form

$$f(C) = l\frac{H}{C - F} + mC \to \min_{C}, \tag{34}$$

$$C \geq F \tag{35}$$

and determines the throughput

$$C = F + \sqrt{\frac{lH}{m}}, \tag{36}$$

at which the average total costs (33) are minimal and amount to

$$f(C) = l\frac{H}{\sqrt{\frac{lH}{m}}} + m\left(F + \sqrt{\frac{lH}{m}}\right) = 2\sqrt{mlH} + mF. \tag{37}$$

Comparison of the Optimal Exponential Ss-Network and the Corresponding Single-Channel QS. To compare the solution (32) to the problem (29), (30) with the solution (36), we rewrite the solution (32) in terms of the total throughput of the Ss-network. Carrying out the substitutions $C_1 = C/(1+\gamma)$, $F_1 = F/(1+\gamma)$, $H_1 = H/(1+\gamma)$ in (32) equivalent for any Ss-network, we obtain the expression

$$\frac{C}{1+\gamma} = \frac{F}{1+\gamma} + \sqrt{\frac{lH}{m(1+\gamma)^2}}, \qquad (38)$$

and, simplifying it, we find the total throughput of the optimal exponential Ss-network

$$C = F + \sqrt{\frac{lH}{m}},$$

coinciding, as we see, with the throughput (35) of the optimal single-channel exponential QS.

Similarly, performing the replacement $H_1 = H/(1+\gamma)$ in (33) equivalent for Ss-networks, we make sure that the costs associated with the use of the optimal exponential Ss-network coincide with the costs associated with the use of the optimal single-channel exponential system. Thus, the solutions to the exponential Ss-network optimization problem, and the optimization problem for the corresponding single-channel exponential QS, obtained in general form, confirm and illustrate Theorem 2 formulated above, proved for any request flow.

5   Networks with Splitting Requests in Constant Proportion

5.1   Fundamentals

Definition. S-networks in which the condition $h_2/h_1 = \gamma = const$ is satisfied when splitting requests, but the condition $C_2 = \gamma C_1$ is not imposed, we will call networks with split requests in equal proportions, or Se-networks. Thus, the Ss-networks considered above are a subset of Se-networks in which both conditions (17), (18) are satisfied.

Optimization of Se-Networks. The optimization problem for Se-networks has certain specific features. It is written, like the general problem (5), (6) of optimization of S-networks, in the form

$$f = lU(C_1, C_2) + mC_1 + mC_2 \rightarrow \min_{C_1, C_2}, \qquad (39)$$

$$\begin{cases} C_1 \geq F_1, \\ C_2 \geq F_2, \end{cases} \qquad (40)$$

(where $U(C_1, C_2) = M\,[\max(u_1, u_2)]$, and inherits the property (as opposed to the network with independent branches, see Fig. 2) that in the general case, in an elementary function, the inexpressible dependence of the mathematical expectation $M\left[\max(u_1, u_2)\right]$ on the variable parameters C_1, C_2 is determined here on stochastically interdependent random variables u_1, u_2 (sojourn time in QS1 and QS2).

In contrast to the optimization problem for Ss-networks, the variable parameters C_1, C_2 in (39), (40) are independent and not related by the condition $C_2 = \gamma C_1$, therefore, both inequalities are preserved in the constraints (40). And so, the progress of requests in the network branches is generally asynchronous here, which makes it difficult to find an explicit formula that accurately expresses time in terms of $U(C_1, C_2) = M\left[\max(u_1, u_2)\right]$ network parameters.

At the same time, it is very important to find the exact solution to the problem (39), (40). This is due to the following considerations. The previously considered problem This is due to the following considerations. The previously considered problem (20), (21) of optimizing Ss-networks is the problem of finding the conditional minimum of the objective function (39), since it connects the arguments of function (39) with an additional condition $C_2 = \gamma C_1$, i.e. limits in the coordinate system $(C_1, 0, C_2)$ the search area for the minimum f to a one-dimensional set of points of the straight line $C_2 = \gamma C_1$. And when solving the problem (20), (21), we found the solution on this straight line that does not increase the costs of a two-channel SSS implementation in comparison with a single-channel implementation. And if in the two-dimensional region of feasible solutions to the problem (39), (40) the only minimum point of the objective function (39) is outside Straight Line $C_2 = \gamma C_1$, then the solution to the problem (39), (40) will be better than the solution to the problem (20), (21). The substantive meaning of such a solution will be to discover the possibility of switching to a two-channel SSS implementation not only without increasing costs (see Theorem 1), but also with their accompanying decrease.

In the next two sections, it is established that such a possibility is excluded in the class of Se-networks: the optimal solutions of the problem (39), (40) with independent throughputs always lie on Straight Line $C_2 = \gamma C_1$.

5.2   Se-Network with Regular Incoming Flow and Fixed Request Size

Theorem 3. The Se-network that is optimal for transmitting a regular flow of fixed-size requests is an Ss-network, i.e., when transmitting a regular flow of requests of a fixed size, the optimal solution to the problem (39), (40) always lies on Straight Line $C_2 = \gamma C_1$.

Proof of the Theorem. Taking into account the condition $h_2/h_1 = \gamma = const$ which defines the Se-network, we represent the domain of feasible solutions to the problem (39), (40) in the form of a union of two domains, R_1 and R_2 (Fig. 3).

Domain R_1 is determined by conditions $C_1 \geq F_1$, $C_2 \geq \gamma C_1$, Domain R_2 is determined by conditions $C_2 \geq F_2$, $C_2 \leq \gamma C_1$. Line $C_2 = \gamma C_1$ for $C_1 \geq F_1$ belongs to both domains. The point (F_1, F_2) lies on this line, since $F_2/F_1 =$

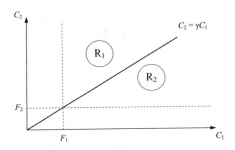

Fig. 3. Domain of feasible solutions to problem (39), (40)

$\lambda h_2/\lambda h_1 = \gamma$ (the coordinates of the point satisfy the equation of the line). Any point (C_1, C_2) belonging to the domain of feasible solutions determines the service time of requests $x_1 = b_1 = h_1/C_1 \leq h_1/F_1 = \tau_1$ for QS1, i.e., the constant service time of requests in the first branch of the network does not exceed the constant period τ_1 of the arrival of requests in this branch. There is no queue to the channel in the first branch, Time u_1 of the sojourn of requests in QS1 is constant and equal to h_1/C_1. Similarly, we find that there is no queue to the channel in the second branch of the network, and Time u_2 of the sojourn of requests in QS2 is equal to constant h_2/C_2.

We use the method of proof by contradiction and assume that the least value of objective function (39) is attained at Point $\overline{C} = (C_1, C_2)$, which does not lie on Straight Line $C_2 = \gamma C_1$. Then the required point lies either in Domain R_1 and above this line, or in Domain R_2 and to the right of this line.

In the first case, the coordinates of Point $\overline{C}$ which lies in the domain of feasible solutions and delivers the smallest value of the objective function, have the form $(C_1, C_2) = (C_1, \gamma C_1 + \varepsilon)$, where $\varepsilon > 0$. In this case, the periodic process of servicing requests in QS1 is characterized by the Service Time $x_1 = h_1/C_1$, in QS2 - by the Service Time $x_2 = h_2/C_2 = \gamma h_1/(\gamma C_1 + \varepsilon) = h_1/(C_1 + \varepsilon/\gamma) < x_1$. Each request in the first branch and its conjugate request in the second branch starts to be served at the same time. The request in the second branch is served earlier and waits for the completion of the service of the conjugate request, which should come from QS1. As a result, Time u spent by the request to the network (until the moment of assembly) becomes equal to $x_1 = h_1/C_1$. But we get the same sojourn time at Point $\overline{C}^* = (C_1, \gamma C_1)$, at which $x_2 = x_1$. And, since at Point $\overline{C}^*$ with the same sojourn time and the same throughput C_1, the throughput of C_2 is lower than at Point $\overline{C}$, we get $f(\overline{C}^*) < f(\overline{C})$ (39). The resulting contradiction excludes the possibility of finding a solution to the problem (39), (40) in Domain R_1 outside Straight Line $C_2 = \gamma C_1$.

A similar contradiction is caused by the assumption that it is possible to find the required minimum point in Domain R_2 outside Straight Line $C_2 = \gamma C_1$ leads to a similar contradiction.

Thus, the solution to the problem (39), (40) always lies on Straight Line $C_2 = \gamma C_1$. With the optimal choice of throughput, the considered Se-network (with a regular flow of fixed-size requests) becomes an Ss-network. The optimal throughput of such a network is found in Sect. 3.2 of the article. It was also shown there that the transmission of such a stream through the optimal Ss-network leads to exactly the same costs as its transmission through the optimal single-channel system. Now the corresponding conclusion also applies to optimal Se-networks.

5.3   Se-Network with Arbitrary Incoming Flow of Requests

Theorem 4. The Se-network that is optimal for transmitting any flow of requests is an Ss-network; Optimal Throughputs C_1, C_2 of the Se-network lie on Straight Line $C_2 = \gamma C_1$.

The Proof of the Theorem is based on the comparison of the Se-network request processes passing through its branches and establishing the fact that if C_1, C_2 are not connected by condition $C_2 = \gamma C_1$, i.e., if the service time of two conjugate requests does not coincide, then the random sojourn time $u^i = \max(u_1^i, u_2^i)$ of the i-th request in the Se-network is determined by the formula

$$u^i = \max(u_1^i, u_2^i) = \begin{cases} u_1^i, & if\ C_2 \geq \gamma C_1, \\ u_2^i, & if\ C_2 \leq \gamma C_1. \end{cases} \tag{41}$$

For $C_2 = \gamma C_1$ for all i we obtain $u_1^i = u_2^i$.

Averaging the sojourn time (41) over all requests, we obtain an expression that is valid for any Se-networks:

$$U = \begin{cases} U_1, & if\ C_2 \geq \gamma C_1, \\ U_2, & if\ C_2 \leq \gamma C_1. \end{cases} \tag{42}$$

And then we complete the establishment of the validity of Theorem 4 by proving it by contradiction (by analogy with the proof of Theorem 3).

We note two important corollaries of Theorem 4:

- for any incoming flow, the Ss-network is the optimal Se-network;
- the optimal two-channel SSS implementation in the form of a Se-network does not increase operating costs in comparison with the optimal (and, therefore, compared to any) single-channel SSS implementation.

Theorem 4 greatly simplifies the solution of the complex problem of nonlinear optimization of Se-networks, since it reduces the search for the optimal values of two variable parameters C_1, C_2, which provide the minimum of the function of two variables, to the search for the optimal value of one variable parameter, e.g., C_1 (the other is determined from the relation $C_2 = \gamma C_1$), that provides the minimum of the objective function. This is especially important when it is necessary to use simulation modeling to calculate the objective function, e.g., when QS in the network branches belongs to G/G/1 class systems for which there are no exact formulas in QT that express the average sojourn time in QS in terms of its parameters.

The average delay in both synchronization queues is zero. Hence it follows that for $C_2 = \gamma C_1$ in Se-networks $u_1 = u_2$ and is determined by the formula (4) $U = U_1$. Consequently, the optimal value of C_1 and the minimum cost (5) can be easily calculated analytically if QS1 is, for example, a system of class M/M /1, M/G/1, etc.

6   Discussion of Research Results (Conclusion)

The article introduces and investigates a mathematical model for the transmission of messages, divided into smaller and larger shares, transmitted over different channels with different throughput. In terms of queuing theory, a network with split requests (S-network) is defined as such a model. This network takes into account the transmission of a split request over two different channels, the formation of queues in front of the channels, and the assembly of split requests on the receiving side of the channel. The mathematical problem of optimizing the throughput of two channels of the S-network is posed in a general form. Methods for solving this problem (methods for optimizing S- network) are formulated and investigated. Analytical methods have solved the problem of optimizing an exponential S-network with independent branches. In practice, the solution to this problem can be used to optimize S-networks, in which the transmission of split requests constitutes a small part of the total traffic transmitted over two channels.

Special methods are used to study optimization problems for such S-networks, in which the transmission of split requests constitutes the main load transmitted over two channels. Four theorems are proved, which makes it possible to reduce two-dimensional optimization problems of such S-networks (i.e.,. problems with two variable parameters) to one-dimensional ones. Several S-networks with sequentially more complex properties are considered, as a result of which the possibility of reducing a two-dimensional optimization problem to a one-dimensional one for S-networks with the most general assumptions regarding the type of flow of requests arriving in the S-network and the distribution laws of the sizes of requests has been established.

All analytical solutions presented in the article have been verified and confirmed by simulation modeling.

References

1. New opportunities in difficult times. https://www.interfax.ru/pressreleases/ 723657, Accessed 10 Feb 2021
2. At home as in the office: it all depends on the correct organization of remote work. https://www.interfax.ru/pressreleases/740431, Accessed 10 Feb 2021
3. Krupin, A.: Steep dive: how the coronavirus pandemic has changed the IT industry. https://3dnews.ru/1007282/krutoe-pike-kak-pandemiya-koronavirusa-izmenila-itotrasl, Accessed 10 Feb 2021
4. Shamir, A.: How to Share a Secret. Commun. ACM **22**(11), 612–613 (1979)

5. Sagaydak, D.A., Faizullin, R.T.: Model secret sharing schemes in systems transmit video. Comput. Opt. **37**(1), 105–112 (2013)
6. Faizullin, R.T., Sagaydak, D.A.: Application of the prefix encoding algorithm for the data array in the video data stream secret sharing scheme. Rep. Tomsk State Univ. Control Syst. Radioelectron. **1**(25), 136–140 (2012)
7. Mikheev, V.V., Sagaidak, D.A., Svench, A.A., Faizullin, R.R.: Information processing algorithms for data transmission and storage based on the secret sharing procedure. Bull. Siberian State Autom. Highw. Acad. **5**(27), 82–87 (2012)
8. Efimov, V.I., Faizullin, R.T.: Multiplexing system of distributed TCP/IP traffic. Math. Struct. Model. **10**, 170–171 (2002)
9. Lavrov, D.N.: Scheme of sharing a secret for data streams of a routed network. Math. Struct. Model. **10**, 192–197 (2002)
10. Lavrov, D.N., Dulkeit, P.I., Mikhailov, P.I., Svench, A.A.: Analysis of the reliability of the secret sharing algorithm in network streams. Math. Struct. Model. **12**, 146–154 (2003)
11. Guss, S.V., Lavrov, D.N.: Approaches to the implementation of a network protocol for ensuring guaranteed delivery in multi-route data transmission. Math. Struct. Model. **2**(46), 95–101 (2018)
12. Open System Interconnection model. ISO standard. https://standards.iso.org/ittf/PubliclyAvailableStandards, Accessed 03 Mar 2021
13. Kleinrock, L.: Computing Systems with Queues. House Mir, Moscow (1979)

Stability of Applied Probability Systems, Their Limit Behavior and Optimization

Ekaterina Bulinskaya[✉] [ID]

Lomonosov Moscow State University, Moscow 119234, Russia
ebulinsk@mech.math.msu.su

Abstract. We consider two input-output applied probability models which can arise in such applications as insurance, inventory, telecommunications, finance, population dynamics and many others. Giving another interpretation to input and output processes one is able to pass from one applied domain to another. For certainty, the models description is given in terms of insurance being the oldest domain among the above mentioned. The first (periodic-review) model treats an insurance company using non-proportional reinsurance and bank loans. In the framework of cost approach we obtain the model optimal control and limit behavior. The model stability with respect to small fluctuations of underlying distribution is also established. The second model is a generalization of the classical Cramér-Lundberg model. The company has several branches of insurance. Premiums, as well as claims, are random. Their flows are described by generalized Poisson processes. Investment in risky- and non-risky assets is also implemented. Analog of the Lundberg inequality for the ultimate ruin probability is obtained using the martingale technique.

Keywords: Limit behavior · Stability · Optimization · Reliability

1 Introduction

In order to study a real process or system one has to choose an appropriate mathematical model. It is well known that the most frequently used models in such applied probability domains as insurance, inventory, telecommunications, finance, population dynamics and many others are of input-output type. They are described by the six-tuple $(T, Z, Y, U, \Psi, \mathcal{L})$ (see, [9]). Here T is the planning horizon, $Z = (Z(t), t \in [0, T])$ is input process and $Y = (Y(t), t \in [0, T])$ is output process. These processes can be not only one-dimensional but multidimensional as well. Moreover, their dimensions are sometimes different. The system may be deterministic, stochastic or mixed. That means, either both processes are deterministic (stochastic) or one of the processes is stochastic and the other one is deterministic. Giving different interpretations to input and output we can study the systems from different applied domains using the same

Supported by the Russian Foundation for Basic Research, project 20-01-00487.

A. Dudin et al. (Eds.): ITMM 2021, CCIS 1605, pp. 238–251, 2022.
https://doi.org/10.1007/978-3-031-09331-9_19

model. The system state $X = \Psi(Z, Y, U)$ depends on input, output and control $U = (U(t), t \in [0, T])$ via functional Ψ describing the system structure and performance mode. The control may be applied to input Z, output Y or system structure Ψ. Control is used to exclude or minimize the risk associated with system performance. According to the choice of the objective function (risk measure, target or criterium) $\mathcal{L}_T(U)$, there exist different approaches, see, e.g., [10, 27]. Our work is employing only two of them. In the framework of the first (cost) approach we calculate either the (expected) costs entailed by system functioning and try to minimize them, or the (expected) profit and maximize it. In the second one (reliability approach) the aim is to minimize the probability of system failure (ruin) or maximize the time of uninterrupted system performance, see, e.g., [2, 29, 31].

Definition 1. *A control $U_T^* = \{U^*(t), t \in [0, T]\}$ is called optimal if*

$$\mathcal{L}_T(U_T^*) = \inf_{U_T \in \mathcal{U}_T} \mathcal{L}_T(U_T), \quad (or \quad \mathcal{L}_T(U_T^*) = \sup_{U_T \in \mathcal{U}_T} \mathcal{L}_T(U_T)), \quad (1)$$

where $\mathcal{U}_T$ is a class of all feasible controls. Furthermore, $U^* = \{U_T^*, T \geq 0\}$ is called an optimal policy (or strategy).

If the extremum in (1) cannot be attained one has to use either the ε-optimal or asymptotic optimal policies.

Below we consider two models in order to take into account different situations. Since insurance is the oldest applied probability domain we describe the models in terms of insurance company functioning. Constructing an insurance company model one has to take into account its two-fold nature. Originally all insurance societies were designed for risk sharing. Hence, their primary task is policyholders indemnification. Nowadays, for the most part they are joint stock companies. Thus, the secondary but very important task is dividend payments to shareholders, see, e.g. [7]. The modern period in actuarial sciences is characterized by investigation of complex systems, including dividends payment, reinsurance, tax, bank loans and investment. Interplay of actuarial and finance methods, in particular, unification of reliability and cost approaches is another feature of the last twenty years, see, e.g., [10, 12]. Discrete-time models recently became popular, because they turned out to be more appropriate for description of some aspects of insurance company performance, see, e.g., [19, 20, 22–24]. Moreover, discrete-time models can be used for approximation of continuous-time ones, see, e.g., [25].

The paper is arranged as follows. Section 2 deals with a mixed-type model, namely, discrete-time model of insurance company using non-proportional reinsurance and bank loans. The premium inflow is deterministic, whereas the claim amounts are stochastic and identically distributed. The models of such type have already been studied however with different additional assumptions. So, in [8] and [11], the asymptotic analysis was carried out for the systems with bank loans. The problem of investment was treated in [15, 16] in the framework of reliability approach, see also [32]. Discrete-time models with dividends and reinsurance are treated in [17]. Optimization of discrete-time insurance model with capital injections and reinsurance is presented in [18]. Non-homogeneous claim flows are considered, e.g., in [1, 3, 4, 19] and [36]. Reviews of discrete-time models one can also find in [9] and [30].

The investigation is carried out as follows. First of all, we prove that the optimal loan policy is characterized by a sequence of critical levels. After that we establish the system stability using the Kantorovich metric, then prove the strong law of large numbers (SLLN) and central limit theorem (CLT) for the company surplus.

In Sect. 3 a generalization of the classical Cramér-Lundberg model is considered. The company is supposed to issue several types (say, n) of policies. Premiums, as well as claims, are random. Their flows are described by generalized Poisson processes. The reliability approach is employed. More precisely, the ruin probability is calculated using the martingale technique. Analog of the Lundberg inequality is also obtained.

2   Non-proportional Reinsurance and Bank Loans

2.1   Model Description

Suppose that the claims arriving to insurance company are described by a sequence of independent identically distributed (i.i.d.) non-negative random variables (r.v.'s) $\{X_i, i \geq 1\}$. Here X_i is the claim amount during the i-th period (year, month or day). Let $F(x)$ be its distribution function (d.f.) having density $\varphi(x)$ and finite expectation. The company uses non-proportional reinsurance with retention a and bank loans. If a loan is taken at the beginning of period (before the claim arrival) the rate is b_1, whereas the emergency loan after the claim arrival is taken at the rate b_2 with $b_2 > b_1$. Our aim is to choose the loans in such a way that the additional payments entailed by loans are minimized. Denote by M the premium acquired by direct insurer (after reinsurance) during each period. Clearly,

$$M = (1 + \beta_1)EX - (1 + \beta_2)E(X - a)^+.$$

Here β_1 and β_2 are the safety loadings of insurer and reinsurer, respectively, usually $\beta_1 < \beta_2$.

If x is the initial capital and y is the capital after the bank loan, then $f_1(x)$, the minimal expected additional cost during one period, is given by

$$f_1(x) = \min_{y \geq x}[b_1(y - x) + b_2 E(\min(X, a) - (y + M))^+]. \qquad (2)$$

Clearly, (2) can be rewritten in the form:

$$f_1(x) = -b_1 x + \min_{y \geq x} G_1(y), \; G_1(y) = b_1 y + b_2 E[\min(X, a) - (y + M)]^+.$$

Now let $f_n(x)$ be the minimal expected costs during n periods and α the discount factor for future expenses. Then, using the dynamic programming, see, e.g., [5], one easily obtains the following relation:

$$f_n(x) = -b_1 x + \min_{y \geq x} G_n(y), \quad G_n(y) = G_1(y) + \alpha E f_{n-1}(y + M - \min(X, a)).$$

It is not difficult to prove by induction the main optimization result.

Theorem 1. *Let $F(M) < 1 - \frac{b_1}{b_2} < F(a)$, then there exists an increasing sequence of critical levels $\{y_n\}_{n \geq 1}$ such that*

$$f_n(x) = -b_1 x + \begin{cases} G_n(y_n), & \text{if } x \leq y_n, \\ G_n(x), & \text{if } x > y_n. \end{cases}$$

The sequence is bounded by $\bar{y}$ satisfying the equation $H(y) = 0$ where $H(y) = G_1'(y) - b_1 \alpha$.

Proof. Obviously, $E[\min(X, a) - z]^+ = \int_z^a \bar{F}(s) \, ds$ with $\bar{F}(s) = 1 - F(s)$. Therefore, $G_1'(y) = b_1 - b_2 \bar{F}(y + M)$ for $y + M < a$ and $G_1'(y) = b_1 > 0$ otherwise. If we assume additionally that $\bar{F}(M) > b_1/b_2 > \bar{F}(a)$ then there exists $y_1 = F^{-1}(1 - b_1/b_2) - M > 0$ such that $G_1'(y_1) = 0$. (The other cases are more intricate and will be treated later.) Thus, we have

$$f_1(x) = -b_1 x + \begin{cases} G_1(y_1), & \text{if } x \leq y_1, \\ G_1(x), & \text{if } x > y_1, \end{cases}$$

and

$$f_1'(x) = \begin{cases} -b_1, & \text{if } x \leq y_1, \\ -b_2 \bar{F}(x + M), & \text{if } x > y_1. \end{cases}$$

In other words, $f_1'(x) < 0$ for all x. Further proof is carried out by induction. Since

$$G_2'(y) = G_1'(y) + \alpha \int_0^\infty f_1'(y + M - \min(s, a)) \varphi(s) \, ds$$

and $G_2''(y) \geq 0$, it is not difficult to get $G_2'(y) \leq G_1'(y)$ for all y, that is, $y_2 \geq y_1$. Moreover,

$$G_2'(y) \geq G_1'(y) - b_1 \alpha = H(y).$$

That entails inequality $y_2 \leq \bar{y}$. It follows immediately that expression of $f_2(x)$ is similar to that of $f_1(x)$ with y_2 instead of y_1 and $f_2'(x) < 0$ for all x. Assuming the same is true for n we can write

$$f_n'(x) - f_{n-1}'(x) = \begin{cases} 0, & x \leq y_{n-1}, \\ -G_{n-1}'(x), & y_{n-1} < x \leq y_n, \\ G_n'(x) - G_{n-1}'(x), & y_n < x. \end{cases}$$

Thus, it is clear that $y_n \leq y_{n+1} \leq \bar{y}$, ending the proof.

Corollary 1. *If $M = 0$ then $\lim_{n \to \infty} y_n = \bar{y}$.*

For $M > 0$ such statement is not valid. To verify this fact the numerical analysis was carried out using Python, for other method see, e.g., [21].

2.2 Stability

In order to establish the model stability to small perturbations of the underlying distribution we are going to use the probability metrics, see, e.g., [33]. Suppose that we have two different sequences of claims $\{X_n\}$ and $\{Y_n\}$ with distribution functions F_X and F_Y, respectively.

The corresponding minimal n-step costs are denoted by $f_{n,X}$ and $f_{n,Y}$, other functions and constants depending on distribution will be marked by subscripts X and Y, as well.

The distance between distributions will be measured in terms of the Kantorovich metric.

Definition 2. *Let random variables X and Y, defined on the same probability space, possess finite expectations. The distance based on the Kantorovich metric is given as follows:*

$$\kappa(X,Y) = \int_{-\infty}^{\infty} |F_X(t) - F_Y(t)| \, dt$$

where F_X and F_Y are the respective distribution functions of X and Y.

The distance between the cost functions is measured in terms of the Kolmogorov uniform metric. Thus, we are going to study

$$\Delta_n = \sup_x |f_{n,X}(x) - f_{n,Y}(x)|.$$

To this end we need the following

Lemma 1. *Let functions $g_i(y)$, $i = 1, 2$, be such that $|g_1(y) - g_2(y)| < \delta$ for some $\delta > 0$ and any y, then $\sup_x |\inf_{y \geq x} g_1(y) - \inf_{y \geq x} g_2(y)| < \delta$.*

Proof. Fix x and put $C_i = \inf_{y \geq x} g_i(y)$. Then, according to definition of infimum, for any $\varepsilon > 0$ there exists such $y_1(\varepsilon) \geq x$ that $g_1(y_1(\varepsilon)) < C_1 + \varepsilon$. Therefore

$$g_2(y_1(\varepsilon)) < g_1(y_1(\varepsilon)) + \delta < C_1 + \varepsilon + \delta$$

implying $C_2 < g_2(y_1(\varepsilon)) < C_1 + \varepsilon + \delta$. Letting $\varepsilon \to 0$ one gets immediately $C_2 < C_1 + \delta$. In a similar way one establishes $C_1 < C_2 + \delta$, thus obtaining the desired result $|C_1 - C_2| < \delta$.

Now we are able to estimate Δ_1.

Lemma 2. *Assume $\kappa(X, Y) = \rho$, then $\Delta_1 \leq b_2 \rho$.*

Proof. According to Lemma 1 we need to estimate $|G_{1,X}(y) - G_{1,Y}(y)|$ for any y. The definition of these functions gives $G_{1,X}(y) - G_{1,Y}(y) = b_2 \int_{y+M}^{a} (\bar{F}_X(s) - \bar{F}_Y(s)) \, ds$. As usually, $\bar{F}_X(s) = 1 - F_X(s)$. This leads immediately to the desired estimate.

To formulate the main result demonstrating the model stability put $D_n = \frac{b_2(1-\alpha^n)}{1-\alpha} + \frac{b_1(\alpha-\alpha^n)}{1-\alpha}$.

Theorem 2. *If $\kappa(X, Y) = \rho$, then $\Delta_n \leq D_n \rho$.*

Proof. As in Lemma 2 we begin by estimating for all y

$$|G_{n,X}(y) - G_{n,Y}(y)| \leq |G_{1,X}(y) - G_{1,Y}(y)| + \alpha \Delta_{n-1}(X, Y)$$

with

$$\Delta_{n-1}(X, Y) = |\int_0^\infty f_{n-1,X}(y + M - \min(s, a)) \, dF_X(s)$$
$$- \int_0^\infty f_{n-1,Y}(y + M - \min(s, a)) \, dF_Y(s)|.$$

Adding and subtracting $\int_0^\infty f_{n-1,X}(y + M - \min(s, a)) \, dF_Y(s)$ in expression under the sign of module and rewriting this integral in the form $f_{n-1,X}(y + M) - \int_0^a f'_{n-1,X}(y + M - s) \bar{F}_X(s) \, ds$, we obtain the recurrent relation

$$\Delta_n \leq \Delta_1 + \alpha(\Delta_{n-1} + b_1 \rho),$$

since $|f'_{n-1}(y)| \leq b_1$ for any y. The desired statement is obvious.

2.3 Limit Theorems

The last problem for this model is the limit behavior of the company surplus as the planning horizon n tends to ∞. Let x be the initial capital. Since we use the reinsurance treaty with retention level a at each step, put $X_k^{(a)} = \min(X_k, a)$ where X_k denotes the claim amount in the k-th period.

According to Theorem 1 the optimal policy of insurer is characterized by the sequence of critical levels y_n as follows. At the first step of n-step process it is necessary to raise the initial capital to level y_n if $x \leq y_n$ and take no loan otherwise. Thus, if $Z_k^{(n)}$ is the surplus at the k-th step of the n-step process then $Z_0^{(n)} = x$ and for $k \geq 1$

$$Z_k^{(n)} = \begin{cases} y_{n+1-k} + M - X_k^{(a)}, & Z_{k-1}^{(n)} \leq y_{n+1-k}, \\ Z_{k-1}^{(n)} + M - X_k^{(a)}, & Z_{k-1}^{(n)} > y_{n+1-k}. \end{cases} \tag{3}$$

Theorem 3 (SLLN for surplus). *For $x > a - M$ with probability 1*

$$\frac{Z_n^{(n)}}{n} \to \delta(a) = M - EX^{(a)}, \quad as \ n \to \infty.$$

Proof. It easily follows from (3) that for $x > a - M$

$$Z_n^{(n)} = x + nM - \sum_{k=1}^n X_k^{(a)}, \tag{4}$$

hence, according to the SLLN for a sequence of i.i.d. r.v's with a finite mean the statement of the theorem is true.

Thus, we can formulate the following

Corollary 2. *The ultimate ruin probability is equal to 1 if* $\delta(a) \leq 0$.

It is also possible to establish the asymptotical normality of surplus. Namely,

Theorem 4 (CLT for surplus). *For* $x > a - M$

$$\frac{Z_n^{(n)} - E Z_n^{(n)}}{\sqrt{Var Z_n^{(n)}}} \xrightarrow{d} \mathcal{N}, \quad as \ n \to \infty,$$

here $\mathcal{N}$ *has Gaussian distribution with parameters* $(0, 1)$ *and* $\xrightarrow{d}$ *signifies convergence in distribution.*

Proof. The assertion easily follows from CLT for i.i.d. r.v.'s. According to (4)

$$\frac{Z_n^{(n)} - E Z_n^{(n)}}{\sqrt{Var Z_n^{(n)}}} = -\frac{\sum_{k=1}^{n} X_k^{(a)} - \sum_{k=1}^{n} E X_k^{(a)}}{\sqrt{\sum_{k=1}^{n} Var X_k^{(a)}}}.$$

So one uses the properties of convergence in distribution (see, e.g., [6]) and the properties of Gaussian distributions to derive the asymptotic normality.

Hence, it is not difficult to obtain the bounds on the size of surplus with probability $1 - \varepsilon$ for small $\varepsilon > 0$ and choose the appropriate values of retention level a and safety loadings β_i, $i = 1, 2$.

3   Generalized Cramér-Lundberg Model

3.1   Model Description

In order to make the model more realistic, we consider n categories (branches) of insurance (not a single one), replace the compound Poisson process by a generalized Poisson process and include the investment. Based on these conditions, a risk model is set up to find (by the martingale methods) the formula of ruin probability and its upper bound. Thus, we put

$$U(t) = u_1 + (u_1 - u_2 - u_3)(b - c + d) + S(t),$$

$$S(t) = u_2 r_1 t + u_3 (r_2 t + a B(t)) + \sum_{j=1}^{n} \sum_{i=1}^{N_j(t)} X_i^{(j)} - \sum_{j=1}^{n} \sum_{i=1}^{M_j(t)} Y_i^{(j)}.$$

Here $U(t)$ is the surplus (capital) of insurance company and $S(t)$ is the gain at time t, u_1 is the initial reserve of insurance company, u_2 is invested in a non-risky asset, u_3 is used in a venture investment, r_1 is rate of return of u_2, $u_3(r_2 t + a B(t))$ is income from investment of u_3 in a Brownian motion with parameters r_2, a (r_2 is drift parameter, a is volatility, $B(t)$ is a standard Brownian motion), b is interest rate, c is inflation rate, d is exchange rate. Clearly, parameters satisfy the following inequalities $t \geq 0$, $u_1 > 0$, $u_2 > 0$, $u_3 > 0$, $r_1 > 0$, $r_2 > 0$, $a > 0$. It is interesting to mention that the models including Brownian motion were studied in [26, 28, 35] for one-dimensional case.

Furthermore, $N_j(t)$ and $M_j(t)$ are Generalized Poisson processes providing inhomogeneity of premium and claim flows, $X_i^{(j)}$ represents the ith premium, while $Y_i^{(j)}$ is the ith claim amount for the jth insurance branch. All r.v.'s possess finite means and variances.

More precisely, $N_j(t) = \sum_{i=1}^{n_j(t)} Z_i^{(j)}$, where $n_j(t)$ is a Poisson process with intensity λ_j and $Z_i^{(j)}$ is the number of insurance policies belonging to the jth insurance branch at the time of the ith jump of $n_j(t)$. Then

$$\sum_{j=1}^{n} \sum_{i=1}^{N_j(t)} X_i^{(j)} = \sum_{j=1}^{n} F^{(j)}(t) \text{ with } F^{(j)}(t) = \sum_{k=1}^{n_j(t)} G_k^{(j)}, \tag{5}$$

where

$$G_k^{(j)} = X_{1+\sum_{i=1}^{k-1} Z_i^{(j)}}^{(j)} + X_{2+\sum_{i=1}^{k-1} Z_i^{(j)}}^{(j)} + \ldots + X_{\sum_{i=1}^{k} Z_i^{(j)}}^{(j)},$$

and its distribution function is $H_j(x)$.

$M_j(t) = \sum_{i=1}^{m_j(t)} Z_i'^{(j)}$, where $m_j(t)$ is a Poisson process with parameter λ_j' and $Z_i'^{(j)}$ is the number of claims of the jth insurance branch at the time of the ith incident, then

$$\sum_{j=1}^{n} \sum_{i=1}^{M_j(t)} Y_i^{(j)} = \sum_{j=1}^{n} F'^{(j)}(t) \text{ with } F'^{(j)}(t) = \sum_{k=1}^{m_j(t)} G_k'^{(j)}, \tag{6}$$

where

$$G_k'^{(j)} = Y_{1+\sum_{i=1}^{k-1} Z_i'^{(j)}}^{(j)} + Y_{2+\sum_{i=1}^{k-1} Z_i'^{(j)}}^{(j)} + \ldots + Y_{\sum_{i=1}^{k} Z_i'^{(j)}}^{(j)},$$

and its distribution function is $H_j'(x)$.

It is also assumed that the sequences $\{Z_i^{(j)}, i \geq 1\}$ and $\{Z_i'^{(j)}, i \geq 1\}$, $j = \overline{1, n}$, are independent, each consisting of integer-valued non-negative i.i.d. r.v.'s.

Next, we put $EX^{(j)} = \mu^{(j)}$, $VarX^{(j)} = (\sigma^{(j)})^2$, $EZ^{(j)} = \mu_1^{(j)}$, $VarZ^{(j)} = (\sigma_1^{(j)})^2$, $EY^{(j)} = \mu'^{(j)}$, $VarY^{(j)} = (\sigma'^{(j)})^2$, $EZ'^{(j)} = \mu_1'^{(j)}$, $VarZ'^{(j)} = (\sigma_1'^{(j)})^2$.

It is not difficult to see that

$$EG_k^{(j)} = \mu^{(j)} \mu_1^{(j)} \text{ and } EG_k'^{(j)} = \mu'^{(j)} \mu_1'^{(j)}.$$

Since the company needs to keep running, an analog of net-profit condition should be satisfied. In other words, we suppose

$$s_3 + \sum_{j=1}^{n} \lambda_j \mu^{(j)} \mu_1^{(j)} - \sum_{j=1}^{n} \lambda_j' \mu'^{(j)} \mu_1'^{(j)} > 0 \tag{7}$$

with $s_3 = u_2 r_1 + u_3 r_2$.

Lemma 3. *If condition (7) holds then* $U(t) \to \infty$ *a.s., as* $t \to \infty$.

Proof. In fact, we need to establish SLLN for the process $\{U(t), t \geq 0\}$. Since

$$\lim_{t\to\infty} \frac{U(t)}{t} = \lim_{t\to\infty} \left[\frac{u_1 + (u_1 - u_2 - u_3)(b - c + d)}{t} + \frac{au_3 B(t)}{t} + s_3 \right.$$
$$\left. + \frac{\sum_{j=1}^{n} \sum_{i=1}^{N_j(t)} X_i^{(j)}}{t} - \frac{\sum_{j=1}^{n} \sum_{i=1}^{M_j(t)} Y_i^{(j)}}{t} \right], \tag{8}$$

we have to obtain the limit (as $t \to \infty$) of each summand on the right-hand side.

First of all, it is clear, that $au_3 B(t)/t \sim N(0, a^2 u_3^2/t)$, hence, for any $\varepsilon > 0$, according to Chebyshev's inequality

$$P(|au_3 B(t)/t| < \varepsilon) \geq 1 - (a^2 u_3^2 / t\varepsilon^2).$$

That means, $au_3 B(t)/t \xrightarrow{p} 0$, furthermore, using the Law of Iterated Logarithm for the Brownian motion, we can obtain

$$\lim_{t\to\infty} au_3 B(t)/t = 0, \text{ a.s. },$$

leading to zero limit for the first two terms. Next, we use the relations (5) and (6). Obviously, representation

$$\frac{1}{t} F^{(j)}(t) = \frac{1}{t} \sum_{k=1}^{n_j(t)} G_k^{(j)} = \frac{n_j(t)}{t} \cdot \frac{1}{n_j(t)} \sum_{k=1}^{n_j(t)} G_k^{(j)}$$

and a similar relation for the last term in (8) leads to conclusion that the last two terms give the limit equal to

$$\sum_{j=1}^{n} \lambda_j \mu^{(j)} \mu_1^{(j)} - \sum_{j=1}^{n} \lambda_j' \mu'^{(j)} \mu_1'^{(j)}.$$

Due to condition (7), it follows immediately that $\lim_{t\to\infty} U(t) = +\infty$ a.s.

Lemma 4. *Surplus process* $\{U(t), t \geq 0\}$ *has stationary independent increments.*

Proof. Taking $0 = t_0 < t_1 < \cdots < t_n$ and denoting

$$s_1(t) = \sum_{j=1}^{n} \sum_{i=1}^{N_j(t)} X_i^{(j)}, \; s_2(t) = \sum_{j=1}^{n} \sum_{i=1}^{M_j(t)} Y_i^{(j)}$$

we can write the following equality $U(t_i) - U(t_{i-1})$

$$= au_3[B(t_i) - B(t_{i-1})] + [s_1(t_i) - s_1(t_{i-1})] - [s_2(t_i) - s_2(t_{i-1})] + s_3(t_i - t_{i-1}).$$

Since the standard Brownian motion has stationary independent increments, as well as the generalized Poisson process, and they are independent of each other, it is clear that $\{S(t), t \geq 0\}$ has also stationary independent increments.

Lemma 5. *There exists a function $h(r)$ satisfying the relation*

$$E[exp(-rS(t))] = exp(th(r)).$$

Here

$$h(r) = -rs_3 + \frac{a^2 u_3^2 r^2}{2} + \sum_{j=1}^{n} \lambda_j [\sum_{m=1}^{\infty} M_{X^{(j)}}^m (-r) P_m^{(j)} - 1] + \sum_{j=1}^{n} \lambda_j' [\sum_{m=1}^{\infty} M_{Y^{(j)}}^m (r) P_m'^{(j)} - 1],$$

$$M_{X^{(j)}}(r) = E \exp(r X^{(j)}), \ M_{Y^{(j)}}(r) = E \exp(r Y^{(j)}),$$

$P_m^{(j)}$ *is the probability that the number of* $M_{X^{(j)}}$ *is* m *and* $P_m'^{(j)}$ *is the probability that the number of* $M_{Y^{(j)}}$ *is* m.

Proof. It is not difficult to obtain the following chain of equalities

$$E \exp[-rS(t)] = E \exp[-rs_3 t - rau_3 B(t) - r \sum_{j=1}^{n} \sum_{i=1}^{N_j(t)} X_i^{(j)} + r \sum_{j=1}^{n} \sum_{i=1}^{M_j(t)} Y_i^{(j)}]$$

$$= \exp[-rs_3 t] \cdot E \exp[-r(au_3 B(t))] \cdot E \exp[-r \sum_{j=1}^{n} \sum_{i=1}^{N_j(t)} X_i^{(j)}] \cdot E \exp[r \sum_{j=1}^{n} \sum_{i=1}^{M_j(t)} Y_i^{(j)}]$$

$$= E \exp\{t[-rs_3 + \frac{a^2 u_3^2 r^2}{2} + \sum_{j=1}^{n} \lambda_j (M_{G^{(j)}}(-r) - 1) + \sum_{j=1}^{n} \lambda_j' (M_{G'^{(j)}}(r) - 1)]\}. \quad (9)$$

Obviously,

$$M_{G^{(j)}}(r) = \sum_{m=1}^{\infty} P_m^{(j)} M_{X^{(j)}}^m (r), \text{ and } M_{G'^{(j)}}(r) = \sum_{m=1}^{\infty} P_m'^{(j)} M_{Y^{(j)}}^m (r),$$

so

$$E \exp[-rS(t)] = \exp[th(r)]$$

with $h(r)$ defined in the statement of lemma.

Lemma 6. *Equation $h(r) = 0$ has only one positive root R.*

Proof. Using (9), we get

$$h'(r) = -s_3 + a^2 u_3^2 r - \sum_{j=1}^{n} \lambda_j E[G^{(j)} \exp(-rG^{(j)})] + \sum_{j=1}^{n} \lambda_j' E[G'^{(j)} \exp(rG'^{(j)})],$$

$$h''(r) = a^2 u_3^2 + \sum_{j=1}^{n} \lambda_j E[(G^{(j)})^2 \exp(-rG^{(j)})] + \sum_{j=1}^{n} \lambda_j' E[(G'^{(j)})^2 \exp(rG'^{(j)})] > 0.$$

Due to (7)

$$h'(0) = -s_3 - \sum_{j=1}^{n} \lambda_j E[G^{(j)}] + \sum_{j=1}^{n} \lambda'_j E[G'^{(j)}]$$

$$= -\{s_3 + \sum_{j=1}^{n} \lambda_j \mu^{(j)} \mu_1^{(j)} - \sum_{j=1}^{n} \lambda'_j \mu'^{(j)} \mu_1'^{(j)}\} < 0,$$

so, equation $h(r) = 0$ has at most two nonnegative roots, moreover, $h(0) = 0$. Thus, there is only one positive root R which is called the *adjustment coefficient*.

3.2   Ruin Probability

Now we are going to choose an objective function. Instead of the additional costs entailed by the system functioning we consider the ruin probability, that is, use the reliability approach. We introduce the following

Definition 3. *Ruin time T of the insurance system under consideration is the first moment when the company capital $U(t)$ becomes negative, that is,*

$$T = \inf\{t > 0 : U(t) < 0\}.$$

We define the ultimate ruin probability as $P(T < \infty)$. *Note* that under assumption (7) this probability is not equal 1 according to Lemma 3.

   In order to use the martingale technique for evaluation of ruin probability we establish the following result.

Lemma 7. *The process $V(t) = \exp[-rU(t) - th(r)]$, $t \geq 0$, with $h(r)$ defined in Lemma 5, is a martingale.*

Proof. Introduce $\mathcal{F}_t = \sigma\{S(s), s \leq t\}$, $t \geq 0$. Since the process $S(t)$ has independent increments, it is obvious that

$$E[V(t)|\mathcal{F}_s] = E[V(s)\exp(-r(S(t) - S(s)) - (t - s)h(r))|\mathcal{F}_s] = V(s).$$

Theorem 5. *For the Generalized Poisson multiple risk model, the upper bound of ultimate ruin probability is given by*

$$P(T < \infty) \leq \exp[-R(u_1 + (u_1 - u_2 - u_3)(b - c + d))].$$

Here R is the unique positive root of the equation $h(r) = 0$.

Proof. By the Stopping-time Theorem, for any fixed $t_0 < \infty$

$$EV(T \wedge t_0) = EV(0) = \exp(-r[u_1 + (u_1 - u_2 - u_3)(b - c + d)]).$$

According to the total expectation formula

$$EV(T \wedge t_0) = E(V(T \wedge t_0)I(T \leq t_0)) + E(V(T \wedge t_0)I(T > t_0))$$
$$\geq E(V(T \wedge t_0)I(T \leq t_0)) = E(V(T)I(T \leq t_0)),$$

where $I(A)$ is the indicator of event A. If $T < \infty$, then

$$U(T) = u_1 + (u_1 - u_2 - u_3)(b - c + d) + S(T) < 0.$$

Thus, we have the following inequalities

$$E(V(T)I(T \leq t_0)) \geq Ee^{-Th(r)}I(T \leq t_0) \geq \inf_{0 \leq t \leq t_0} e^{-th(r)} \cdot P(T \leq t_0).$$

Hence

$$P(T \leq t_0) \leq \exp(-r[u_1 + (u_1 - u_2 - u_3)(b - c + d)]) \cdot \sup_{0 \leq t \leq t_0} e^{th(r)}.$$

Letting $t_0 \to \infty$ we get

$$P(T < \infty) \leq \exp(-r[u_1 + (u_1 - u_2 - u_3)(b - c + d)]) \cdot \sup_{t \geq 0} \exp[th(r)].$$

Since the adjustment coefficient $R = \sup_{r>0}\{r : h(r) \leq 0\}$, we finally obtain

$$P(T < \infty) \leq \exp(-R[u_1 + (u_1 - u_2 - u_3)(b - c + d)]).$$

4 Conclusion

Two new insurance models were studied. The first one is a discrete-time model of insurance company using non-proportional reinsurance and bank loans. The aim is minimization of expected additional costs associated with loans during n periods. It is proved that the optimal policy is determined by an increasing bounded sequence of critical levels. The model stability is established in terms of Kantorovich metric. For the optimal company surplus SLLN and CLT are proved. Further investigation directions are treatment of incomplete information, see, [13], and non-homogeneous flows, as well as, the choice of optimal reinsurance treaty.

The second model describes a company having several business lines. The model is doubly stochastic and uses investment in risky and non-risky assets. Moreover, the input (premiums flow) and output (claims flow) are generalized Poisson processes (not compound ones). The properties of company surplus are studied. This model illustrates the reliability approach, providing the ruin probability. Next step is investigation of system stability and choice of optimal parameters, as in [14], see also [8,34].

References

1. Alfa, A.-S., Drekic, S.: Algorithmic analysis of the Sparre Andersen model in discrete time. ASTIN Bull. **37**, 293–317 (2007)
2. Asmussen, S., Albrecher, H.: Ruin Probabilities, 2nd edn. World Scientific, New Jersey (2010)

3. Bao, Z., Liu, H.: On the discounted factorial moments of the deficit in discrete time renewal risk model. Int. J. Pure Appl. Math. **79**(2), 329–341 (2012)

4. Blaževičius, K., Bieliauskienė, E., Šiaulys, J.: Finite-time ruin probability in the nonhomogeneous claim case. Lithuanian Math. J. **50**(3), 260–270 (2010)

5. Bellman, R.: Dynamic Programming. Princeton University Press, Princeton (1957)

6. Billingsley, P.: Convergence of Probability Measures. John Wiley and Sons, New York (1968)

7. Bulinskaya, E.: New dividend strategies. In: Dimotikalis, Y., et al. (eds.) Applied Modeling Techniques and Data Analysis 2, vol. 3, pp. 39–52. ISTE Ltd., London (2021)

8. Bulinskaya, E.: Asymptotic analysis and optimization of some insurance models. Appl. Stochastic Models Bus. Ind. **34**(6), 762–773 (2018)

9. Bulinskaya, E.: New research directions in modern actuarial sciences. In: Panov, V. (ed.) MPSAS 2016. SPMS, vol. 208, pp. 349–408. Springer, Cham (2017). https://doi.org/10.1007/978-3-319-65313-6_15

10. Bulinskaya, E.-V.: Cost approach versus reliability. In: Proceedings of International Conference DCCN-2017, Technosphera, Moscow, pp. 382–389 (2017)

11. Bulinskaya, E.: Asymptotic analysis of insurance models with bank loans. In: Bozeman, J.-R., Girardin, V., Skiadas, Ch. (eds.) New Perspectives on Stochastic Modeling and Data Analysis, pp. 255–270. ISAST, Athens (2014)

12. Bulinskaya, E.: On the cost approach in insurance. Revue Appl. Ind. Math. **10**(2), 276–286 (2003). In Russian

13. Bulinskaya, E., Gusak, J.: Insurance models under incomplete information. In: Pilz, J., Rasch, D., Melas, V.B., Moder, K. (eds.) IWS 2015. SPMS, vol. 231, pp. 171–185. Springer, Cham (2018). https://doi.org/10.1007/978-3-319-76035-3_12

14. Bulinskaya, E., Gusak, J.: Optimal control and sensitivity analysis for two risk models. Commun. Stat. Simul. Comput. **45**(5), 1451–1466 (2016)

15. Bulinskaya, E., Kolesnik, A.: Reliability of a discrete-time system with investment. In: Vishnevskiy, V.M., Kozyrev, D.V. (eds.) DCCN 2018. CCIS, vol. 919, pp. 365–376. Springer, Cham (2018). https://doi.org/10.1007/978-3-319-99447-5_31

16. Bulinskaya, E., Shigida, B.: Discrete-time model of company capital dynamics with investment of a certain part of surplus in a non-risky asset for a fixed period. Methodol. Comput. Appl. Prob. **23**(1), 103–121 (2021). https://doi.org/10.1007/s11009-020-09843-5

17. Bulinskaya, E., Yartseva, D.: Discrete-time models with dividends and reinsurance. In: Proceedings of SMTDA 2010, Chania, Greece, 8–11 June 2010, pp. 155–162 (2010)

18. Bulinskaya, E., Gusak, J., Muromskaya, A.: Discrete-time insurance model with capital injections and reinsurance. Methodol. Comput. Appl. Prob. **17**, 899–914 (2015)

19. Castañer, A., Claramunt, M.-M., Lefèvre, C., Gathy, M., Mármol, M.: Ruin probabilities for a discrete-time risk model with non-homogeneous conditions. Scand. Actuarial J. **2013**(2), 83–102 (2013)

20. Chan, W., Zhang, L.: Direct derivation of finite-time ruin probabilities in the discrete risk model with exponential or geometric claims. North Am. Actuarial J. **10**(4), 269–279 (2006)

21. Coulibaly, I., Lefèvre, C.: On a simple quasi-Monte Carlo approach for classical ultimate ruin probabilities. Insur. Math. Econ. **42**, 935–942 (2008)

22. Czarna, I., Palmovsky, Z., Swiątek, P.: Discrete time ruin probability with Parisian delay(2017). arXiv:1403.7761v2 [math.PR], Accessed 14 June 2017

23. Damarackas, J., Šiaulys. J.: Bi-seasonal discrete time risk model. Appl. Math. Comput. 247, 930–940 (2014)
24. Diasparra, M., Romera, R.: Inequalities for the ruin probability in a controlled discrete-time risk process. Eur. J. Oper. Res. **204**(3), 496–504 (2010)
25. Dickson, D.-C.-M., Waters, H.-R.: Some optimal dividends problems. ASTIN Bull. **34**, 49–74 (2004)
26. Dufresne, F., Gerber, H.-U.: Risk theory for the compound Poisson process that is perturbed by diffusion. Insur. Math. Econ. **10**, 51–59 (1991)
27. Gallager, R.-G.: Stochastic Processes: Theory for Application. Cambridge University Press, Cambridge (2013)
28. Gerber, H.-U., Landry, B.: On the discounted penalty at ruin in a jump-diffusion and the perpetual put option. Insur. Math. Econ. **22**, 263–276 (1998)
29. Grandell, J.: Aspects of Risk Theory. Springer, New York (1991). https://doi.org/10.1007/978-1-4613-9058-9
30. Li, S., Lu, Y., Garrido, J.: A review of discrete-time risk models. Revista de la Real Academia de Ciencias Naturales. Serie A Matemàticas **103**, 321–337 (2009)
31. Mikosch, T.: Non-Life Insurance Mathematics. U, Springer, Heidelberg (2009). https://doi.org/10.1007/978-3-540-88233-6
32. Paulsen, J.: Ruin models with investement income (2008). arXiv:0806.4125v1 (math. PR), Accessed 25 June 2008
33. Rachev, S.T., Klebanov, L., Stoyanov, S.V., Fabozzi, F.: The Methods of Distances in the Theory of Probability. and Statistics. Springer-Verlag, New York (2013). https://doi.org/10.1007/978-1-4614-4869-3
34. Saltelli, A., Tarantola, S., Campolongo, F.: Sensitivity analysis as an ingredient of modeling. Stat. Sci. **15**(4), 377–395 (2000)
35. Wang, G., Wu, R.: Some distribution for classical risk process that is perturbed by diffusion. Insur. Math. Econ. **26**(1), 15–24 (2000)
36. Zhou, Q., Sakhanenko, A., Gou, J.: Exponential bounds of ruin probabilities for non-homogeneous risk models. Prob. Math. Stat. **41**(2), 217–235 (2021)

Retrial Queue MMPP/M/N Under Heavy Load Condition

Ekaterina Fedorova[✉][iD]

National Research Tomsk State University, Lenina Avenue, 36, Tomsk, Russia
moiskate@mail.ru

Abstract. In the paper, a multi-server retrial queueing system with MMPP arrivals is considered. The service and retrial times are exponentially distributed. The two-dimension stochastic process of number of calls in the orbit and states of service unit is analyzed. The system of Kolmogorov differential equations is composed. The matrix form of the equations in steady-state regime for partial characteristic functions is written. The method of asymptotic analysis under the heavy load condition for its solving is proposed. It is proved that the asymptotic characteristic function of the number of calls in the orbit has the gamma distribution with obtained parameters. Some numerical examples of comparison asymptotic and simulate distributions are presented.

Keywords: Retrial queue · MMPP · Heavy load · Asymptotic analysis

1 Introduction

Retrial queueing systems are mathematical models widely used in telecommunication networks, computer systems, call centers, etc. [1–5]. The distinguishing feature of such models is that an arriving call, which can not be served, does not join a queue and does not leave the system immediately (as in classical queueing systems). It joins to an orbit (virtual place), where a call waits some random time and then it tries to be served. Now a large number of publications are devoted to retrial queues. The most detailed description, the comparison of classical queueing systems and retrial queues and detailed overviews up to 2008 are contained in monographs of J. Artalejo and A. Gómez-Corral [6], G. Falin and J. Templeton [7].

In most papers devoted retrial queues with MAP (or MMPP), authors use truncation methods [6,8–11] or matrix methods [12–14] and further numerical analysis. While explicit formulas for probability distributions or performance characteristic of complex retrial queues (e.g. with MMPP arrivals, several orbits, non-exponential retrial or service times) cannot be usually obtained. But some approximations or asymptotic solutions can be proposed. One of approximate

The reported study was funded by RFBR and Tomsk region according to the research project № 19-41-703002.

methods is the method of diffusion approximation of retrial queue proposed in [15,16], etc.

In this paper, the asymptotic analysis method [17,18] is used for the multi-server retrial queue with MMPP arrivals. This method is developed in Tomsk and has different modifications for different types of queueing models and queueing networks. It consists of a derivation of some asymptotic equations determining models characteristics and further getting formulas for asymptotic functions under some limit condition. In previous papers [19,20], we have obtained asymptotic solutions under the heavy load condition for different types of single-server retrial queues: $M/M/1$, $M/GI/1$ and even $MMPP/M/1$, $MMPP/GI/1$. So here, we are going to generalize our results to more complex RQ: the multi-server system with MMPP arrivals. Retrial queues with non-Poisson arrival processes are also studied in [1,10,12,21].

The paper is organized as follows. In Sect. 2, the considered mathematical model is described and the stochastic process under study is defined. Section 3 is devoted to method of asymptotic analysis and study of the retrial queue under a limit condition of heavy load. The theorem about the gamma form of the asymptotic characteristic function is proved and parameters of the distribution are obtained. In Sect. 4, numerical examples of the comparison of the asymptotic distributions with simulation ones are shown. The last section contains conclusions.

2   Mathematical Model

Let us describe the model under study. We consider a multi-server retrial queueing system $MMPP/M/N$. Primary calls arrive at the system according to Markovian Modulated Poisson Process (MMPP) defined by matrices $\mathbf{D_0}$ and $\mathbf{D_1}$ [22,23]. If a primary call finds a server free, it starts service with exponentially distributed service time with rate μ'. If all servers is busy, the call goes to an orbit, where it stays during random time distributed by the exponential law with rate σ. After the delay, the call makes an attempt to get service again. If any server is free, the call gets the service, otherwise, the call instantly returns to the orbit. The arrival process, the service times, the retrial times are assumed to be mutually independent. The system structure is presented in Fig. 1.

The MMPP underlying process $n(t)$ is a Markov chain with continuous time and finite set of states $n = 1, 2, \ldots, W$. Matrix $\mathbf{Q} = \mathbf{D_0} + \mathbf{D_1} = (q_{mv})$ is a generator of the process $n(t)$, where $m, v = 1, 2, \ldots, W$. Matrix $\mathbf{D_1}$ is diagonal with elements λ_n $(n = 1, 2, \ldots, W)$. Further, we will use denotation $\mathbf{D_1} = \mathbf{\Lambda} = \mathtt{diag}\{\lambda_n\}$.

Let us denote a stationary probability distribution of $n(t)$ by $\mathbf{r}$, which is row-vector uniquely determined by the following system

$$\begin{cases} \mathbf{rQ} = \mathbf{0}, \\ \mathbf{re} = 1, \end{cases} \tag{1}$$

where $\mathbf{e} = \{1, 1, \ldots, 1\}^T$ and $\mathbf{0} = \{0, 0, \ldots, 0\}$.

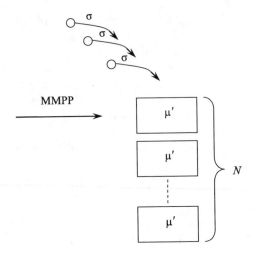

Fig. 1. Retrial queueing system $MMPP/M/N$

Obviously, that the fundamental rate of the arrival process is $\lambda = \mathbf{r} \cdot \mathbf{\Lambda} \cdot \mathbf{e}$.

Let process $i(t)$ define the number of calls in the orbit and $k(t)$ define the service unit state in the following way

$$
k(t) = \begin{cases} 0, & \text{if all servers are free,} \\ 1, & \text{if one server is busy,} \\ ..., & \\ N, & \text{if all servers are busy.} \end{cases}
$$

The aim of the study is to obtain the stationary probability distribution of the number of calls in the orbit.

Because of process $i(t)$ is not Markovian, we consider the multi-dimensional process $\{k(t), n(t), i(t)\}$, which is a continuous time Markov chain.

Denote $P(k, n, i, t) = P\{k(t) = k, n(t) = n, i(t) = i\}$. The system of Kolmogorov equations is written for $i > 0$, $n = \overline{1, W}$ as follows:

$$\begin{cases} \dfrac{\partial P(0,n,i,t)}{\partial t} = -(\lambda_n + i\sigma - q_{nn})P(0,n,i,t) + \mu'P(1,n,i,t) + \sum_{v \neq n} P(0,v,i,t)q_{vn}, \\[2mm] \dfrac{\partial P(k,n,i,t)}{\partial t} = -(\lambda_n + k\mu' + i\sigma - q_{nn})P(k,n,i,t) + \lambda_n P(k,n,i-1,t) \\[1mm] + \lambda_n P(k-1,n,i,t) + (i+1)\sigma P(k-1,n,i+1,t) \\[1mm] + (k+1)\mu'P(k+1,n,i,t) + \sum_{v \neq n} P(k,v,i,t)q_{vn} \quad \text{for } 1 \leq k \leq N-1, \\[2mm] \dfrac{\partial P(N,n,i,t)}{\partial t} = -(\lambda_n + N\mu' - q_{nn})P(N,n,i,t) + \lambda_n P(N,n,i-1,t) \\[1mm] + \lambda_n P(N-1,n,i,t) + (i+1)\sigma P(N-1,n,i+1,t) + \sum_{v \neq n} P(N,v,i,t)q_{vn}. \end{cases}$$

$$(2)$$

In steady-state regime, we have

$$\begin{cases} -(\lambda_n + i\sigma - q_{nn})P(0,n,i) + \mu'P(1,n,i) + \sum_{v \neq n} P(0,v,i)q_{vn} = 0, \\[2mm] -(\lambda_n + k\mu' + i\sigma - q_{nn})P(k,n,i) + \lambda_n P(k,n,i-1) \\[1mm] + \lambda_n P(k-1,n,i) + (i+1)\sigma P(k-1,n,i+1) \\[1mm] + (k+1)\mu'P(k+1,n,i) + \sum_{v \neq n} P(k,v,i)q_{vn} = 0 \quad \text{for } 1 \leq k \leq N-1, \\[2mm] -(\lambda_n + N\mu' - q_{nn})P(N,n,i) + \lambda_n P(N,n,i-1) \\[1mm] + \lambda_n P(N-1,n,i) + (i+1)\sigma P(N-1,n,i+1) + \sum_{v \neq n} P(N,v,i)q_{vn} = 0, \end{cases}$$

$$(3)$$

where $P(k,n,i) = \lim_{t \to \infty} P(k,n,i,t)$.

Let us introduce row-vectors $\mathbf{P}_k(i) = \{P(k,1,i), P(k,2,i), \ldots, P(k,W,i)\}$. Then System (3) can be written in matrix form as follows:

$$\begin{cases} -\mathbf{P}_0(i)(\mathbf{\Lambda} + i\sigma\mathbf{I} - \mathbf{Q}) + \mu'\mathbf{P}_1(i) = \mathbf{0}, \\[1mm] -\mathbf{P}_k(i)(\mathbf{\Lambda} + k\mu'\mathbf{I} + i\sigma\mathbf{I} - \mathbf{Q}) + \mathbf{P}_{k-1}(i)\mathbf{\Lambda} \\[1mm] + \sigma(i+1)\mathbf{P}_{k-1}(i+1) + (k+1)\mu'\mathbf{P}_{k+1}(i) = \mathbf{0} \quad \text{for } 1 \leq k \leq N-1, \quad (4) \\[1mm] -\mathbf{P}_N(i)(\mathbf{\Lambda} + N\mu'\mathbf{I} - \mathbf{Q}) + \mathbf{P}_{N-1}(i)\mathbf{\Lambda} \\[1mm] + \sigma(i+1)\mathbf{P}_{N-1}(i+1) + \mathbf{P}_N(i-1)\mathbf{\Lambda} = \mathbf{0}. \end{cases}$$

where $\mathbf{I}$ is the identity matrix.

Denoting partial characteristic functions by $\mathbf{H}_k(u) = \sum_i e^{jui}\mathbf{P}_k(i)$, where $k = 0, 1, \ldots, N$ and $j = \sqrt{-1}$, System (4) is rewritten as follows

$$\begin{cases} \mathbf{H}_0(u)(\mathbf{Q} - \mathbf{\Lambda}) + j\sigma\mathbf{H}_0'(u) + \mu'\mathbf{H}_1(u) = \mathbf{0}, \\[1mm] \mathbf{H}_k(u)(\mathbf{Q} - \mathbf{\Lambda} - k\mu'\mathbf{I}) + j\sigma\mathbf{H}_0'(u) + \mathbf{H}_{k-1}(u)\mathbf{\Lambda} \\[1mm] -j\sigma e^{-ju}\mathbf{H}_{k-1}'(u) + (k+1)\mu'\mathbf{H}_{k+1}(u) = \mathbf{0} \quad \text{for } 1 \leq k \leq N-1, \quad (5) \\[1mm] \mathbf{H}_N(u)(\mathbf{Q} - \mathbf{\Lambda} - N\mu'\mathbf{I}) + \mathbf{H}_{N-1}(u)\mathbf{\Lambda} + \mathbf{H}_N(u)\mathbf{\Lambda}e^{ju} \\[1mm] -j\sigma e^{-ju}\mathbf{H}_{N-1}'(u) = \mathbf{0}, \end{cases}$$

System (5) can not be exactly solved. Thus, we propose the method of asymptotic analysis under the heavy load condition [19,20] for its solution.

3   Asymptotic Analysis Under Heavy Load Condition

Let us introduce load parameter $\rho = (\mathbf{r}\mathbf{\Lambda}\mathbf{e})/(N\mu')$. Denoting $\mu = \rho N \mu'$ (such as $\mu = \mathbf{r}\mathbf{\Lambda}\mathbf{e}$), System (5) is rewritten as

$$
\begin{cases}
\mathbf{H}_0(u)(\mathbf{Q} - \mathbf{\Lambda}) + j\sigma\mathbf{H}_0'(u) + \dfrac{\mu}{\rho N}\mathbf{H}_1(u) = \mathbf{0}, \\[2mm]
\mathbf{H}_k(u)\left(\mathbf{Q} - \mathbf{\Lambda} - \dfrac{k\mu}{\rho N}\mathbf{I}\right) + j\sigma\mathbf{H}_0'(u) + \mathbf{H}_{k-1}(u)\mathbf{\Lambda} \\[2mm]
\quad -j\sigma e^{-ju}\mathbf{H}_{k-1}'(u) + \dfrac{(k+1)\mu}{\rho N}\mathbf{H}_{k+1}(u) = \mathbf{0} \quad \text{for } 1 \le k \le N-1, \\[2mm]
\mathbf{H}_N(u)\left(\mathbf{Q} - \mathbf{\Lambda} - \dfrac{N\mu}{\rho N}\mathbf{I}\right) + \mathbf{H}_{N-1}(u)\mathbf{\Lambda} + \mathbf{H}_N(u)\mathbf{\Lambda}e^{ju} \\[2mm]
\quad -j\sigma e^{-ju}\mathbf{H}_{N-1}'(u) = \mathbf{0},
\end{cases} \tag{6}
$$

Let us prove the following theorem.

Theorem 1. *The limit characteristic function $h(u)$ of the process of the number of calls in the orbit in the $MMPP/M/N$ retrial queueing system in the steady-state regime under the heavy load condition has the gamma distribution form*

$$
h(u) = \lim_{\rho \to 1} E\left\{e^{jw(1-\rho)i(t)}\right\} = \left(1 - \frac{ju}{(1-\rho)\beta}\right)^{-\gamma},
$$

with parameters

$$
\beta = \frac{\mu}{\mathbf{v}\mathbf{\Lambda}\mathbf{e} + \mu}, \quad \gamma = 1 + \frac{\mu}{N\sigma}\beta, \tag{7}
$$

where vector $\mathbf{v}$ is a solution of the following system

$$
\begin{cases}
\mathbf{v}\mathbf{Q} = \mathbf{r}(\mu\mathbf{I} - \mathbf{\Lambda}), \\
\mathbf{v}\mathbf{e} = 0.
\end{cases}
$$

Proof. The proof consists of two parts: deriving of asymptotic equations and its solving.

Derivation of Asymptotic Equations
First of all, we introduce the notations:

$$
\varepsilon = 1 - \rho,\, u = \varepsilon w,
$$

$$
\mathbf{H}_0(u) = \varepsilon^N \mathbf{F}_0(w, \varepsilon),\, \mathbf{H}_1(u) = \varepsilon^{N-1}\mathbf{F}_1(w, \varepsilon), ..., \mathbf{H}_N(u) = \mathbf{F}_N(w, \varepsilon). \tag{8}
$$

The condition of heavy load is defined as $\rho \uparrow 1$ (or $\varepsilon \downarrow 0$).

System of Eqs. (6) can be rewritten in Notations (8) as follows

$$
\begin{cases}
\varepsilon^N \mathbf{F}_0(w,\varepsilon)(\mathbf{Q}-\boldsymbol{\Lambda}) + j\sigma\varepsilon^{N-1}\dfrac{\partial \mathbf{F}_0(w,\varepsilon)}{\partial w} + \dfrac{\mu}{(1-\varepsilon)N}\varepsilon^{N-1}\mathbf{F}_1(w,\varepsilon) = \mathbf{0}, \\[2mm]
\varepsilon^{N-k}\mathbf{F}_k(w,\varepsilon)\left(\mathbf{Q}-\boldsymbol{\Lambda}-\dfrac{k\mu}{(1-\varepsilon)N}\mathbf{I}\right) + j\sigma\varepsilon^{N-k-1}\dfrac{\partial \mathbf{F}_k(w,\varepsilon)}{\partial w} \\[2mm]
\quad + \varepsilon^{N-(k-1)}\mathbf{F}_{k-1}(w,\varepsilon)\boldsymbol{\Lambda} - j\sigma e^{-j\varepsilon w}\varepsilon^{N-(k-1)-1}\dfrac{\partial \mathbf{F}_{k-1}(w,\varepsilon)}{\partial w} \\[2mm]
\quad + \dfrac{(k+1)\mu}{(1-\varepsilon)N}\varepsilon^{N-(k+1)}\mathbf{F}_{k+1}(w,\varepsilon) = \mathbf{0} \quad \text{for } 1 \le k \le N-1, \\[2mm]
\mathbf{F}_N(w,\varepsilon)\left(\mathbf{Q}-\boldsymbol{\Lambda}-\dfrac{\mu}{(1-\varepsilon)}\mathbf{I}\right) + \varepsilon\mathbf{F}_{N-1}(w,\varepsilon)\boldsymbol{\Lambda} \\[2mm]
\quad + \mathbf{F}_N(w,\varepsilon)\boldsymbol{\Lambda}e^{j\varepsilon w} - j\sigma e^{-j\varepsilon w}\dfrac{\partial \mathbf{F}_{N-1}(w,\varepsilon)}{\partial w} = \mathbf{0},
\end{cases}
$$

After some transformations, we obtain

$$
\begin{cases}
\varepsilon(1-\varepsilon)\mathbf{F}_0(w,\varepsilon)(\mathbf{Q}-\boldsymbol{\Lambda}) + j\sigma(1-\varepsilon)\dfrac{\partial \mathbf{F}_0(w,\varepsilon)}{\partial w} + \dfrac{\mu}{N}\mathbf{F}_1(w,\varepsilon) = \mathbf{0}, \\[2mm]
\varepsilon\mathbf{F}_k(w,\varepsilon)\left((\mathbf{Q}-\boldsymbol{\Lambda})(1-\varepsilon)-\dfrac{k\mu}{N}\mathbf{I}\right) + j\sigma(1-\varepsilon)\dfrac{\partial \mathbf{F}_k(w,\varepsilon)}{\partial w} \\[2mm]
\quad + \varepsilon^2(1-\varepsilon)\mathbf{F}_{k-1}(w,\varepsilon)\boldsymbol{\Lambda} - j\sigma e^{-j\varepsilon w}\varepsilon(1-\varepsilon)\dfrac{\partial \mathbf{F}_{k-1}(w,\varepsilon)}{\partial w} \\[2mm]
\quad + \dfrac{(k+1)\mu}{N}\mathbf{F}_{k+1}(w,\varepsilon) = \mathbf{0} \quad \text{for } 1 \le k \le N-1, \\[2mm]
\mathbf{F}_N(w,\varepsilon)\left((\mathbf{Q}-\boldsymbol{\Lambda})(1-\varepsilon)-\mu\mathbf{I}\right) + \varepsilon(1-\varepsilon)\mathbf{F}_{N-1}(w,\varepsilon)\boldsymbol{\Lambda} \\[2mm]
\quad + (1-\varepsilon)e^{j\varepsilon w}\mathbf{F}_N(w,\varepsilon)\boldsymbol{\Lambda} - j\sigma(1-\varepsilon)e^{-j\varepsilon w}\dfrac{\partial \mathbf{F}_{N-1}(w,\varepsilon)}{\partial w} = \mathbf{0},
\end{cases}
\tag{9}
$$

First of all, in System (9) we make limit $\varepsilon \to 0$.

$$
\begin{cases}
j\sigma\mathbf{F}_0'(w) + \dfrac{\mu}{N}\mathbf{F}_1(w) = \mathbf{0}, \\[2mm]
j\sigma\mathbf{F}_k'(w) + \dfrac{(k+1)\mu}{N}\mathbf{F}_{k+1}(w) = \mathbf{0}, \\[2mm]
\mathbf{F}_N(w)\left(\mathbf{Q}-\mu\mathbf{I}\right) - j\sigma\mathbf{F}_{N-1}'(w) = \mathbf{0},
\end{cases}
\tag{10}
$$

where $\mathbf{F}_k(w) = \lim\limits_{\varepsilon \to 0} \mathbf{F}_k(w,\varepsilon)$.

Let us consider expansions of functions $\mathbf{F}_k(w,\varepsilon)$ in the form

$$
\mathbf{F}_k(w,\varepsilon) = \mathbf{F}_k(w) + \varepsilon\mathbf{f}_k(w) + \mathbf{O}(\varepsilon^2),
\tag{11}
$$

where $\mathbf{O}(\varepsilon^2)$ is an infinitesimal value of order ε^2.

Substituting Expansions (11) into System (6) and making some transformations, we obtain the following system of equations in limit $\varepsilon \to 0$

$$
\begin{cases}
\mathbf{F}_0(w)(\mathbf{Q}-\boldsymbol{\Lambda}) - j\sigma\mathbf{F}_0'(w) + j\sigma\mathbf{f}_0'(w) + \dfrac{\mu}{N}\mathbf{f}_1(w) = \mathbf{0}, \\[2mm]
\mathbf{F}_k(w)\left(\mathbf{Q}-\boldsymbol{\Lambda}-\dfrac{k\mu}{N}\mathbf{I}\right) - j\sigma\mathbf{F}_k'(w) \\[2mm]
\quad + j\sigma\mathbf{f}_k'(w) - j\sigma\mathbf{F}_{k-1}'(w) + \dfrac{(k+1)\mu}{N}\mathbf{f}_{k+1}(w) = \mathbf{0} \quad \text{for } 1 \le k \le N-1, \\[2mm]
-\mathbf{F}_N(w)\mathbf{Q} + \mathbf{f}_N(w)\left(\mathbf{Q}-\mu\mathbf{I}\right) + \mathbf{F}_{N-1}(w)\boldsymbol{\Lambda} \\[2mm]
\quad + jw\mathbf{F}_N(w)\boldsymbol{\Lambda} + j\sigma(1+jw)\mathbf{F}_{N-1}'(w) - j\sigma\mathbf{f}_{N-1}'(w) = \mathbf{0}.
\end{cases}
\tag{12}
$$

In addition, we sum up all equations of System (6) and multiply the result by vector **e**.

$$\mathbf{F}_N(w,\varepsilon)e^{jw\varepsilon}\boldsymbol{\Lambda}\mathbf{e} + j\sigma \sum_{k=0}^{N-1} \varepsilon^{N-k-1}\frac{\partial \mathbf{F}_k(w,\varepsilon)}{\partial w}\mathbf{e} = 0.$$

Substituting Expansions (11) and writing equalities for members with equal powers of ε, we obtain two additional scalar equations

$$\begin{cases} \mathbf{F}_N(w)\boldsymbol{\Lambda}\mathbf{e} + j\sigma\mathbf{F}'_{N-1}(w)\mathbf{e} = 0, \\ jw\mathbf{F}_N(w)\boldsymbol{\Lambda}\mathbf{e} + \mathbf{f}_N(w)\boldsymbol{\Lambda}\mathbf{e} + j\sigma\mathbf{F}'_{N-2}(w)\mathbf{e} + j\sigma\mathbf{f}'_{N-1}(w)\mathbf{e} = 0. \end{cases} \tag{13}$$

Thus, we have System (10), (12), (13) of $2(N+1)$ matrix and two scalar differential equations.

Analysis of the Equations

The partial characteristic function of the number of calls in the orbit is calculated as follows

$$H(u) = \mathrm{E}\left\{e^{jui(t)}\right\} = \sum_{k=0}^{N} \mathbf{H}_k(u)\mathbf{e}.$$

Under the heavy load condition, the asymptotic characteristic function $h(u)$ can be written as

$$h(u) = \lim_{\rho \to 1} \mathrm{E}\left\{e^{jw(1-\rho)i(t)}\right\} = \mathbf{F}_N\left(\frac{u}{1-\rho}\right)\mathbf{e} + O(\varepsilon). \tag{14}$$

Therefore, it is necessary to find only scalar function $\mathbf{F}_N(w)\mathbf{e}$ from Equations (10), (12), (13). We make it in three steps.

Step 1. By using Equations (10), we obtain that

$$-j\sigma\mathbf{F}'_k(w) = \frac{(k+1)\mu}{N}\mathbf{F}_{k+1}(w) \quad \text{for } k < N. \tag{15}$$

Comparing the equation for $k = N$ in (10) and the equation for $k = N - 1$ of (15), we get

$$\mathbf{F}_N(w)\mathbf{Q} = \mathbf{0}.$$

Taking into account (1), function $\mathbf{F}_N(w)$ can be written as the following product:

$$\mathbf{F}_N(w) = \mathbf{r} \cdot \Phi(w), \tag{16}$$

where $\Phi(w)$ is an unknown scalar function.

Step 2. From Eqs. (12) and Equalities (15), it can be written that

$$
\begin{cases}
j\sigma\mathbf{f}_0'(w) = -\mathbf{F}_0(w)(\mathbf{Q}-\mathbf{\Lambda}) - \dfrac{\mu}{N}\mathbf{F}_1(w) - \dfrac{\mu}{N}\mathbf{f}_1(w), \\[2mm]
j\sigma\mathbf{f}_k'(w) = -\mathbf{F}_k(w)\left(\mathbf{Q}-\mathbf{\Lambda}-\dfrac{k\mu}{N}\mathbf{I}\right) \\[1mm]
\quad -\dfrac{(k+1)\mu}{N}\mathbf{F}_{k+1}(w) - \dfrac{k\mu}{N}\mathbf{F}_k(w) - \dfrac{(k+1)\mu}{N}\mathbf{f}_{k+1}(w) \text{ for } 1 \le k \le N-2, \\[2mm]
j\sigma\mathbf{f}_{N-1}'(w) = -\mathbf{F}_{N-1}(w)\left(\mathbf{Q}-\mathbf{\Lambda}-\dfrac{N-1\mu}{N}\mathbf{I}\right) \\[1mm]
\quad -\dfrac{(N)\mu}{N}\mathbf{F}_N(w) - \dfrac{N-1\mu}{N}\mathbf{F}_{(N-1)}(w) - \dfrac{(N)\mu}{N}\mathbf{f}_N(w), \\[2mm]
j\sigma\mathbf{f}_{N-1}'(w) = -\mathbf{f}_N(w)(\mathbf{Q}-\mu\mathbf{I}) + \mathbf{F}_{N-1}(w)\mathbf{\Lambda} + jw\mathbf{F}_N(w)\mathbf{\Lambda} - (1+jw)\mu\mathbf{F}_N(w).
\end{cases}
$$
$$\tag{17}$$

Subtracting the two last equations of System (17), we obtain

$$
(\mathbf{F}_{N-1}(w) + \mathbf{f}_N(w))\mathbf{Q} = \mathbf{F}_N(w)\mathbf{Q} + jw\mathbf{F}_N(w)(\mathbf{\Lambda}-\mu\mathbf{I}).
$$

Substituting Formula (16), we have the following equation

$$
(\mathbf{F}_{N-1}(w) + \mathbf{f}_N(w))\mathbf{Q} = -jw\Phi(w)\mathbf{r}(\mathbf{\Lambda}-\mu\mathbf{I}).
\tag{18}
$$

Let us introduce the following notation:

$$
\mathbf{F}_{N-1}(w) + \mathbf{f}_N(w) = -jw\Phi(w)\mathbf{v},
\tag{19}
$$

where vector $\mathbf{v}$ is a solution of the equation

$$
\mathbf{v}\mathbf{Q} = \mathbf{r}(\mu\mathbf{I}-\mathbf{\Lambda}).
\tag{20}
$$

For Eq. (20) solution existence, it is necessary that ranks of the system matrix and augmented one will be equal. Because $\mathbf{r}(\mu\mathbf{I}-\mathbf{\Lambda})\mathbf{e} = 0$, that it is true.

Matrix Eq. (20) has infinitely many solutions. We can present the general solution as follows

$$
\mathbf{v} = C\mathbf{r} + \mathbf{v}_0,
$$

where $C = const$ and $\mathbf{v}_0$ is a particular solution, for example, $\mathbf{v}_0\mathbf{e} = 0$.

Step 3. Substituting (10), (15), (17) into the last equation of System (13), we obtain the following equation:

$$
2jw\mathbf{F}_N(w)\mathbf{\Lambda}\mathbf{e} + \mathbf{f}_N(w)\mathbf{\Lambda}\mathbf{e} - \dfrac{(N-1)\mu}{N}\mathbf{F}_{N-1}(w)\mathbf{e} \\
-\mu\mathbf{f}_N(w)\mathbf{e} + \mathbf{F}_{N-1}(w)\mathbf{\Lambda}\mathbf{e} - (1+jw)\mu\mathbf{F}_N(w)\mathbf{e} = 0.
$$

Taking into account Equality (19), we have

$$
jw\Phi(w)(2\mathbf{r}\mathbf{\Lambda}\mathbf{e} + \mathbf{v}(\mathbf{\Lambda}\mathbf{e}-\mu\mathbf{e}) - \mu) - \mu\Phi(w) + \dfrac{\mu}{N}\mathbf{F}_{N-1}(w)\mathbf{e} = 0.
\tag{21}
$$

The we differentiate this equation. Taking into account (15), we obtain the following differential equation

$$
j\Phi(w)\left(\mathbf{v}(\mathbf{\Lambda}\mathbf{e}-\mu\mathbf{e}) + \mu + \dfrac{\mu^2}{N\sigma}\right) - \Phi'(w)(\mu - jw(\mathbf{v}(\mathbf{\Lambda}\mathbf{e}-\mu\mathbf{e}) + \mu)) = 0.
$$
$$\tag{22}$$

Let us divide (22) by $(\mathbf{v}\Lambda\mathbf{e} - \mu\mathbf{v}\mathbf{e} + \mu)$ and introduce denotations

$$\beta = \frac{\mu}{\mathbf{v}\Lambda\mathbf{e} - \mu\mathbf{v}\mathbf{e} + \mu}, \quad \gamma = 1 + \frac{\mu}{N\sigma}\beta.$$

Thus Eq. (22) is rewritten as

$$\Phi'(w)(\beta - jw) = j\gamma\Phi(w).$$

Clearly, the solution of this equation has the form

$$\Phi(w) = C_0\left(1 - \frac{jw}{\beta}\right)^{-\gamma}.$$

From formula (16), we obtain

$$\mathbf{F}_N(w) = \mathbf{r} \cdot C_0\left(1 - \frac{jw}{\beta}\right)^{-\gamma}.$$

Taking into account $\mathbf{v} = C\mathbf{r} + \mathbf{v}_0$, it is easy to show that the parameters β and γ do not depend on C. Choosing a solution $\mathbf{v}_0$ such as

$$\begin{cases} \mathbf{v}_0 Q = \mathbf{r}(\mu\mathbf{I} - \Lambda), \\ \mathbf{v}_0\mathbf{e} = 0, \end{cases}$$

we can write that

$$\beta = \frac{\mu}{\mathbf{v}_0\Lambda\mathbf{e} + \mu}, \quad \gamma = 1 + \frac{\mu}{N\sigma}\beta,$$

Returning to characteristic function (14), we can write that

$$h(u) = C_0\left(1 - \frac{ju}{\beta}\right)^{-\gamma},$$

where $C_0 = 1$ due to the normalisation requirement.

Thus, we have proved that the asymptotic characteristic function of the probability distribution of the number of calls in the orbit under the heavy load condition has the gamma distribution form.

4   Numerical Analysis

In this section, we present some numerical examples and make conclusions about the asymptotic method applicability area. First of all, we denote the probability distribution function of the gamma distribution with parameters (7) as $\Gamma(x)$. We will calculate of the discrete probability distribution of the number of calls in the orbit $p(i)$ as follows

$$p(i) = \Gamma(i+1) - \Gamma(i).$$

Further, we present the comparison of asymptotic and simulated distributions for different values of the retrial queuing system parameters.

In the first example, let the retrial queue have three server ($N = 3$), and the arrival MMPP have three states and be defined by following matrices

$$\Lambda = \begin{bmatrix} 1 & 0 & 0 \\ 0 & 2 & 0 \\ 0 & 0 & 3 \end{bmatrix}, \qquad Q = \begin{bmatrix} -0.5 & 0.2 & 0.3 \\ 0.1 & -0.3 & 0.2 \\ 0.3 & 0.6 & -0.9 \end{bmatrix}.$$

The retrial rate is $\sigma = 1$, the service rate equals $\mu = \dfrac{r\Lambda e}{N\rho}$, then the load parameter ρ has values $0 < \rho < 1$.

In Fig. 2, the comparison of the asymptotic and simulated distributions is presented for $\rho = 0.90$ and $\rho = 0.95$, where dashed lines are the asymptotic distributions and solid lines are simulated ones.

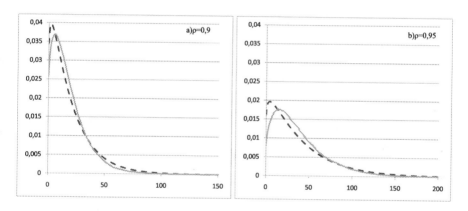

Fig. 2. Comparison of the asymptotic and the simulated distributions for $MMPP/M/3$ with a) $\rho = 0.90$ and b) $\rho = 0.95$

In the second example, let us consider a particular cases of the retrial queue - the single-server retrial queue with following values of parameters

$$N = 1, \qquad \sigma = 1, \qquad \mu = \frac{r\Lambda e}{\rho},$$

the comparison of the asymptotic and simulated distributions is presented in Fig. 3 and 4.

Also we demonstrate a numerical example for multi-server retrial queue with Poisson arrival process (Fig. 4), where $\lambda = 1, N = 10, \sigma = 1, \mu = \dfrac{\lambda}{N\rho}$ (Fig. 5).

In this example, the main difference between asymptotic and the simulation distributions is in point $i = 0$.

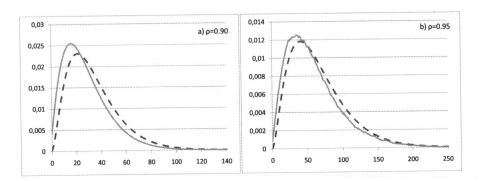

Fig. 3. Comparison of the asymptotic and the simulated distributions for the single-server retrial queue with a) $\rho = 0.90$ and b) $\rho = 0.95$

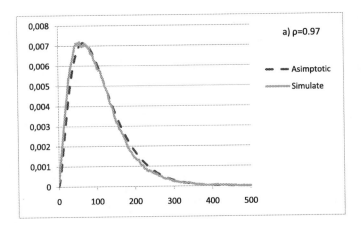

Fig. 4. Comparison of the asymptotic and the simulated distributions for the single-server retrial queue $\rho = 0.97$

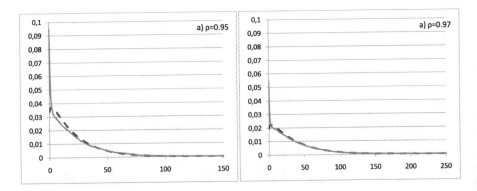

Fig. 5. Comparison of the asymptotic and the simulated distributions for Poisson arrival process with a) $\rho = 0.95$ and b) $\rho = 0.97$

For the method accuracy estimation, we use Kolmogorov distance between respective distribution functions:

$$d = \max_{i \geq 0} \left| \sum_{l=0}^{i} [\tilde{p}(l) - p(l)] \right|,$$

where $p(l)$ is an asymptotic probability distribution and $\tilde{p}(l)$ is a simulated one. In Table 1, there are values of the Kolmogorov distance for all presented numerical examples.

Table 1. Kolmogorov distances d for various values of the parameter ρ

	$N = 1$	$N = 3$	$N = 10, \lambda = 1$
$\rho = 0.90$	0.070	0.070	0.068
$\rho = 0.95$	0.043	0.043	0.040
$\rho = 0.97$	0.036	0.038	0.035

Note, we have obtained the same results of the numerical analysis for different arrivals and number of servers. For our purpose, the asymptotic analysis method under the heavy load condition can be applied for $\rho \geq 0.95$, where the Kolmogorov distance between asymptotic and the simulation distributions $d \leq 0.05$.

5   Conclusions

In the paper, the multi-server retrial queueing system with MMPP arrivals has been studied by the asymptotic analysis method under the heavy load condition. We have proved that the asymptotic characteristic function of the number of calls in the orbit has the gamma distribution form, as for single-server retrial queue. In this way, we generalize our results for more complex model. By means of the numerical analysis, we have shown a good accuracy of the proposed approximation in the applicability area $\rho \geq 0.95$.

References

1. Choi, B.D., Chang, Y., Kim, B.: MAP 1, MAP2 /M/c retrial queue with guard channels and its application to cellular networks. Top **7**, 231–248 (1999). https://doi.org/10.1007/BF02564724
2. Zhu, D.B., Choi, B.D.: Performance analysis of CSMA in an unslotted cognitive radio network with licensed channels and unlicensed channels. J. Wirel. Com. Netw. **2012**, 12 (2012). https://doi.org/10.1186/1687-1499-2012-12
3. Phung-Duc, T., Kawanishi, K.: Performance analysis of call centers with abandonment, retrial and after-call work. Perf. Eval. **80**, 43–62 (2014). https://doi.org/10.1016/j.peva.2014.03.001

4. Phung-Duc, T., Kawanishi, K.: Multiserver retrial queue with setup time and its application to data centers. J. Ind. Manag. Optim. **15**(1), 15–35 (2019). https://doi.org/10.3934/jimo.2018030

5. Dudin, A.N., Lee, M.H., Dudina, O., Lee, S.K.: Analysis of priority retrial queue with many types of customers and servers reservation as a model of cognitive radio system. IEEE Trans. Commun. **65**(1), 186–199 (2017). https://doi.org/10.1109/TCOMM.2016.2606379

6. Artalejo, J.R., Gómez-Corral, A.: Retrial Queueing Systems: A Computational Approach. Springer, Stockholm (2008). https://doi.org/10.1007/978-3-540-78725-9

7. Falin, G.I., Templeton, J.G.C.: Retrial Queues. Chapman & Hall, London (1997)

8. Artalejo, J.R., Pozo, M.: Numerical calculation of the stationary distribution of the main multiserver retrial queue. Ann. Oper. Res. **116**, 41–56 (2002). https://doi.org/10.1023/A:1021359709489

9. Neuts, M.F., Rao, B.M.: Numerical investigation of a multiserver retrial model. Queue. Syst. **7**(2), 169–189 (1990). https://doi.org/10.1007/BF01158473

10. Artalejo, J.R., Chakravarthy, S.R.: Algorithmic analysis of the MAP/PH/1 retrial queue. TOP **14**, 293–332 (2006). https://doi.org/10.1007/BF02837565

11. Chakravarthy, S.R.: Busy period analysis of multi-server retrial queueing systems. In: Joshua, V.C., Varadhan, S.R.S., Vishnevsky, V.M. (eds.) Applied Probability and Stochastic Processes. ISFS, pp. 61–76. Springer, Singapore (2020). https://doi.org/10.1007/978-981-15-5951-8_5

12. Dudin, A.N., Klimenok, V.I., Vishnevsky, V.M.: The Theory of Queuing Systems with Correlated Flows. Springer, Cham (2020). https://doi.org/10.1007/978-3-030-32072-0

13. Gómez-Corral, A.G.: A bibliographical guide to the analysis of retrial queues through matrix analytic techniques. Ann. Oper. Res. **141**, 163–191 (2006). https://doi.org/10.1007/s10479-006-5298-4

14. Breuer, L., Dudin, A.N., Klimenok, V.I.: A retrial BMAP/PN/N system. Queue. Syst. **40**, 433–457 (2002). https://doi.org/10.1023/A:1015041602946

15. Nazarov, A., Moiseev, A., Phung-Duc, T., Paul, S.: Diffusion limit of multi-server retrial queue with setup time. Mathematics **8**, 2232 (2020). https://doi.org/10.3390/math8122232

16. Nazarov, A., Phung-Duc, T., Paul, S., Lizyura, O.: Diffusion approximation for multiserver retrial queue with two-way communication. In: Vishnevskiy, V.M., Samouylov, K.E., Kozyrev, D.V. (eds.) DCCN 2020. LNCS, vol. 12563, pp. 567–578. Springer, Cham (2020). https://doi.org/10.1007/978-3-030-66471-8_43

17. Danilyuk, E.Y., Fedorova, E.A., Moiseeva, S.P.: Asymptotic analysis of an retrial queueing system M—M—1 with collisions and impatient calls. Autom. Rem. Control **79**(12), 2136–2146 (2018). https://doi.org/10.1134/S0005117918120044

18. Danilyuk, E., Vygoskaya, O., Moiseeva, S.: Retrial queue M/M/N with impatient customer in the orbit. In: Vishnevskiy, V.M., Kozyrev, D.V. (eds.) DCCN 2018. CCIS, vol. 919, pp. 493–504. Springer, Cham (2018). https://doi.org/10.1007/978-3-319-99447-5_42

19. Fedorova, E.: The second order asymptotic analysis under heavy load condition for retrial queueing system MMPP/M/1. In: Dudin, A., Nazarov, A., Yakupov, R. (eds.) ITMM 2015. CCIS, vol. 564, pp. 344–357. Springer, Cham (2015). https://doi.org/10.1007/978-3-319-25861-4_29

20. Fedorova, E.A., Nazarov, A.A., Farkhadov, M.P. Asymptotic analysis of the $MMPP/M/1$ retrial queue with negative calls under the heavy load condition. Izvestiya of Saratov Univ. Math. Mech. Inf. **20**(4), 534–547 (2020).https://doi.org/10.18500/1816-9791-2020-20-4-534-547

21. Dudin, A., Deepak, T.G., Joshua, V.C., Krishnamoorthy, A., Vishnevsky, V.: On a $BMAP/G/1$ retrial system with two types of search of customers from the orbit. In: Dudin, A., Nazarov, A., Kirpichnikov, A. (eds.) ITMM 2017. CCIS, vol. 800, pp. 1–12. Springer, Cham (2017). https://doi.org/10.1007/978-3-319-68069-9_1
22. Neuts, M.F.: A Versatile Markovian point process. J. Appl. Prob. **16**(4), 764–779 (1979). https://doi.org/10.2307/3213143
23. Lucantoni, D.M.: New results on the single server queue with a batch Markovian arrival process. Stochastic Models **7**, 1–46 (1991). https://doi.org/10.1080/15326349108807174

On the Application of Queuing Theory in the Analysis of Transients in the Operation of a Freight Railway Station

Maxim Zharkov$^{(\boxtimes)}$, Alexander Kazakov , and Anna Lempert

Matrosov Institute for System Dynamics and Control Theory of Siberian Branch of Russian Academy of Sciences (IDSTU SB RAS), Irkutsk, Russia
zharkm@mail.ru
http://idstu.irk.ru

Abstract. The paper continues our research carried out earlier, in which we applied the queuing theory for modeling the operation of railway stations. The mathematical model is constructed as a multiphase queuing system with Batch Markovian Arrival Process. Its stationary characteristics are determined numerically. It is known that there exist cases when stationary characteristics are not enough for a profound study of the operation of those technical systems. Therefore, it is required to consider the properties of the transient processes that occur in railway stations. In the paper, we apply the proposed approach to construct the model of the operation of the typical freight railway station. We compose and study the Kolmogorov ordinary differential equations system that describes the dependence of the probabilities of system states on time. Its solution allows us to determine the transients' behavior and convergence rate to the stationary mode.

Keywords: Queuing system · $BMAP$ · Kolmogorov equations · Freight railway station · Transient process

1 Introduction

Queuing theory is an effective tool in studying technical device operations in fields of information and telecommunication systems [1–4], mass manufacture [5, 6], and trade [6]. The distinctive features of models based on queuing theory are the random nature of the request arrivals and their non-deterministic processing. These features make it possible to apply the developed mathematical apparatus in the field of transport, in particular, railway. Here, queuing theory is one of the relevant scientific directions. It allows evaluating the efficiency, stability, and reliability of railway stations, taking into account the influence of random factors [7].

The study was funded by the Ministry of Science and Education of the Russian Federation in the framework of the basic part, project No 121041300065-9, and by RFBR, project No 20-010-00724.

Models based on queuing theory have been used in the field of railway transport since the 70s. As a rule, single-phase Markov and semi-Markov queuing systems describe the operation of railway stations and other elements of railway infrastructure [8,9]. In particular, the researchers use a queuing system (QS) with batch arrival of requests or service [10]. Today, we need more complex and accurate models due to the development and complexity of transport systems. In paper [11], the authors use queuing systems with a loss and a finite queue to describe the process of the train disbanding by a hump. The paper [12] presents the stochastic model that allows determining train delays on railway sections without regard to the schedule. The model is based on a special type semi-Markov QS. In [13,14], the reliability models of railway stations in which the arrival, service, and repair times have a phase-type distribution are presented. The researchers also use queuing theory to model larger systems that are railway sections [15–17]. They use queueing networks with infinite queues.

However, all the models have shortcomings associated either with a coarsening of the system structure or an insufficiently adequate and accurate description of the incoming traffic flow. Previously, we proposed the approach for modeling freight and marshaling railway stations (RS) to overcome the shortcomings [18–21]. It uses of queueing networks [19,20] and multiphase queuing systems [18,21] with batch service of requests and final queues. The distinctive feature of the approach is the use of the Batch Markovian Arrival Process $(BMAP)$, which is usually applied to describe data flow in telecommunication systems [4,22]. This allows us to describe a few traffic flows with different characteristics as an integral structure in detail.

We studied the operation of the RS in the stationary mode in the previous papers [18,21]. However, RS are dynamic systems, and their parameters change over time. In particular, incoming traffic flows have daily or seasonal deviations. As a result, transients occur in the operation of RS. In this case, we need to determine the following properties of the processes. The first property is the duration of transients. That is, how long will it last before the system reaches a stationary mode? The second property is the value of the deviation of a system's performance measures in a transient mode from similar parameters in a stationary mode. In this paper, we design the model of the freight railway operation in the form of the three-phase QS and find the probabilities of system states as a function of time. Then, based on results obtained, we draw conclusions about the properties of transients in the considered railway system.

2   The Object of the Study

Freight railroads carry out loading, unloading, and shunting work to bring cars to the cargo fronts and withdrawal them. The typical scheme of the freight RS includes a receiving and departure yard (RDY), a sorting bowl (SB), a cargo yard (CY), and a departure yard (DY). RDY can be separated into two subsystems at large stations. At the same time, DY can not exist on a station. The neighboring RS can perform functions of DY if it is required.

Trains arrive at RS from two or more directions and can be accepted only at an RDY or a receiving yard. Then the trains are served at the SB and transferred to the CY for loading or unloading. After that, cars return to the RDY for departure from the station. The subsystems have different parameters, particularly the capacity of the cars, the number of service devices, and the type of operations performed. In addition, subsystems have a batch service of cars. Batch sizes are determined by the type of operations performed.

In the RDY the main device is a hump; in the SB are diesel locomotives, which rearrange the cars to the next subsystem; in the CY are loading or unloading fronts. Incoming train traffic and the operation of subsystems are affected by many random factors, as a rule, negative. For example, they can be equipment breakdowns, personnel errors, and weather conditions. This leads to the operation rhythm of a station being a disruption. Therefore, we need to use probabilistic models.

3   Mathematical Model of the Transients at a Freight Railway Station

The study of transients at a freight railway station includes four stages:

1. designing the model in the form of a QS;
2. composing the ordinary differential equations system (Kolmogorov equations);
3. performing an numerical analysis of the Kolmogorov equations;
4. interpretation of the results obtained in terms of the object studied.

Stage 1. According to the proposed approach [19–21], we use models of different types to describe the flow of the arriving trains and the process of station operation. We assume that a car is a request and a train is a batch. We described an arrival flow by $BMAP$ [4,22].

$BMAP$ is a generalization of the batch Poisson process, allowing the change in the intensity of the arrival of request batches. The intensity of arrival request batches λ_v depends on the state number of the Markov chain v_t with continuous time and state space $\{0, 1, \ldots, W\}$. The residence time in each state is exponentially distributed with parameter λ_v. With probability $p_k(v, v')$ the chain can go to state v'. This generates a batch of random size $k \geq 0$. The normalization condition is satisfied:

$$\sum_{k=0}^{\infty} \sum_{v=0}^{W} p_k(v, v') = 1.$$

The transitions intensities are written in matrix form

$$\begin{aligned}
(D_0)_{v,v} &= -\lambda_v, & v &= \overline{0, W}, \\
(D_0)_{v,v'} &= \lambda_v p_0(v, v'), & v, v' &= \overline{0, W}, \\
(D_k)_{v,v'} &= \lambda_v p_k(v, v'), & v, v' &= \overline{0, W}, k \geq 1.
\end{aligned}$$

We use a multiphase QS for modeling car service at the railway station. It allows us to take into account the structural features of the object considered. A separate QS describes each subsystem and form a phase. The number of service devices determines the number of channels, and the queue length is the total capacity of the cars in the subsystem. Each QS serves requests in batches. Next, it is assumed that at each phase, requests are selected from the queue according to FIFO (first in, first out). We use channel locks to avoid the loss of requests between phases.

We consider only one direction of cars moving: form the RDY to the SB, then to the CY. Such the direction includes the main operations with trains: receiving, processing in the SB, and loading or unloading. The operations for the departure of trains are performed much faster and do not affect the operations of other subsystems. Thus, in terms of queuing theory, the model of a freight railway station is a three-phase QS $BMAP/M^{X1}/n_1/m_1 \rightarrow */M^{X2}/n_2/m_2 \rightarrow */M^{X3}/n_3/m_3$, where Xi is the distribution of requests in batches which selecting for service in the channel in phase i.

Stage 2. To study transient processes in the QS, we consider the Kolmogorov equations [5,23]. The unknown functions are the probabilities of the states of the system $p_i(t), i = 0, ..., K$, where K is the maximum number of requests, and the independent variable is time. The initial data is the state of the system at zero time. Earlier, we constructed formulas that allow us to compose the Kolmogorov equations for a three-phase QS with $BMAP$ [21]. Note that it can be applied to a QS, in which the flows between phases are Poisson flows. Here we consider a special type of a QS. First, the intensity of incoming flow is lower than the intensity of service at all phases. It means that almost all requests will be served. Second, the time between the arrival of requests, the service time in phases, and the time of channels locking obeys an exponential distribution. In this case, we can simplify model and use formulas from [21] to obtain the Kolmogorov equations.

The number of channels in subsystems can be measured in tens and the queue length in hundreds of units. Therefore, Kolmogorov equations can consist of one or more million equations. To reduce the dimension of the Kolmogorov equations, we assume that the arriving train is one service request.

Stage 3. The Kolmogorov equations include more than a hundred equations even after simplifying of the model. Therefore, we use numerical methods to solve it. The most popular method is the Runge-Kutta method [24]. The solution to the Kolmogorov equations is the transition probabilities of states, that is, the probabilities of the system states depending on time. Base on these probabilities, we can determine the performance measures for the considered QS [1,5], that is the purpose of modeling.

Stage 4. We interpret the results obtained in terms of the object studied. The most important for railway stations is an assessment of the duration of such processes and their deviations from stationary parameters.

4   Study of a Freight Railway Station Operation

We consider the freight railway station Sukhovskaya, which is located on the East Siberian Railway (Russia). It was selected for the following reasons. First, Sukhovskaya is a standard station on the Russian Railway. Second, we managed to obtain statistics on arriving train traffic and information about its structure. Note that we have already considered this station [18,21]. However, firstly, we have not studied the properties of transient processes when the intensity of the incoming train traffic changes. Secondly, the data on the operation of the subsystems and incoming train traffic was updated at the end of 2021.

Figure 1 shows the station scheme, where the arrows indicate the directions of train movement. Sukhovskaya includes the receiving yard with the hump, the sorting bowl, the cargo yard, and the receiving and departure yard.

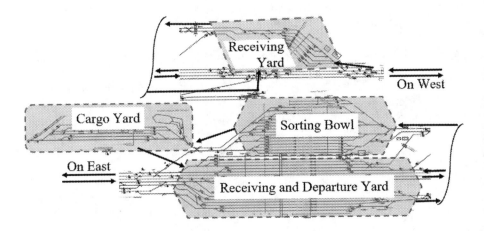

Fig. 1. Scheme of freight Sukhovskaya

The receiving yard has one hump and a capacity of 5 trains. The sorting bowl has two diesel locomotives and a capacity of 8 trains. The cargo yard has ten fronts, which can process three trains in parallel. In the cargo yard, we assume that ten service devices and their tracks is three channels. It allows us to fit the whole train into the channel. Therefore, we have three channels working in the cargo yard, and there is no queue. In terms of queuing theory, we have the three-phase QS $BMAP/M/1/5 \rightarrow */M/2/8 \rightarrow */M/3/0$. Figure 2 shows its scheme, where the arrows indicate the directions of requests movement, and dotted arrows are feedback.

We obtained the service time distributions from field observations. The data include statistics on 124 trains that arrived at the station in the period from 10 June 2021 to 4 July 2021. The service parameters of the channels of each phase are following: $\mu_1 = 0.74, \mu_2 = 0.222, \mu_3 = 0.083$ trains per hour, respectively. We do not consider the receiving and departure yard in the model.

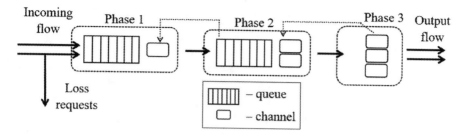

Fig. 2. Scheme of the three-phase QS

The station receives on average three trains per day from even and two trains from odd directions. The $BMAP$ matrices have the form

$$D_0 = \begin{pmatrix} -0.204 & 0 \\ 0 & -0.212 \end{pmatrix}, \ D_1 = \begin{pmatrix} 0.114 & 0.08 \\ 0.083 & 0.129 \end{pmatrix}, \ D_2 = \begin{pmatrix} 0.01 & 0 \\ 0 & 0 \end{pmatrix}. \quad (1)$$

The Kolmogorov system for the three-phase QS includes

$$(K1 + 1)(K2 + 2)(K3 + 3) - K2 = 408$$

equations. Here Ki is the maximum number of requests on phase i. We present only a part of the resulting system since it is cumbersome.

$$
\begin{cases}
P'_{000}(t) = D_0 P_{000}(t) + I\mu_3 P_{001}(t), \\
P'_{001}(t) = (D_0 - I\mu_3)P_{001}(t) + I\mu_2 P_{010}(t) + 2I\mu_3 P_{002}(t), \\
P'_{002}(t) = (D_0 - 2I\mu_1)P_{002}(t) + I\mu_2 P_{011}(t) + 3I\mu_3 P_{003}(t), \\
P'_{003}(t) = (D_0 - 3I\mu_1)P_{003}(t) + I\mu_2 P_{012}(t) + 3I\mu_3 P_{004}(t), \\
\quad \dots \\
P'_{ijz}(t) = \left(D_0 + \displaystyle\sum_{k=1}^{\min(1,\, 6-i)} D_k - I\mu_1 - \min(j,1)I\mu_2 - \min(z,1)I\mu_3 \right) P_{ijz}(t) + \\
\quad + \displaystyle\sum_{k=1}^{\min(1,\, i)} D_k P_{i-k,j,z}(t) + I\mu_1 P_{i+1,j-1,z}(t) + \min(j+1,1)I\mu_2 P_{i,j+1,z-1}(t) + \\
\quad + \min(z+1,1)I\mu_3 P_{i,j,z+1}(t), i = \overline{0,6}, j = \overline{0,11}, z = \overline{0,4}.
\end{cases}
$$
$$(2)$$

System (2) has a high dimension to be analytically analyzed.

5   Numerical Experiment

We use the fourth-order Runge-Kutta method with the step $h = 0.05$ in the numerical experiment.

Experiment 1. We consider the Kolmogorov Eqs. (2) with parameters $\mu_1 = 0.74, \mu_2 = 0.222, \mu_3 = 0.083$ and incoming $BMAP$ (1). The initial probability distribution gives the Cauchy conditions $p_{000}(0) = 1, p_{i,j,z}(0) = 0, i = \overline{0,6}, j = \overline{0,11}, z = \overline{0,4}$, where i, j and z are the number of requests in phases 1, 2, and 3, respectively. States $p_{i,11,z}(t)$ and $p_{i,j,4}(t)$ mean locking of channels of phases 1 and 2. As a result, we obtain the transition probabilities of states and, on its basis, the transient of performance measures for the QS considered.

The following figures show graphs of changes in the loss probability depending on t and the initial state of the system. Lines 1–6 correspond to $p_{1,9,3}(0) = 1$ (the phase 2 and 3 is full loaded), $p_{000}(0) = 1$ (requests are absent), $p_{2,5,0}(0) = 1$ (the system is half loaded, the phase 1 and 2 work), $p_{3,5,1}(0) = 1$ (the system is half loaded, the third phase is empty), $p_{5,11,3}(0) = 1$ (locking phase 1 channels), and $p_{6,10,1}(0) = 1$ (the system is full loaded without locking), respectively.

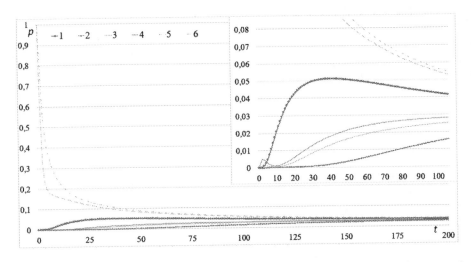

Fig. 3. Results of experiment 1. Loss probability depending on the initial state and $t = [0, 200]$

Table 1 shows the limiting or stationary performance measures of the considered QS. Here P_{loss} is a loss probability, k is an average number of working channels, L is a queue length, P_{lock} is a probability of locking.

Table 1. The stationary performance measures of the three-phase QS

P_{loss}		k	L	P_{lock}	P_{lock} (Phase 1 and 2)
0.0319	**Phase 1**	0.453	0.576	0.067	0.108
	Phase 2	1.678	3.628	0.294	
	Phase 3	2.491	–	–	

Analysis of transition probabilities shows the following. Its behavior and the rate of convergence to stationary parameters significantly depend on the initial state of the system. First, the minimum time required for the transition probabilities to converge to the stationary ones with a deviation of $\pm 5\%$ is achieved at the average system load and the maximum time at the lowest or the highest. For example, the convergence time is $t = 97.5$ (hours) for the initial state $p_{3,5,0}(0) = 1$. For $p_{0,0,0}(0) = 1$ or $p_{6,10,1}(0) = 1$, the time $t = 200$. Second, there are local extremes on the graph of the loss probability with an average load of the system (see Fig. 3, lines 3 and 4). Besides, there is a special case. The global maximum appears if phase 1 is free, and phase 2 and 3 are loaded at $t = 0$, in particular if $p_{1,9,3}(0) = 1$. Its value can be twice the stationary parameters (line 1).

It is known that fluctuations of various types can appear in railway transport traffic flows. We can mention daily, seasonal, and other deviations [7, 8]. Sukhovskaya is mainly subject to daily deviations. We registered the case when six trains arrived in 25 h. It is 15% above the daily average value. The $BMAP$ matrices for the period of increased intensity have the form

$$D_0 = \begin{pmatrix} -0.235 & 0 \\ 0 & -0.244 \end{pmatrix}, \; D_1 = \begin{pmatrix} 0.132 & 0.092 \\ 0.096 & 0.148 \end{pmatrix}, \; D_2 = \begin{pmatrix} 0.011 & 0 \\ 0 & 0 \end{pmatrix}. \quad (3)$$

The location of this period on the time axis has a significant impact on the properties of transients. We show below two of the most illustrative cases.

Experiment 2. The incoming $BMAP$ is set by matrices (3) on the interval $t = [5; 29]$, and by matrices (1) on the rest of the time in contrast to experiment 1. Figure 4 shows graphs of changes in the loss probability depending on $t = [0, 250]$ and the initial state of the system.

An increase in the intensity of the arrival of requests at the initial time moment leads to the following changes. The local extrema of the loss probability graph arise in contrast to experiment 1 (see Fig. 4 lines 1, 3, and 4). The extremes observed in experiment 1 increase their values by 59% on average. In particular, the global maximum increases 1.7 times in comparison with the same extremum in experiment 1 for the special case $p_{1,9,3}(0) = 1$ (line 1).

Experiment 3. The incoming $BMAP$ is set by matrices (3) on the interval $t = [50; 74]$ in contrast to experiment 1. Figure 5 shows graphs of changes in the loss probability depending on $t = [0, 250]$ and the initial state of the system.

The transient processes have the following features in this experiment (see Fig. 5). First, the global extrema appear on the graph (lines 1, 3, and 4) in contrast to experiment 1, where the loss probability is monotonic and increasing. Their values are on average 31% higher than stationary ones (lines 3 and 4). Second, special cases are critical to system performance. The global maximum for the initial state $p_{1,9,3}(0) = 1$ (line 1) increased by 39% compared to experiment 1.

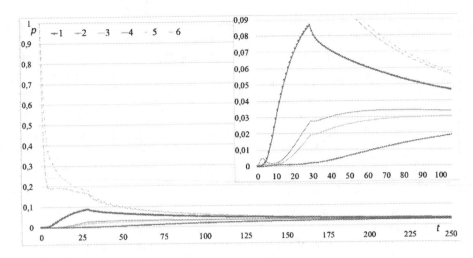

Fig. 4. Results of experiment 2. Loss probability depending on the initial state and $t = [0, 250]$

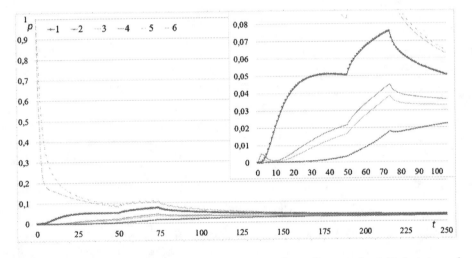

Fig. 5. Results of experiment 3. Loss probability depending on the initial state and $t = [0, 250]$

In experiments 2 and 3, the average time that requires the transition probabilities to converge to the stationary ones with the deviation ±5% increased by 12.5% an average. The increase of the intensity of the requests flow by 15% for $t \geq 85$ leads to results similar to experiment 3. Thus, the loss probability on average exceeds the stationary parameters by 31%.

6   Interpretation of Numerical Results

This section interprets the results obtained in terms of the Sukhovskaya freight station. First, the duration of transients is a significant factor for this type of station. More than four days required the station to transit to stationary mode. Secondly, we need to take into account the system load at the initial moment to estimate its capacity. In particular, we considered the operation of the station after the removal of an accident in the cargo yard. It corresponds to the total loading of phases 2 and 3 in the model. In this case, the recovery period for the station operation is more than five days, and the loss probability is doubles compared to the stationary ones. Third, as a rule, an increase in the volume of train traffic can be compensated by the system capacity if it is insignificant and has a relatively short duration. However, such an increase after accidents on the station leads to a rapid overflow of the receiving yard and the system as a whole. We present this case in experiments 2 and 3, where the maximum value of the loss probability is more than three times higher than the stationary parameter. Obviously, we see that an increase in the volume of train traffic leads to an increase in the time to converge the system to stationary mode.

Thus, we can ignore the transient processes in the following cases. The first is when a railway station has been just established or restarted; that is, the system is almost empty. The second is during periods of unbroken operating with minimal fluctuations in the volume of incoming traffic. In other cases, we should analyze transients for completeness of the study.

7   Conclusions

We have performed the analysis of transient processes for the three-phase QS that describes the operation of the freight railway station. As a result, we have found that the initial states of the systems substantially determine the rate of convergence and behavior of transients to stationary mode. Moreover, it is necessary to take into account the influence of transients at an average or maximum station load at the initial moment of time. The analysis of transients can be neglected when predicting the system operation in the case of minimal load.

The proposed approach has the following advantages. First, models in the form of a QS can be adapted to different railway stations with minimal effort. Researchers can use such models to study transient processes at other types of stations, in particular, marshaling stations. Second, we can use the approach to predict the operation of stations with a lack of information. For example, to simulate an increase in the traffic volume for the future, when the train schedule is unknown.

We point out the following possible directions for further research. First, the development of a procedure that makes it possible to compose the Kolmogorov equations for a queueing network with four nodes. Second, the adaptation and applying the proposed approach to the study of other types of transport systems, in particular, ports and transport hubs.

References

1. Vishnevsky, V., Semenova, O.: Polling systems and their application to telecommunication networks. Mathematics **9**(2), 117 (2021). https://doi.org/10.3390/math9020117
2. Nazarov, A., et al.: Multi-level MMPP as a model of fractal traffic. In: Dudin, A., Nazarov, A., Moiseev, A. (eds.) ITMM 2020. CCIS, vol. 1391, pp. 61–77. Springer, Cham (2021). https://doi.org/10.1007/978-3-030-72247-0_5
3. Bushkova, T., Danilyuk, E., Moiseeva, S., Pavlova, E.: Resource queueing system with dual requests and their parallel service. In: Vishnevskiy, V.M., Samouylov, K.E., Kozyrev, D.V. (eds.) DCCN 2019. CCIS, vol. 1141, pp. 364–374. Springer, Cham (2019). https://doi.org/10.1007/978-3-030-36625-4_29
4. Dudin, A.N., Klimenok, V.I., Vishnevsky, V.M.: Tandem queues with correlated arrivals and their application to system structure performance evaluation. In: The Theory of Queuing Systems with Correlated Flows, pp. 307–392. Springer, Cham (2020). https://doi.org/10.1007/978-3-030-32072-0_6
5. Medhi, J.: Stochastic Models in Queuing Theory, 2nd edn. Academic Press, San Diego (2002)
6. Vuuren, M.: Performance analysis of manufacturing systems: queueing approximations and algorithms. Technische Universiteit Eindhoven, Eindhoven (2007). https://doi.org/10.6100/IR625074
7. Milenkovic, M., Bojovic, N.: Optimization Models for Rail Car Fleet Management, 1st edn. Elsevier Inc., Amsterdam (2019)
8. Potthoff, G.: Verkehrs Stromungs Lehre. TRANSPRESS-VEB, Berlin (1970)
9. Daganzo, C., Dowling, R., Hall, R.: Railroad classification yard throughput: the case of multistage triangular sorting. Transp. Res. **17**(2), 95–106 (1986). https://doi.org/10.1016/0191-2607(83)90063-8
10. Turnquist, M.A., Daskin, M.S.: Queuing models of classification and connection delay in railyards. Transp. Sci. **16**(2), 207–230 (1982). https://doi.org/10.1287/trsc.16.2.207
11. De Kort, A., Heidergott, B., Van Egmand, R.J., Hooghiemstra, G.: Train Movement Analysis at Railway Stations: Procedures & Evaluation of Wakob's Approach, 1st edn. Delft University Press, Delft (1999)
12. Huisman, T., Boucherie, R.J.: Running times on railway sections with heterogeneous train traffic. Transp. Res. Part B: Methodol. **35**, 271–292 (2001). https://doi.org/10.1016/S0191-2615(99)00051-X
13. Meester, L.E., Muns, S.: Stochastic delay propagation in railway networks and phase-type distributions. Transp. Res. Part B: Methodol. **41**(2), 218–230 (2007). https://doi.org/10.1016/j.trb.2006.02.007
14. Weik, N., Nießen, N.: A quasi-birth-and-death process approach for integrated capacity and reliability modeling of railway systems. J. Rail Transp. Plan. Manag. **7**, 114–126 (2017). https://doi.org/10.1016/j.jrtpm.2017.06.001
15. Huisman, T., Boucherie, R.J., Van Dijk, N.M.: A solvable queueing network model for railway networks and its validation and applications for the Netherlands. Eur. J. Oper. Res. **142**, 30–51 (2002). https://doi.org/10.1016/S0377-2217(01)00269-7
16. Schmitz, C., et al.: Markov models for the performance analysis of railway networks. In: Proceeding of 7th International Conference on Railway Operations Modelling and Analysis 2017, Lille, pp. 1–23. https://raillille2017.sciencesconf.org, Accessed 24 Dec 2021

17. Weik, N., Nießen, N.: Quantifying the effects of running time variability on the capacity of rail corridors. J. Rail Transp. Plan. Manag. **15**, 100203 (2020). https://doi.org/10.1016/j.jrtpm.2020.100203

18. Bychkov, I.V., Kazakov, A.L., Lempert, A.A., Bukharov, D.S., Stolbov, A.B.: An intelligent management system for the development of a regional transport logistics infrastructure. Autom. Rem. Control **77**(2), 332–343 (2016). https://doi.org/10.1134/S0005117916020090

19. Bychkov, I., Kazakov, A., Lempert, A., Zharkov, M.: Modeling of railway stations based on queuing networks. Appl. Sci. **11**(5), 2425 (2021). https://doi.org/10.3390/app11052425

20. Zharkov, M., Lempert, A., Pavidis, M.: Simulation of railway marshalling yards based on four-phase queuing systems. In: Dudin, A., Nazarov, A., Moiseev, A. (eds.) ITMM 2020. CCIS, vol. 1391, pp. 143–154. Springer, Cham (2021). https://doi.org/10.1007/978-3-030-72247-0_11

21. Zharkov, M.L., Kazakov, A.L., Lempert, A.A.: Transient process modeling in micrologistic transport systems. In: IOP Conference Series: Earth and Environmental Science 2021, vol. 629, pp. 012023 (2021). https://doi.org/10.1088/1755-1315/629/1/012023

22. Lucantoni, D.-M.: New results on single server queue with a Batch Markovian Arrival Process. Commun. Statist. Stochastic Models **7**, 1–46 (1991). https://doi.org/10.1080/15326349108807174

23. Bolch, G., et al.: Queueing Networks and Markov Chains: Modeling and Performance Evaluation with Computer Science Applications. John Wiley & Sons, Hoboken (2006)

24. Shampine, L.F.: Tolerance proportionality in ODE codes. In: Bellen, A., Gear, C.W., Russo, E. (eds.) Numerical Methods for Ordinary Differential Equations. LNM, vol. 1386, pp. 118–136. Springer, Heidelberg (1989). https://doi.org/10.1007/BFb0089235

Optimization of the Bartlett Conflict Traffic Flows Control Process in the Class of Algorithms with a Pedestrian Mode

Stepan Lembrikov[✉] and Elena Kuvykina

Lobachevsky State University of Nizhny Novgorod, Nizhny Novgorod, Russia
stepan.lembrikov.ru@mail.ru

Abstract. This article deals with a special class of algorithms with feedback. Such algorithms are designed to control the traffic and the pedestrian flows at the intersection. A specific feature of the traffic control system is changing the probability structure of input flows under the influence of random factors such as accident or worsening weather conditions. The purpose of this research is optimization of the system work for the different types of flows such as Poisson and Bartlett input flows. Using the simulation method, in this work is studied the influence of the probability structure of input flows and the average batch size on work of the system. The influence of the Bartlett flow parameters on the quasi – optimal control parameters and the minimal value of the average sojourn time of a random customer at the intersection is considered. Using of the control parameters allows to regulate traffic at the intersection and reduce the traffic load.

Keywords: Traffic control · Conflict flows · Bartlett flow · Poisson flow · Traffic flow · Pedestrian flow · Intensity of the input flow · Average batch size · Optimal control parameters · Transport batch

1 Introduction

Everyday traffic seems to get worse on the roads. The traffic load is increasing in cities throughout the world. Study of the transport systems is becoming increasingly relevant in that regard. At the same time many problems and tasks are considered, such as road safety improvement [1], congestion combating that could reduce the negative environmental impact [2]. Among others, the solution of the traffic and pedestrian flows control problem at the intersection is of considerable interest [3–5]. By applying various algorithms for the transport system control it is possible to reduce the traffic load, which in turn has a positive effect on both safety and the environmental situation.

2 Problem Statement

The article is based on a queuing system with a wait. It's a T – junction intersection that schematically depicted in Fig. 1.

© Springer Nature Switzerland AG 2022
A. Dudin et al. (Eds.): ITMM 2021, CCIS 1605, pp. 278–290, 2022.
https://doi.org/10.1007/978-3-031-09331-9_22

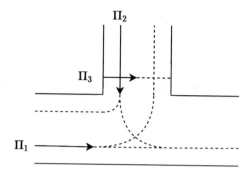

Fig. 1. The intersection diagram.

The system receives three independent input flows Π_1, Π_2, Π_3 where Π_1 and Π_2 are the traffic flows, and Π_3 is the flow of pedestrians. Their simultaneous service is impossible so the flows are pairwise conflict [6].

The automatic traffic light acts as the server. It has eight modes of working $\Gamma^{(1)}, \Gamma^{(2)}, \ldots, \Gamma^{(8)}$. The traffic flows Π_1 and Π_2 are serviced in the modes $\Gamma^{(1)}$ and $\Gamma^{(3)}$ respectively, $\Gamma^{(5)}$ and $\Gamma^{(7)}$ are responsible for servicing the pedestrian flow Π_3. The modes $\Gamma^{(2)}, \Gamma^{(4)}, \Gamma^{(6)}$ and $\Gamma^{(8)}$ correspond to the yellow traffic light. They are called the orientation – changeover modes, and the flows are not serviced in them. These modes are necessary for the road safety. Each mode $\Gamma^{(r)}$ lasting for a fixed time $T_r, r = \overline{1,8}$. Herewith, the duration of yellow traffic light is selected based on the road safety. Parameters T_5 and T_7 are also considered as constant. The duration of the traffic service modes T_1 and T_3 are control parameters of the system.

The scheme of the server operation modes is shown in the Fig. 2. The modes switching $\Gamma^{(1)} \rightarrow \Gamma^{(2)}, \Gamma^{(3)} \rightarrow \Gamma^{(4)}, \Gamma^{(5)} \rightarrow \Gamma^{(6)}, \Gamma^{(6)} \rightarrow \Gamma^{(3)}, \Gamma^{(7)} \rightarrow \Gamma^{(8)},$ $\Gamma^{(8)} \rightarrow \Gamma^{(1)}$ are uniquely determined. Let's consider the switch from modes $\Gamma^{(2)}$ and $\Gamma^{(4)}$. At the end of $\Gamma^{(2)}$ the server switches to $\Gamma^{(3)}$ if there are no more than N_1 pedestrians at the intersection. Otherwise, the $\Gamma^{(5)}$ mode is activated. Similarly, at the end of $\Gamma^{(4)}$ mode the system switches to $\Gamma^{(1)}$ if there are no more than N_2 pedestrians at the intersection. Otherwise, $\Gamma^{(7)}$ mode is activated.

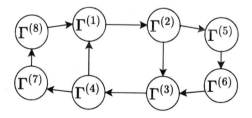

Fig. 2. Graph of the server operation modes.

The described algorithm is close to cyclic, so it's easy to implement it in practice. At the same time, it is a feedback algorithm because it takes into account the entrance and the service of the pedestrian flow Π_3 customers.

One of the most important criteria of the work of the system with a wait is the average sojourn time γ of a random customer at the intersection. This article focuses on the looking for the quasi – optimal control parameters. That is those parameters in which the average sojourn time γ is minimal. The optimization problem is achieved numerically using the simulation method. Two types of the input flows are considered: flows with independent movement of customers and the batch flows.

The mathematical model of the flow with independent customer movement is the Poisson flow with the intensity λ. In this article the pedestrian flow Π_3 is always assumed Poisson. This is due to the peculiarity of pedestrian movements. Their speed is relatively small and the external factors have small impact on their margin for manoeuvre.

The second type of the flows are the flows that have a batch structure. The emergence of batches is caused by limited overtaking capacities when the fast cars are built up after the slow one. The mathematical model of the such struc-ture flows is the Bartlett flow. These flows are described non-locally using a vector random sequence $\{(\tau_i, \eta_i), i \geq 0\}$ [7,8]. The observe moments $\tau_i, i \geq 0$ are considered to be the entering moments of the first customers in the batches. Random variables η_i describe the number of customers in the i-th batch and have the following distribution:

$$\begin{cases} P(\eta_i = 1) = 1 - r \\ \\ P(\eta_i = k) = r \cdot (1 - g) \cdot g^{k-2}, k \geq 2, \end{cases} \tag{1}$$

where r, g—parameters of the distribution, $0 < r, g < 1$.

The average customers count in the i-th batch is calculated by the following way:

$$M_{\eta_i} = 1 + r/(1 - g) \tag{2}$$

In the Formula 1, the parameter r is the probability that a batch of two or more customers will enter the system. The parameter g is related to the average size of the batch (see Formula 2). By converting the Formula 2, the following equation can be obtained:

$$r = (M_{\eta_i} - 1) \cdot (1 - g) \tag{3}$$

It is easy to see that if the average size of the batch is fixed, then as the value of the parameter r grows, the value of the parameter g decreases.

3   Numerical Study

In this research is numerically studied the influence of the probability structure of the input flows and the average size of the batch on the work of the system.

It is established that as the mathematical expectation of customers count in a batch grows, the quasi – optimal control parameters increases. The minimal value of the average sojourn time γ also increases. In addition, it has been able to establish that the change of the probability structure of the input flows also affects the quasi – optimal control parameters and the minimal value of the average sojourn time γ.

To illustrate this conclusion, consider the following example. Let the intensity of the flows Π_1, Π_2 and Π_3 coincide: $\lambda_1 = \lambda_2 = \lambda_3 = 0,1$ (customers per second). Critical values for queue lengths of the pedestrian flow Π_3 are assumed to be zero: $N_1 = N_2 = 0$ (customers). The duration of the pedestrian service modes $T_5 = T_7 = 30$ (s). In case of the Poisson input flows Π_1, Π_2 and Π_3 the following quasi – optimal control parameters are obtained: $T_1 = 16$ (s), $T_3 = 15$ (s). The minimal value of the average sojourn time γ is $33,72$ (s). Now let's consider two cases of the Bartlett traffic flows Π_1 and Π_2. For the average size of the batch $M_{\eta_{j,i}} = 2$ (customers) the following quasi – optimal parameters are obtained: $T_1 = 41$ (s), $T_3 = 42$ (s). The minimal value of the average sojourn time γ of a random customer at the intersection is 52.51 (s). For the value $M_{\eta_{j,i}} = 11$ (customers) the following quasi – optimal control parameters are obtained: $T_1 = 49$ (s), $T_3 = 51$ (s). The minimal value of the average sojourn time γ is 60.62 (s). To compare the results obtained, see Table 1.

Table 1. The influence of the parameter $M_{\eta_{j,i}}, j = 1, 2$ on the system work.

All input flows are Poisson	$M_{\eta_{j,i}} = 2$ (customers) $g_j = 0,91$ $r_j = 0,09, j = 1,2$	$M_{\eta_{j,i}} = 11$ (customers) $g_j = 0,91$ $r_j = 0,9, j = 1,2$
$T_1 = 16$ (s)	$T_1 = 41$ (s)	$T_1 = 49$ (s)
$T_3 = 15$ (s)	$T_3 = 42$ (s)	$T_3 = 51$ (s)
$\gamma = 33,72$ (s)	$\gamma = 52,51$ (s)	$\gamma = 60,62$ (s)

In Table 1, the top row is responsible for the input flows parameters. The first column represents the case of the Poisson input flows and the next two columns represent the Bartlett traffic flows Π_1 and Π_2. The bottom row is responsible for the results obtained. Table 1 shows that in the case of the Poisson input flows the system copes much better with customers service. Compared the results of the first two columns, it can be seen that the difference between the minimal values of the average sojourn time γ is $18,79$ (s). The quasi – optimal control parameters have more than doubled. A comparison of the second and the third columns confirms that increase of the average batch size for the traffic flows affects the system work. The minimal value of the average sojourn time γ increased — the difference is $8,11$ (s). In doing so, the quasi – optimal control parameters are also increased. These conclusions about the relationship between the system work and the probability structure of the input flows and the average batch size are consistent with the results obtained in the article [9] for another control algorithm.

Further research is carried out to find out the influence of the Bartlett distribution parameters g_j and $r_j, j = 1, 2$ on the quasi – optimal control parameters and the minimal value of the average sojourn time γ of a random customer at the intersection. It's found out that changing the ratio between the parameters g_j and $r_j, j = 1, 2$ significantly affects the system work. It's established that as the parameter $g_j, j = 1, 2$ grows, the quasi – optimal control parameters increases. The minimal value of the average sojourn time γ also increases.

The following example illustrates this fact. Let the intensity of the flows Π_1, Π_2 and Π_3 coincide: $\lambda_1 = \lambda_2 = \lambda_3 = 0, 1$ (customers per second). Critical values for queue lengths of the pedestrian flow Π_3 are assumed to be zero: $N_1 = N_2 = 0$ (customers). The duration of the pedestrian service modes $T_5 = T_7 = 30$ (s). The pedestrian flow is Poisson, and the traffic flows Π_1 and Π_2 are Bartlett flows with the average batch size $M_{\eta_{j,i}} = 2$ (customers).

Table 2. The influence of the g_j and r_j parameters on the system work, $j = 1, 2$.

$g_j = 0, 91$ $r_j = 0, 09, j = 1, 2$	$g_j = 0, 95$ $r_j = 0, 05, j = 1, 2$
$T_1 = 40$ (s)	$T_1 = 52$ (s)
$T_3 = 45$ (s)	$T_3 = 54$ (s)
$\gamma = 52, 47$ (s)	$\gamma = 65, 38$ (s)

Table 2 shows that as the parameter $g_j, j = 1, 2$ for the traffic flows grows, the minimal value of the average sojourn time γ increases from $52, 47$ (s) to $65, 38$ (s), so the difference is $12, 91$ (s). At the same time, the quasi – optimal control parameters also increases.

Based on results of the two examples (Table 1 and Table 2), the smallest values of the control parameters and the average sojourn time γ were obtained in the case of the Poisson input flows. In addition to this, increasing of the average batch size from 2 to 11 customers does not have such effect on the system work as changing the ratio between the parameters g_j and $r_j, j = 1, 2$ with the fixed mathematical expectation of customers count in a batch.

To understand why changing the ratio between the parameters g_j and $r_j, j = 1, 2$ affects the system work so much, let's consider two figures. The first one is the figure of the minimal value of the average sojourn time γ against the value of the parameter $g_j, j = 1, 2$. The second one is the figure of the average value of the control parameters T_1 and T_3 against the value of the parameter $g_j, j = 1, 2$. Let the intensity of the flows Π_1, Π_2 and Π_3 coincide: $\lambda_1 = \lambda_2 = \lambda_3 = 0, 1$ (customers per second). Critical values for queue lengths of the pedestrian flow Π_3 are assumed to be zero: $N_1 = N_2 = 0$ (customers). The duration of the pedestrian service modes $T_5 = T_7 = 30$ (s). The pedestrian flow is Poisson, and the traffic flows Π_1 and Π_2 are Bartlett flows with the average batch size $M_{\eta_{j,i}} = 2$ (customers). The value of the parameter $g_j, j = 1, 2$ will be varied from $0, 1$ to $0, 99$. At the interval $[0, 1; 0, 9)$ it will be increased to $0, 1$ by one step, and at the interval $[0,9;0,99)$—to $0,01$. The parameter $r_j, j = 1,2$ will be varied according to Formula 2. Also for comparison there is a plot for the case of the Poisson input flows in the Fig. 3.

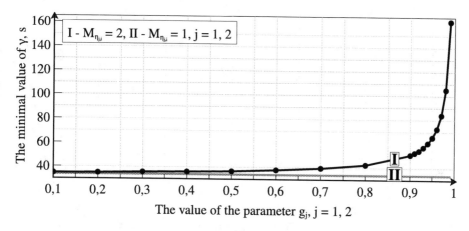

Fig. 3. The plots of the average sojourn time γ against the value of the parameter $g_j, j = 1,2$.

The Fig. 3 shows that in the case of the Poisson input flows (line II), the minimal value of the average sojourn time γ of a random customer at the intersection is $33,72$ (s). In this case 100% of customers are entered the system one at a time. In doing so, the curve I shows that despite the fact that the average batch size remains constant, the ratio between the parameters g_j and $r_j, j = 1,2$ significantly affects the minimal value of the average sojourn time γ. Let's consider the leftmost point of the curve I. In this point parameter $g_j = 0,1$ and $r_j = 0,9, j = 1,2$. The value of the parameter $r_j = 0,9, j = 1,2$ indicates that on average only 10% of the entering the system batches consists of one customer. Calculations show that the minimal value of the average sojourn time γ is very close to the case of the Poisson input flows and equal to $34,72$ (s). For the rightmost point of the curve I, the parameters take the following values: $g_j = 0,99, r_j = 0,01, j = 1,2$. That is, the probability of a single customer entering the system is equal to $0,99$. In doing so, the average sojourn time γ is $161,41$ (s). Despite the fact, that in the second case the flows Π_1 and Π_2 are almost identical with the Poisson ones, there is a sharp increase of the average sojourn time γ. The results shows that as the parameter $g_j, j = 1,2$ grows, the minimal value of the average sojourn time γ of a random customer at the intersection increases. And starting from the level $g_j = 0,8, j = 1,2$ there is a significant difference between cases of the Bartlett and the Poisson input flows. Based on this, it can be assumed that the main parameter, that affects the minimal value of the average sojourn time γ of a random customer at the intersection is $g_j, j = 1,2$.

The following figure represents of the average value of the control parameters T_1 and T_3 against the value of the parameter $g_j, j = 1, 2$ with the same input parameters. The average value of the control parameters T_1 and T_3 are considered, because the system is symmetric under the input flows Π_1 and Π_2.

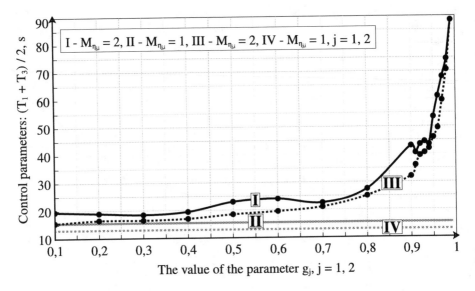

Fig. 4. The plots of the average value of the control parameters T_1 and T_3 against the value of the parameter $g_j, j = 1, 2$.

The Fig. 4 shows that the ratio between the parameters g_j and $r_j, j = 1, 2$ affects not only the minimal value of the average sojourn time γ of a random customer at the intersection, but also the average value of the control parameters T_1 and T_3. For curve I and line II the quasi – optimal control parameters are selected as the parameters T_1 and T_3. With the same input parameters, depending on the specific implementation the different quasi – optimal control parameters T_1 and T_3 can be obtained in modeling. Nevertheless, the curve I shows that as the parameter $g_j, j = 1, 2$ grows, the average value of the quasi – optimal control parameters T_1 and T_3 tends to increase. For example, if we compare the opposite points of the curve I, when the parameter $g_j, j = 1, 2$ takes the values $0, 1$ and $0, 99$ respectively, we can see that the average value of the quasi – optimal control parameters T_1 and T_3 increases from $19, 5$ (s) to $88, 5$ (s). Let's choose the parameters T_1 and T_3 another way to reduce the fluctuations and get a more illustrative picture. Let's find the range of the control parameters at which the average sojourn time γ differs from the minimal one by no more than one second. The parameters T_1 and T_3 are taken from this area, so that their sum is minimal. These values are called close to optimal. Thus, in the Fig. 4 curve III and line IV are plotted. Obviously, the two given plots lies no higher than curve

I and line II respectively. Choosing the parameters T_1 and T_3 close to optimal allows to minimize random deviations and see that as the parameter $g_j, j = 1, 2$ grows, the average value of the close to optimal control parameters T_1 and T_3 increases. And starting from the level $g_j = 0, 8, j = 1, 2$ there is a significant difference between cases of the Bartlett and the Poisson input flows. Based on this, it can be assumed that the main parameter, that affects the average value of the quasi – optimal and close to them control parameters T_1 and T_3, is $g_j, j = 1, 2$.

Consider the second equality of the Formula 1. It describes the probability of a batch with a k size entering the system, $k \geq 2$. Let $M_{\eta_{j,i}} = 11$ (customers), $j = 1, 2$. Let's plot two curves describing the values of the customers count in a batch probability for two following sets of the Bartlett parameters: $g_j = 0, 91, r_j = 0, 9$ in the first case and $g_j = 0, 99, r_j = 0, 1$ in the second, $j = 1, 2$.

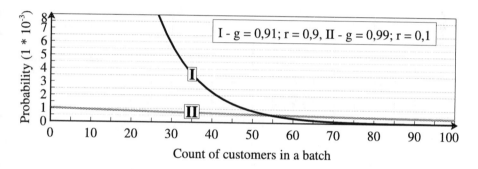

Fig. 5. The plots of a batch entering probability against its size.

The Fig. 5 shows that initially curve I lies higher than curve II. This is due to the value of the parameter $r_j, j = 1, 2$, because it's the probability of a batch of two or more customers entering the system. Indeed, for the curve I the probability of a single customer entering the system is equal to 0, 1 and for the curve II this value is equal to 0, 9. This shows that the batches count with a size greater than one is higher in the first case. For example, for the first parameters set the probability $P(\eta_{j,i} = 2) = r_j \cdot (1 - g_j) \cdot g_j^{(0)} = 0, 081$, and for the second— $P(\eta_{j,i} = 2) = 0, 001, j = 1, 2$. But according to the Formula 1, the probability of a batch with a k size, $k \geq 2$ entering the system is the power function of g_j parameter, $0 < g_j < 1, j = 1, 2$. That's why with an increasing k, the first curve decreases much faster than the second one. In this regard, for large values of k, the probability that a k size batch will enter the system is higher for the second parameters set. This leads to a situation where the system starts working worse: the minimal value of the average sojourn time γ and the quasi – optimal control parameters T_1 and T_3 increases.

It follows from the Formula 1 that the parameters $r_j, g_j \in (0; 1), j = 1, 2$. But changing the average batch size leads to changing the range of the g_j parameter, $j = 1, 2$. For example, when $M_{\eta_{j,i}} = 2$ (customers), the parameter $g_j, j = 1, 2$ is really ranged from zero to one. But if $M_{\eta_{j,i}} = 11$ (customers) the value of

$g_j \in (0,9;1), j = 1,2$. Thus, the large values of the $g_j, j = 1,2$ parameter corresponds to the big batch size. That's why increasing the average batch size leads to rising of the minimal value of the average sojourn time γ and the quasi – optimal parameters T_1 and T_3. Of course, not only the parameter $g_j, j = 1,2$ affects the result, but nevertheless it's a fundamental factor.

Consider the following example. Let the intensity of the flows Π_1, Π_2 and Π_3 coincide: $\lambda_1 = \lambda_2 = \lambda_3 = 0,1$ (customers per second). Critical values for queue lengths of the pedestrian flow Π_3 are assumed to be zero: $N_1 = N_2 = 0$ (customers). The duration of the pedestrian service modes $T_5 = T_7 = 30$ (s). The pedestrian flow is Poisson. And the traffic flows Π_1 and Π_2 are the Bartlett flows with the same average batch size: $M_{\eta_{1,i}} = M_{\eta_{2,i}}$. To study the dependence of the average sojourn time γ on the average batch size and the parameter g_j consider the following values $M_{\eta_{j,i}}$: $M_{\eta_{j,i}} = 2$, $M_{\eta_{j,i}} = 3$, $M_{\eta_{j,i}} = 5$ and $M_{\eta_{j,i}} = 11$ (customers), $j = 1,2$. Also for each of these values the parameter $g_j, j = 1,2$ will be ranged from $0,91$ to $0,99$ with a step $0,01$.

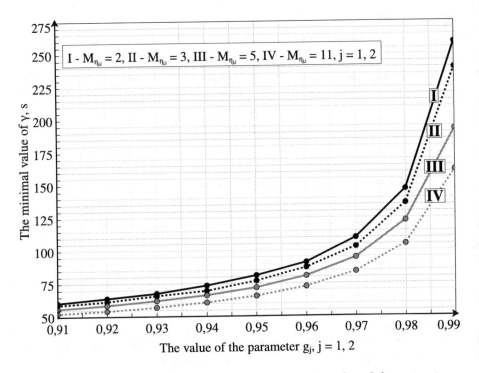

Fig. 6. The plots of the average sojourn time γ against the value of the parameter g_j for a different $M_{\eta_{j,i}}, j = 1,2$.

For each value of the average batch size, as the parameter $g_j, j = 1,2$ grows, the minimal value of the average sojourn time γ increases. For example, when $M_{\eta_{j,i}} = 2, j = 1,2$ (curve IV), the average sojourn time increases more than 100 seconds. In addition, if the parameter $g_j, j = 1, 2$ is fixed, then as the mathe-matical expectation of customers count in a batch grows, the minimal value of the average sojourn time γ of a random customer at the intersection increases. For example, when $g_j = 0, 91, j = 1, 2$, the average sojourn time γ increases by about 10 s. Thus, Fig. 6 shows that by changing the parameter $g_j, j = 1, 2$ with a smaller batch size, a much larger minimal value of the average sojourn time γ of a random customer at the intersection can be got.

Let's plot the influence of the average values of the quasi – optimal control parameters to the parameter $g_j, j = 1, 2$.

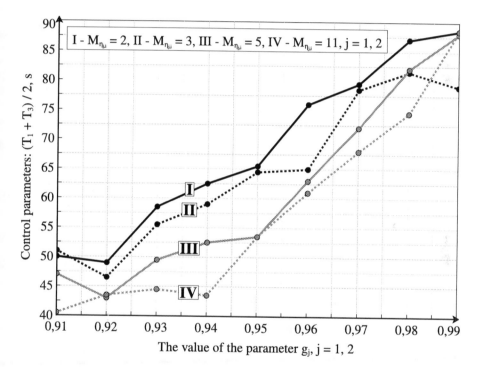

Fig. 7. The plots of the average value of the control parameters T_1 and T_3 against the value of the parameter $g_j, j = 1, 2$ for a different $M_{\eta_{j,i}}, j = 1, 2$.

The plots in Fig. 7 allow to confirm that as the parameter $g_j, j = 1, 2$ grows, the average value of the quasi – optimal parameters T_1 and T_3 increases. For example, when $g_j = 0, 91, j = 1, 2$, the average value of the parameters T_1 and T_3 varies from 40, 5 (s) to 51 (s). In this way, the average batch size increases from 2 to 11 (customers). Let's consider the curve IV. Calculations shows that the average value of the parameters T_1 and T_3 varies from 40, 5 (s) to 88, 5 (s). Thus, by changing the parameter $g_j, j = 1, 2$ with a smaller batch size a much larger average value of the quasi – optimal control parameters T_1 and T_3 can be got.

Similar results are obtained if the bigger values of the traffic input flows intensities are considered. Let's plot two figures similar to graphics in Fig. 6 and Fig. 7 with $\lambda_j = 0,2$ (customers per second), $j = 1,2$ using the same other parameters as in the previous example ($\lambda_3 = 0,1$ (customers per second), $N_1 = N_2 = 0$ (customers), $T_5 = T_7 = 30$ (s)). The first one is the figure of the minimal value of the average sojourn time γ against the value of the parameter $g_j, j = 1,2$. The second one is the figure of the average value of the control parameters T_1 and T_3 against the value of the parameter $g_j, j = 1,2$.

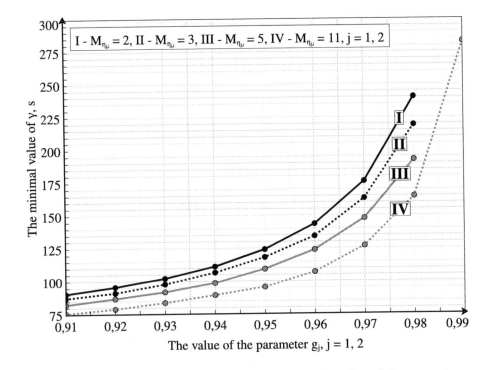

Fig. 8. The plots of the average sojourn time γ against the value of the parameter g_j for a different $M_{\eta_{j,i}}, j = 1,2$.

Comparing Fig. 6 and Fig. 8, it can be seen that with a higher value of the intensity the spread of the minimal values of the parameter γ becomes stronger with changing the average batch size $M_{\eta_{j,i}}, j = 1,2$. For example, when $g_j = 0,91$, in the case of lower intensity $\lambda_j = 0,1$ (customers per second) the spread of the average sojourn time γ is $10,11$ (s), $j = 1,2$. And in the case of higher intensity $\lambda_j = 0,2$ (customers per second), this spread is increased to 15 (s), $j = 1,2$. In the second case when the value of the parameter $g_j = 0,99$, the area of the stationary mode existence could be found only for $M_{\eta_{j,i}} = 2, j = 1,2$ (curve IV). For other values of the average batch size for traffic flows the system not worked with any control parameters. At the same time, it's easy to see that the parameter $g_j, j = 1, 2$ affects the minimal value of the average sojourn time γ of a random customer at the intersection mainly.

The same results are observed for the influence of the average value of the quasi – optimal control parameters T_1 and T_3 to the value of the parameter $g_j, j = 1, 2$.

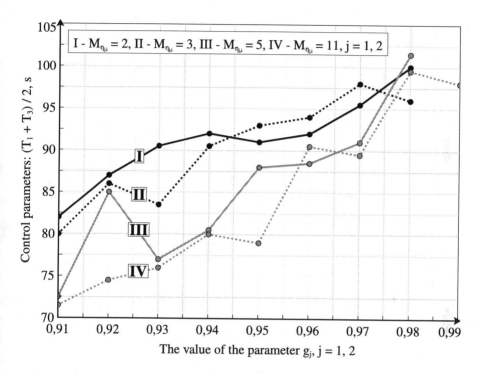

Fig. 9. The plots of the average value of the control parameters T_1 and T_3 against the value of the parameter $g_j, j = 1, 2$ for a different $M_{\eta_{j,i}}, j = 1, 2$.

Comparing Fig. 7 and Fig. 9, it can be seen that with a higher value of the intensity the spread of the average values of the quasi – optimal control parameters T_1 and T_3 becomes stronger with changing the average batch size $M_{\eta_{j,i}}, j = 1, 2$, and the random fluctuations between nearby points increases. For example, when $M_{\eta_{j,i}} = 3$ (customers), $j = 1, 2$ with increasing the parameter $g_j, j = 1, 2$ from $0, 92$ to $0, 93$ the average value of the quasi – optimal control parameters T_1 and T_3 decrease from 85 (s) to 77 (s). Nevertheless, the general trend of increasing the average value of the quasi – optimal control parameters T_1 and T_3, as the parameter $g_j, j = 1, 2$ grows, still remains unchanged.

4   Conclusion

The paper considers a transport system of the Bartlett conflict traffic flows control process in the class of algorithms with a pedestrian mode. The problem of determining the optimal control parameters in which the average sojourn time γ of a random customer at the intersection is minimal is achieved numerically using the simulation method for cases of Poisson and Bartlett input flows. The results showed that the probability structure of the input flows significantly affects the system work. The influence of the Bartlett flow parameters with a batch structure on the optimal system control is also studied. In practice, such indicators as the average batch size and the probability of a single customer entering the system are often used to characterize the batch flows. However, the conducted studies allows to infer that despite the fact these indicators affect the optimal system control this effect is relatively small compared to the influence of the Bartlett distribution parameter g. It is established that the optimal control parameters mostly depend on g. As the parameter g grows there is a significant increase in both the optimal control parameters and the minimal value of the average sojourn time γ of a random customer at the intersection. That's why with the growth of g the system starts working worse.

References

1. Popov, A.V., Chernova, G.A., Baranov, A.A.: Assessment of traffic safety on the Trade - Union boulevard of the Volzhsky city. Mod. Prob. Transp. Comp. Russia **3**(1), 36–42 (2013)
2. Vladimirov, S.N.: Traffic congestion in the metropolis. Izvestiya MGTU MAMI **3**(1), 77–84 (2014)
3. Kudryavtsev, E.V., Fedotkin, M.A.: Analysis of discrete models of adaptive control system conflict nonhomogeneous flows. Bull. Moscow Univ. **1**, 19–26 (2019)
4. Zagutin, D. S., Skudina, A. A., Bakhteev, S. A., Mironov, S. A.: Investigation of parameters of installation of transport and pedestrian traffic lights. Eng. Bull. Don (1) (2019)
5. Zagutin, T.S.: Methods of queuing theory in the study and optimization of traffic at controlled intersections. Proc. MIPT **7**(2), 119–130 (2015)
6. Zorin, A.V.: Optimization of conflict flow control parameters in the class of cyclic algorithms. Bull. Tomsk State Univ. Manag. Comput. Eng. Comput. Sci. (3), 70–77 (2011)
7. Fedotkin, M.A.: Incomplete Description of Flows of Heterogeneous Requirements: Theory of Queuing. MSU, VNIISI, Moscow (1981)
8. Kuvykina, E. V.: General statement of the problem of algorithmic management of conflicting flows of customers with their non-local description. Modeling of dynamic systems. In: Collection of Scientific Papers of the Nizhny Novgorod Branch of the Institute of Machine Science of the Russian Academy of Sciences, pp. 65–71 (2007)
9. Kuvykina, E.V.: Studying the management system of conflict traffic flows in the class of algorithms with maintenance. Sci. Prob. Water Transp. **48**, 56–61 (2016)

PASTA for Remaining Service Time in Stable and Unstable Queues

Evsey Morozov[1,2,3] and Taisia Morozova[1,2,4(✉)]

[1] Petrozavodsk State University, Petrozavodsk, Russia
tiamorozova@gmail.com
[2] Institute of Applied Mathematical Research of the Karelian research
Centre of RAS, Petrozavodsk, Russia
[3] Moscow Center for Fundamental and Applied Mathematics,
Moscow State University, Moscow, Russia
[4] Uppsala University, Uppsala, Sweden

Abstract. In this paper, we establish the PASTA property for the limiting distribution of the remaining service time in stable and non-stable multiclass $M/G/1$ queueing systems. Our asymptotic analysis heavily exploits the regenerative property of the queueing system. The proof is first given for the stable system. A key observation for non-stable systems is that while the basic processes diverge to infinity, a proper time-average limit of the remaining service time exists. Some numerical results demonstrating the PASTA property are included as well.

Keywords: PASTA · multiclass system · stability · remaining service time · regenerative approach

1 Introduction

In this section, we describe the problem we address in the paper and present in brief the existing results in the area. The remaining service time plays quite important role in the regenerative stability analysis of a wide class of queueing systems [5,11]. More exactly, the proof of the tightness of the remaining service time process is an important stage in the stability analysis both single-server and multiserver classic queueing systems as well in the stability analysis of the retrial systems, see [4–6,10].

By an evident reason, the basic queueing processes such as queue size and workload (remaining work), as a rule, are the main object of the corresponding research, while the research of the remaining service time attracted much less attention. In this regard, we mention a few following papers [2,7–9,12,13] where the remaining service time is the main object of research. In particular, the *tightness* of the remaining service time in a wide class of classic queueing

This work is supported by the Russian Science Foundation, Project No. 21-71-10135, https://rscf.ru/en/project/21-71-10135/.

systems has been proved in the paper [9] based on the construction given in [4]. It is worth mentioning that the tightness property of the remaining service time holds regardless of whether the underlying queueing system is stable (positive recurrent) or not. (Below we give exact definition of the stable/unstable system.) At that, while the proof for the stable system is straightforward and based on the tightness of the remaining regeneration time, the proof of the tightness for the unstable system is the most challenging case [10, 11].

In this paper, we continue to study the remaining service time process with focus on the property PASTA. This research complements our previous work [12] where some explicit results related to the stationary remaining service time in classic and retrial queueing systems have been obtained. Moreover, the limiting distribution of the remaining service time in the unstable systems has also been found in [12] by the regenerative method. We emphasize the importance the PASTA property in the discrete-event simulation and also in the analysis of the multiclass systems, because this property allows to show that some important performance indexes are independent of the customer's class.

The main contribution of this paper is a direct proof of the property PASTA for the remaining service time both in stable and unstable $M/G/1$ queueing system. To the best of our knowledge, the proof of PASTA for the unstable system is performed for the first time because in the known works it is based on the stationary distribution of a basic process describing the system [1]. Some numerical results demonstrating the PASTA property are included as well.

The paper is organized as follows. In Sect. 1 we summarize the main previous results directly related to the subject of this research. In particular, we give the explicit expressions for the stationary remaining service time distribution in the stable systems (both classic multiclass multiserver and retrial), and also for the limiting distribution (being the corresponding time-average limit) in the unstable system. In Sect. 2, we present the direct proof of the property PASTA for the stable multiclass system $M/G/1$. This proof has an independent interest and also can be used in other analogous settings. In Sect. 3, the proof of PASTA for the unstable system is given. An important observation is that in this case the mean fraction of the idle time of server is asymptotically negligible. As a result, PASTA in this case holds in the form of the "convergence in mean". In Sect. 4, we give a few numerical examples which illustrate the obtained theoretical results.

2   Description of the System and Previous Results

In this section, we first consider a classical multiclass $M/G/1$-type queueing system with N classes of customers which follow independent stationary Poisson input processes with rates λ_i, $i = 1, \ldots, N$. Denote by $\lambda = \sum_i \lambda_i$ the rate of the superposed (Poisson) input, and by $\{t_n, n \geq 1\}$ the instants of this input, with $t_1 := 0$. Also let τ denote the generic interarrival time in the merged Poisson input. The service times $\{S_n^{(i)}, n \geq 1\}$ of class-i customers are assumed to be iid with generic element $S^{(i)}$, service rate $\mu_i = 1/\mathsf{E}S^{(i)}$ and distribution function F_i, $i = 1, \ldots, N$. To describe the regenerative structure of the system, we denote by $Q(t)$ the total number of customers in the system at instant t^-, and let $Q(t_k) = Q_k$ be the number of customers just before the kth arrival. (All continuous-time processes are assumed to be right-continuous with left-hand limits [5].) Then the regeneration instants $\{T_n\}$ of the process $\{Q(t), t \geq 0\}$ are recursively defined as

$$T_{n+1} = \inf_k(t_k > T_n : Q_k = 0), \ n \geq 0, \tag{1}$$

where we put $T_0 := 0$. We call the case $Q_1 = 0$, $t_1 = 0$ zero initial state when the 1st customer arrives at the empty system at instant $t_1 = 0$. The generic regeneration period, the distance between two regeneration points, is denoted by T. We call the regenerative process $\{Q(t)\}$ (and the queueing system) *positive recurrent* (stable) if the mean regeneration period is finite, $ET < \infty$. (If $ET = \infty$ then the system is called *null-recurrent* or unstable.) It is well-known that positive recurrence of this system implies *stability*, that is the existence of the stationary distribution of $Q(t)$ as $t \to \infty$. (The underlying theory of the regenerative processes can be found in [1,3].)

Denote by $S_i(t)$ the remaining service time of class-i customer at instant t, where by definition, $S_i(t) = 0$ if the server either empty or serves class-k customer, $k \neq i$. Also denote, when exists, the weak limit (in distribution)

$$S_i(t) \Rightarrow \mathbb{S}_i, \ t \to \infty, \ i = 1, \ldots, N,$$

where $\mathbb{S}_i$ is the stationary remaining service time of class-i customer, and let

$$\rho_i = \lambda_i E S^{(i)} \text{ and } \rho = \sum_i \rho_i.$$

It is well-known that if $\rho < 1$ then the system is positive recurrent (for instance, see [10,11]). The following result has been proven in [12]:

$$P(\mathbb{S}_i > x) = P_B^{(i)} \mu_i \int_x^\infty (1 - F_i(u))du, \tag{2}$$

where the limit with probability 1 (w.p.1)

$$\lim_{t \to \infty} \frac{1}{t} \int_0^t 1(S_i(u) > 0)du =: P_B^{(i)} = \rho_i, \ i = 1, \ldots, N, \tag{3}$$

is the stationary probability that the server is occupied by a class-i customer and $1(\cdot)$ denotes indicator function. Thus, expression (2) becomes

$$P(\mathbb{S}_i > x) = \lambda_i \int_x^\infty (1 - F_i(u))du, \tag{4}$$

in turn implying the expression (3) for the stationary busy probability if $x = 0$. (Indeed, the analysis in [12] has been performed for each server in a far more general multiserver $M/G/m$ system with the identical servers.)

Below we give another proofs of these results and also the proof of the corresponding versions of PASTA property. In turn this allows further to establish PASTA for the remaining service time is an unstable queueing system.

3   PASTA in Positive Recurrent Case

In this section, we give a simple proof of the PASTA property for the above described $M/G/1$ system, under assumption $\rho < 1$. Alternatively, we could refer to the established property PASTA [1]. However the independent proof is useful because then we apply the same idea to prove PASTA in the *null-recurrent* systems, in which case we can not appeal to the stationary distributions at the arbitrary and 'embedded' instants.

First of all, we have the following balance equation, connecting the class-i work $V_i(t)$ arrived in the interval $[0, t]$ with the remaining class-i work $W_i(t)$ at instant t and the busy time $B_i(t)$ of server in $[0, t]$ when it is occupied by class-i customers:

$$V_i(t) = W_i(t) + B_i(t), \ t \geq 0, \tag{5}$$

where, by the positive recurrence, $W_i(t) = o(t)$ w.p.1 as $t \to \infty$ [11]. Denote by $A_i(t)$ the number of class-i arrivals in the interval $[0, t]$, then

$$V_i(t) = \sum_{k=1}^{A_i(t)} S_k^{(i)}, \ t \geq 0,$$

and, it follows from (5) by the Strong Law of Large Numbers that, w.p.1,

$$\lim_{t \to \infty} \frac{V_i(t)}{t} = \lim_{t \to \infty} \frac{V_i(t)}{A_i(t)} \frac{A_i(t)}{t} = \lambda_i \mathsf{E} S^{(i)} = \rho_i. \tag{6}$$

It now follows in the limit from the balance equation (5) that, w.p.1,

$$\lim_{t \to \infty} \frac{B_i(t)}{t} = \mathsf{P}_B^{(i)} = \rho_i,$$

and it corresponds to expression (3). We stress that the probability $\mathsf{P}_B^{(i)}$ is obtained as the *time-average* limit which is also a weak limit, that is

$$\mathsf{P}_B^{(i)} = \lim_{t \to \infty} \mathsf{P}(S_i(t) > 0) = \mathsf{P}(\mathbb{S}_i > 0).$$

Now we obtain this probability as the *customer-average* limit, establishing the PASTA property for $\mathsf{P}_B^{(i)}$. Denote by θ the number of arrivals during a regeneration cycle of the system, and let θ_i be the number of class-i customers arrived within a cycle, that is the (stochastic) equality $\theta =_{st} \sum_{i=1}^{N} \theta_i$ holds. We note that each customer entering server belongs to class i with the probability λ_i/λ. It then easy to see that, by the property of (the assumed) FIFO service discipline,

$$\frac{\mathsf{E}\theta_i}{\mathsf{E}\theta} = \frac{\lambda_i}{\lambda}, \ i = 1, \ldots, N. \tag{7}$$

Denote by τ_i the interarrival time between i-class arrivals. Because the mean regeneration cycle length $\mathsf{E}T = \mathsf{E}\theta\mathsf{E}\tau$, then equality (7) can be also written in a more intuitive form as

$$\mathsf{E}\theta_i\mathsf{E}\tau_i = \mathsf{E}\theta\mathsf{E}\tau = \mathsf{E}T,$$

meaning that the mean regeneration cycle length can be also expressed, for each i, as the mean sum of interarrival times between θ_i class-i arrivals within the cycle. Also we denote by $\alpha_k^{(i)}$ the number of arrivals (of all classes) during service of the kth class-i customer (within a cycle), so all such customers meet server busy. Note that $\{\alpha_n^{(i)}, n \geq 1\}$ are iid variables with generic element $\alpha^{(i)}$. Then it is easy to find from the regenerative theory and Wald's identity [1] that

$$\lim_{n \to \infty} \frac{1}{n} \sum_{k=1}^{n} 1(S_i(t_k) > 0) = \frac{E \sum_{k=1}^{\theta_i} \alpha_k^{(i)}}{E\theta} = \frac{E\theta_i E\alpha^{(i)}}{E\theta} = \frac{\lambda_i E\alpha^{(i)}}{\lambda}. \tag{8}$$

Note that we can apply Wald's identity to the sum $E \sum_{k=1}^{\theta_i} \alpha_k^{(i)}$ because θ_i is a *stopping time* with respect to the summands [4]. It is left to find $E\alpha^{(i)}$. Because the superposed Poisson input has rate λ, we easily find that the mean number of arrivals during a class-i service time satisfies

$$E\alpha^{(i)} = \int_0^\infty \lambda x dF_i(x) = \lambda E S^{(i)}.$$

Substituting this result in (8), we arrive to the relation

$$\lim_{n \to \infty} \frac{1}{n} \sum_{k=1}^{n} 1(S_i(t_k) > 0) = \rho_i = P_B^{(i)}, \tag{9}$$

which, together with the time-average limit (3), establishes PASTA property for the busy probability $P_B^{(i)}$, $i = 1, \ldots, N$.

Now we obtain the PASTA for the *entire stationary distribution* of the remaining service time. First of all we calculate the limiting distribution as the time-average limit using construction from [12]: collecting together all θ_i service times $\{S_k^{(i)}, k = 1, \ldots, \theta_i\}$ of class-i customers served within a regeneration cycle and then shifting the obtained busy period $B_i := \sum_{k=1}^{\theta_i} S_k^{(i)}$ to the beginning of the cycle. Combining approach based on a regenerative argument and previous analysis we obtain the following relations:

$$\lim_{t \to \infty} \frac{1}{t} \int_0^t 1(S_i(u) > x) du = \frac{E \int_0^{B_i} 1(S_i(u) > x) du}{ET}$$

$$= \frac{E \sum_{k=1}^{\theta_i} \int_0^{S_k^{(i)}} 1(S_i(u) > x) du}{E\theta \, E\tau} = \lambda \frac{E\theta_i E \int_0^{S^{(i)}} 1(S_i(u) > x) du}{E\theta}. \tag{10}$$

where also the Wald's identity $ET = E\theta \, E\tau$ is applied. Now we calculate the generic term

$$E \int_0^{S^{(i)}} 1(S_i(u) > x) du = \int_0^\infty P(S_i(u) > x, S^{(i)} > u) du$$

$$= \int_0^\infty P(S^{(i)} > u + x) du = \int_x^\infty (1 - F_i(u)) du, \tag{11}$$

where we use the equality $S_i(u) = S^{(i)} - u$, provided $u \leq S^{(i)}$. Substituting (11) in (10) implies

$$\lim_{t \to \infty} \frac{1}{t} \int_0^t 1(S_i(u) > x)du = \frac{\mathsf{E}\theta_i}{\mathsf{E}\theta} \lambda \int_x^\infty (1 - F_i(u))du = \lambda_i \int_x^\infty (1 - F_i(u))du,$$

and hence the stationary remaining service time $\mathbb{S}_i$ has the (tail) distribution (4) obtained in [12].

To prove PASTA, we must obtain the same result considering the embedded process at the arrival instants $\{t_n\}$ (and also at the arrival instants $\{t_n^{(i)}\}$ of class-i customers only). Denote by $\mathcal{B}_{ik}$ the set of numbers of arrivals during service of the kth class-i customer. Note that the capacity $|\mathcal{B}_{ik}| = \alpha_k^{(i)}$. By analogy with (8), for an arbitrary fixed $x \geq 0$, we can write

$$\lim_{n \to \infty} \frac{1}{n} \sum_{k=1}^n 1(S_i(t_k) > x) = \frac{\mathsf{E}\sum_{k=1}^{\theta_i} \sum_{j \in \mathcal{B}_{ik}} 1(S_i(t_j) > x)}{\mathsf{E}\theta}$$

$$= \frac{\mathsf{E}\theta_i \mathsf{E}\sum_{j=1}^{\alpha^{(i)}} 1(S_i(t_j) > x)}{\mathsf{E}\theta}, \qquad (12)$$

where $\sum_{j=1}^{\alpha^{(i)}} 1(S_i(t_j) > x)$ denotes the generic number of arrivals, during service time of a class-i customer, which observe the remaining service time bigger than x. It can be realized by taking an independent Poisson process of arrivals during a generic class-i service time. It is easy to calculate, using integration by parts, that

$$\mathsf{E}\sum_{j=1}^{\alpha^{(i)}} 1(S_i(t_j) > x) = \int_x^\infty \lambda(u - x)dF_i(u) = \lambda \int_x^\infty (1 - F_i(u))du. \qquad (13)$$

Inserting (13) into (12) we arrive to (4), and it proves PASTA.

Now denote by $\widehat{\theta}_i$ the number of *class-i* arrivals during a type-i regeneration cycle which is started by a class-i customer arriving in an empty system. (The corresponding regeneration instants can be easily constructed by analogy with (1).) We stress that now we only observe the remaining class-i service time which a *class-i arrival* meets. Denote by β_{ik} the set of numbers of i-arrivals during service of the kth class-i customer. (Notice a difference between β_{ik} and $\mathcal{B}_{ik}$.) We note that the capacity of these sets are iid with generic element denoted by β_i. Then, with a minor modification of the analysis above, we obtain

$$\lim_{n \to \infty} \frac{1}{n} \sum_{k=1}^n 1(S_i(t_k^{(i)}) > x) = \frac{\mathsf{E}\sum_{k=1}^{\widehat{\theta}_i} \sum_{j \in \beta_{ik}} 1(S_i(t_j^{(i)}) > x)}{\mathsf{E}\widehat{\theta}_i}$$

$$= \mathsf{E}\sum_{j=1}^{\beta_i} 1(S_i(t_j^{(i)}) > x), \qquad (14)$$

where $\sum_{j=1}^{\beta_i} 1(S_i(t_j^{(i)}) > x)$ is a generic number of class-i arrivals, during a generic class-i service time, which observe the remaining service time bigger than x. As above we obtain

$$\mathsf{E} \sum_{j=1}^{\beta_i} 1(S_i(t_j^{(i)}) > x) = \int_x^\infty \lambda_i(u - x)dF_i(u) = \lambda_i \int_x^\infty (1 - F_i(u))du, \quad (15)$$

proving PASTA in this case as well. It is straightforward to obtain the same result for the *attained service time* of a class-i customer at instant t, denoted by $\hat{S}_i(t)$. Note only that $\hat{S}_i(t_k) \leq t_k$ for each t_k. Then, by analogy with (14), (15) and for arbitrary fixed x we again obtain that,

$$\lim_{n \to \infty} \frac{1}{n} \sum_{k=1}^n 1(\hat{S}_i(t_k) > x)du = \lambda_i \int_x^\infty (1 - F_i(u))du.$$

This result expresses the well-known fact that the attained and remaining (renewal) times have the same equilibrium limit.

The analysis developed above can be easily extended to some other positive recurrent systems, for instance, to retrial systems with constant retrial rate studied in [12]. In these systems specific of structure of a system is reflected in the value of stationary busy probability (3), while the integral term of the stationary distribution of the remaining service time and the proof of PASTA remain the same.

Remark 1. The results given above can be directly extended to m-server $M/G/m$ system with identical servers, because, by FCFS discipline, the structure of the input in each server remains the same as in the single-server case, only with the rates λ, λ_i replaced by λ/m, λ_i/m, respectively.

4   Null-Recurrent System

Assume now that the N-class system $M/G/1$ is *non-positive recurrent* (or *null-recurrent*), that is $\mathsf{E}T = \infty$. Because, by Wald's identity, $\lambda \mathsf{E}T = \mathsf{E}\theta$, then we also have $\mathsf{E}\theta = \infty$. (In this identity, both sides either finite or infinite simultaneously, see [4].) Then, by assumption, the number of customers $Q(t) \Rightarrow \infty$ (in probability) and hence,

$$\lim_{t \to \infty} \mathsf{P}(Q(t) > k) = 1,$$

for each $k \geq 0$, implying in particular, that (unconditional) remaining service time $S(t)$ satisfies

$$\lim_{t \to \infty} \mathsf{P}(S(t) > 0) = 1, \quad (16)$$

see for instance, [11]. Then the idle time of the server in the time interval $[0, t]$,

$$I(t) = \int_0^t 1(S(u) = 0)du,$$

satisfies

$$\lim_{t \to \infty} \frac{\mathsf{E}I(t)}{t} = \lim_{t \to \infty} \frac{1}{t} \int_0^t \mathsf{P}(S(u) = 0)du = 0. \tag{17}$$

Denote by $\tilde{S}_i(t)$ the remaining *class-i service time* at instant t in the process generated by the service times realized in the server. We denote such a process (being a combination of differently distributed renewal intervals) as $\tilde{\mathbb{Z}}$. As above, $\tilde{S}_i(t) = 0$ if the server serves class-k customer, $k \neq i$. We stress that this service process has no idle periods. Denote by $V(t) = \sum_{i=1}^N V_i(t)$ the total workload arrived in the system in $[0, t]$. It has been proved in [12] (also see [5]) that the limiting probability $\widehat{\mathsf{P}}_B^{(i)}$ that the server is occupied by a class-i customer in the service process $\tilde{\mathbb{Z}}$ satisfies

$$\widehat{\mathsf{P}}_B^{(i)} = \lim_{t \to \infty} \frac{V_i(t)}{V(t)} = \frac{\rho_i}{\rho}, \tag{18}$$

and that, w.p.1,

$$\begin{aligned}
\lim_{t \to \infty} \frac{1}{t} \int_0^t 1(\tilde{S}_i(u) \leq x)du &= \frac{\rho_i}{\rho}\mu_i \int_0^x (1 - F_i(u))du \\
&= \frac{\lambda_i}{\rho} \int_0^x (1 - F_i(u))du
\end{aligned} \tag{19}$$

Recall that $S_i(t)$ is the remaining service time (at instant t) in the real service process (in which the mean idle time satisfies (17)). It then easy to see that the following inequalities hold for all t:

$$\int_0^t 1(\tilde{S}_i(u) \leq x)du \leq \int_0^t 1(S_i(u) \leq x)du \leq I(t) + \int_0^t 1(\tilde{S}_i(u) \leq x)du. \tag{20}$$

Because, for each i,

$$\frac{1}{t} \int_0^t 1(\tilde{S}_i(u) \leq x)du \leq 1,$$

then by the dominated convergence theorem [4], we obtain from (19) that

$$\lim_{t \to \infty} \frac{1}{t}\mathsf{E} \int_0^t 1(\tilde{S}_i(u) \leq x)du = \frac{\lambda_i}{\rho} \int_0^x (1 - F_i(u))du.$$

Now (20) and (17) imply

$$\begin{aligned}
\lim_{t \to \infty} \frac{1}{t}\mathsf{E} \int_0^t 1(S_i(u) \leq x)du &= \frac{\lambda_i}{\rho} \int_0^x (1 - F_i(u))du \\
&= \frac{\rho_i}{\rho}\mu_i \int_0^x (1 - F_i(u))du.
\end{aligned} \tag{21}$$

This result has a clear intuitive interpretation and states that in the null-recurrent case the time-average limit of the remaining service time distribution, with probability $\widehat{\mathsf{P}}_B^{(i)} = \rho_i/\rho$, equals the distribution of the *stationary remaining renewal time* in the renewal process generated by class-i service times. Note that we can not claim the weak convergence $S_i(t) \Rightarrow \mathbb{S}_i$ in the non-positive recurrent case.

Remark 2. In the *single-class* $M/G/1$ system, with service time distribution F and service rate $\mu = 1/\mathsf{E}S$, expression (21) becomes

$$\lim_{t \to \infty} \frac{1}{t} \int_0^t \mathsf{P}(S(u) \le x) du = \mu \int_0^x (1 - F(u)) du, \ x \ge 0 \ ,$$

and is the distribution of the stationary remaining renewal time in the corresponding renewal process.

Now we establish PASTA property for this system considering the remaining service time in the process $\tilde{\mathbb{Z}}$ at the arrival instants of the original input process. To this end, we denote by $T_n^{(i)}$ the departure time of the nth class-i customer, and define *i-type regeneration cycle* as the interval $(T_n^{(i)}, T_n^{(i)}]$, with generic length $T^{(i)}$. In other words, this regeneration period contains exactly one class-i service time, and a new cycle starts just after class-i customer leaves the system. First of all, it is easy to understand, that, for each i, the lengths of such constructed periods indeed form an iid sequence. This definition first allows easily to calculate the mean regeneration period length in continuous-time setting. Namely, it follows by construction and by regenerative argument, that w.p.1,

$$\lim_{t \to \infty} \frac{1}{t} \int_0^t 1(\tilde{S}_i(u) > 0) du = \frac{\mathsf{E}S^{(i)}}{\mathsf{E}T^{(i)}}. \tag{22}$$

On the other hand, it follows from (18) that the limit (22) must also be equal to ρ_i/ρ, implying equality $\mathsf{E}T^{(i)} = \rho/\lambda_i$. Moreover, denoting by D_i the number of all arrivals during i-type regeneration cycle, we have from Wald's identity

$$\mathsf{E}D_i = \lambda \mathsf{E}T^{(i)} = \lambda\rho/\lambda_i. \tag{23}$$

(We note that in general $D_i \ne \theta_i$ because these quantities relate to differently defined regeneration cycles.) Using a coupling, we remain distributions of all processes unchanged if take the arrival instants during each service time in the original system the same as the arrival instants during the *same service time* in the process $\tilde{\mathbb{Z}}$. (It is easy to do using memoryless property of exponential distribution.) Denote by $\mathcal{Z}_n$ the number of arrivals meeting server idle within interval $[0, t_n]$. The coupling of service times allows to obtain the following *stochastic inequalities*:

$$\sum_{k=1}^n 1(\tilde{S}_i(t_k) \le x) \le_{st} \sum_{k=1}^n 1(S_i(t_k) \le x) \le_{st} \sum_{k=1}^n 1(\tilde{S}_i(t_k) \le x) + \mathcal{Z}_n. \tag{24}$$

Note that

$$\mathcal{Z}_n = \sum_{k=1}^{n} 1(S(t_k) = 0),$$

and that $P(S(t_k) = 0) \to 0$ as $k \to \infty$ by assumption $E\theta = \infty$. (In the opposite case, assuming $P(S(t_k) = 0) \nrightarrow 0$, we obtain a contradiction: $E\theta < \infty$, see [11].) It immediately implies that $E\mathcal{Z}_n = o(n)$. On the other hand, using (13) we obtain from regenerative argument applied to the process $\tilde{\mathbb{Z}}$ (with regeneration periods with length D_i) that, w.p.1,

$$\frac{1}{n} \sum_{k=1}^{n} 1(\tilde{S}_i(t_k) > x) \to \frac{\lambda}{ED_i} \int_x^\infty (u - x) dF_i(u) = \frac{\lambda}{ED_i} \int_x^\infty (1 - F_i(u)) du.$$

By the dominated convergence theorem, the convergence in mean in the latter relation holds as well. Because, from (21), $ED_i = \lambda\rho/\lambda_i$ then, after taking expectation in all inequalities in (24), dividing by n, we obtain

$$\lim_{n\to\infty} \frac{1}{n} E \sum_{k=1}^{n} 1(S_i(t_k) \le x) = \frac{\lambda_i}{\rho} \int_0^x (1 - F_i(u)) du.$$

Now comparing the latter result with (21), we conclude that PASTA *in mean* holds in the non-positive recurrent system.

5   Simulation Results

In this section we verify the theoretical results obtained above, using stochastic (discrete-event) simulation and software R. We model a queueing system with $N = 2$ classes of Poisson customers with rate $\lambda_1 = \lambda_2 = 1$. We consider both positive recurrent case and null-recurrent case, and in both scenarios we use Weibull and Pareto service time, denoted by $Weibull(k, \lambda), Pareto(x_0, \alpha)$, with distribution function,

$$F_i(x) = 1 - e^{(-\lambda x)^k}, \ x \ge 0,$$
$$F_i(x) = 1 - \left(\frac{x_0}{x}\right)^\alpha, \ x \ge x_0,$$

respectively.

Positive recurrent case. In the first two experiments, we use service times $Weibull(2, 1/2)$ and $Weibull(3, 1/3)$ for classes 1,2, respectively. It is easy to calculate that

$$\rho_1 + \rho_2 = 0.443 + 0.298 = 0.74 < 1,$$

and stability criteria is fulfilled.

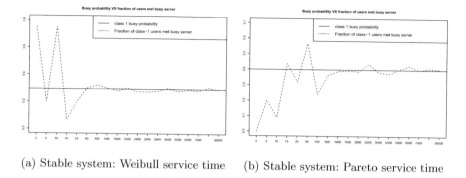

(a) Stable system: Weibull service time   (b) Stable system: Pareto service time

Fig. 1. An illustration of PASTA for the 1st class customers: $Weibull(2, 1/3)$ (left) and $Pareto(1/5, 2)$ (right).

Figure 1 shows how fast the fraction of the 1st class arrivals which meet server busy approaches the line expressing the 1st class busy probability ρ_1 (vs. the number of arrivals K), for the Weibull and Pareto service time distributions, respectively. (The plots for the 2nd class are quite similar and omitted.)

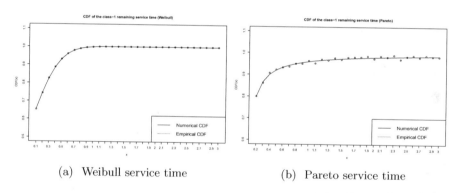

(a) Weibull service time   (b) Pareto service time

Fig. 2. Comparison of the actual cumulative distribution function (CDF), obtained numerically, and ECDF of the class-1 remaining service time observed by all customers; $K = 1000$ arrivals, averaging over $N = 100$ simulation runs.

In the next experiment we again consider a two-class stable system. It is worth mentioning that in modelling, instead of the actual remaining service time, we compute the attained service time which converges to the same limit [5]. By this reason and because we discuss the limiting results, we remain the term 'remaining time' (instead of 'attained' time). More specifically, to construct empirical distribution function (ECDF) of the 1st class remaining service time, we record the attained service time at the moments of all arrivals which meet server occupied by class-1 arrival. We run the simulation with Weibull and Pareto distributions, with the same parameters as above, and with $K = 1000$. Figure 2 shows a remarkable proximity between the theoretical distribution and ECDF in both cases. (The results for the 2nd class remaining service time are similar.)

Null-recurrent case. In the final experiments, we verify the results obtained in Sect. 4 for the null-recurrent (unstable) system in which case $\rho \geq 1$. We compute the ECDF of the remaining service time and compare it with the theoretical distribution function (CDF) constructed numerically. In these experiments we choose the following parameters: $Weibull(2,1)$ and $Weibull(1/2,1)$ implying $\rho = 2.9 > 1$, and $Pareto(1,2)$ and $Pareto(1/2,3)$ implying $\rho = 2.75 > 1$. CDF and ECDF are shown on the Fig. 3 again demonstrate a perfect agreement between theoretical and empirical curves for the 1st class remaining service time observed by all arrivals. (The results for the 2nd class remaining service time are quite similar.)

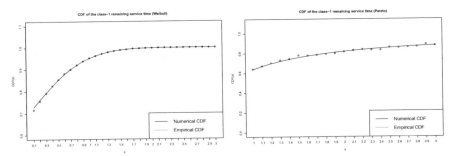

(a) CDF and ECDF of the 1st class remaining (Weibull) service time

(b) CDF and ECDF of the 1st class remaining (Pareto) service time

Fig. 3. Unstable system: comparison of CDF and ECDF of the 1st class remaining service time observed by all arrivals: (a) $Weibull(2,1)$ service time and (b) $Pareto(1,2)$ service time; $K = 1000$, averaging over $N = 100$ simulation runs.

6   Conclusion

In this research we consider a multiclass queuing system with Poisson inputs and give a direct proof of the PASTA property for the remaining service time. The proof first is applied to the positive recurrent (stable) system, when PASTA means the equality w.p.1 limits of the continuous-time sample mean and discrete-time sample mean. Then this result is extended to the *null-recurrent* system, in which case the corresponding limits *in mean* coincide. Some numerical results are included as well.

References

1. Asmussen, S.: Applied Probability and Queues, 2nd edn. Springer, New York (2003). https://doi.org/10.1007/b97236

2. Boer, P.T.D., Nicola, V.F., van Ommeren, J.K.C.: The remaining service time upon reaching a high level in $M/G/1$ queues. Questa **39**, 55–7823 (2001). https://doi.org/10.1023/A:1017935616446

3. Serfozo, R.F.: Basics of Applied Stochastic Processes. Springer-Verlag, Heidelberg (2009). https://doi.org/10.1007/978-3-540-89332-5

4. Borovkov, A.A.: Probability Theory. Springer Science and Business Media, London (2013). https://doi.org/10.1007/978-1-4471-5201-9

5. Morozov, E., Steyaert, B.: Stability Analysis of Regenerative Queueing Models. Springer, Cham (2021). https://doi.org/10.1007/978-3-030-82438-9

6. Iglehart, D., Whitt, W.: Multiple channel queues in heavy traffic. Adv. Appl. Prob. **2**, 150–170 (1970)

7. Kerner, Y.: The conditional distribution of the residual service time in the $M_n/G/1$ queue. Stochast. Models **24**(3), 364–375 (2008)

8. Kerner, Y.: Equilibrium joining probabilities for an $M/G/1$ queue. Games Econ. Behav. **71**(2), 521–526 (2011)

9. Morozov, E.: The tightness in the ergodic analysis of regenerative queueing processes. Queueing Syst. **27**, 179–203 (1997). https://doi.org/10.1023/A:1019114131583

10. Morozov, E., Delgado, R.: Stability analysis of regenerative queues. Autom. Remote Contr. **70**, 1977–1991 (2009). https://doi.org/10.1134/S0005117909120066

11. Morozov, E., Steyaert, B.: A stability analysis method of regenerative queueing systems. In: Anisimov, V., Limnios, N. (eds.) Queueing Theory 2. Advanced Trends, Wiley/ISTE, New York (2021)

12. Morozov, E., Morozova, T.: On the stationary remaining service time in the queueing systems. CEUR Workshop Proc. **2792**, 140–149 (2020)

13. Ross, S.M., Seshadri, S.: Hitting time in an $M/G/1$ queue. J. Appl. Prob. **36**, 934–940 (1999)

Application of Machine Learning Methods to Solving Problems of Queuing Theory

Vladimir Vishnevsky[ID] and Anastasia V. Gorbunova[✉][ID]

V.A. Trapeznikov Institute of Control Sciences of Russian Academy of Sciences, 65, Profsoyuznaya Street, Moscow 117997, Russia
avgorbunova@list.ru

Abstract. This review is the first to propose the systematic presentation of a new approach to the study of queuing systems and networks. The concept of the new approach is based on a combination of traditional methods of queuing theory with various machine learning algorithms. The detailed description and justification of the possibility of applying the approach are given on the example of a combination of simulation with artificial neural networks. The analysis of publications allows us to conclude that the application of machine learning methods is highly effective, promising for further research, as well as for the possible separation of this new approach into an independent direction in the field of solving complex problems of the theory of queues.

Keywords: queuing theory · queuing system · simulation · data mining · machine learning · artificial neural networks

1 Introduction

Mathematical models of networks and systems based on the queuing theory are widely used in the design of modern telecommunication networks, including analysis of the characteristics of existing and future network protocols, optimization of routing algorithms and network topological structure, etc. Starting with the pioneering works of A. Erlang, A. Hinchin, L. Kleinrock, and up to the present, a huge number of articles have been published on the study of various systems and queuing networks and their application in telecommunication networks.

In recent years, the interest of researchers in the field of queuing theory has shifted towards the analysis of queuing systems with correlated input flows (MAP, BMAP, MMAP) since such flows are typical for modern computer networks [1,2]. The formation and development of this scientific direction were facilitated by the work of N. Newts, G.P. Basharin, D. Lucantoni, A.N. Dudin, V.I. Klimenok, and others.

Traditionally, the primary attention in the study of complex queuing systems (QS) is directed to the analysis of the stationary mode of operation, the construction of a multidimensional Markov chain, the derivation of the Kolmogorov

The reported study was funded by RFBR, project number 19-29-06043.

A. Dudin et al. (Eds.): ITMM 2021, CCIS 1605, pp. 304–316, 2022.
https://doi.org/10.1007/978-3-031-09331-9_24

system of differential equations, and their solution using the Laplace-Stieltjes transformation apparatus and/or generating functions. However, a significant number of unsolved problems in the queuing theory (QT) still remain, the analytical and numerical solution of which is either difficult or completely impossible by traditional methods. Examples of such tasks are the investigation of multi-server priority queuing systems with correlated input flows (MMAP) and a buffer of limited size; analysis of the characteristics of networks and of multiphase QS of large dimension with incoming BMAP-flow and general functions of service distribution on phases; the study of adaptive dynamic polling systems and of the fork-join QS, as well as a number of problems of the theory of reliability of the k-out-of-n-type, etc. The lack of new methods and approaches to the analytical and numerical solution of such problems hinders the practical application of queuing theory models in the performance evaluation and design of telecommunication networks.

In this paper, we propose one of the new approaches based on a combination of machine learning and simulation methods. A brief description of the new approach, the features of its application on specific examples, and, consequently, the possible prospects for its separation into an independent direction in further research in the field of queuing theory are given. Although the development and implementation of this new approach began relatively recently, this review also considers early publications in which this topic was only indirectly touched upon. These publications deserve attention since they allow tracing the development of a new method from its starting point to the current state. At the same time, the strict formulation of the proposed approach and the demonstration of its effectiveness in the study of various models of queuing are reflected mainly in the recent works of the authors of this review.

2   Artificial Neural Networks and Queuing Theory

There are three main types of problems that can be solved by using various machine learning algorithms: classification, clustering, and forecasting. The last task is to predict the behavior of the system from its previous reactions, which actually reduces to the problem of approximating a function of several variables. In what follows, we will be interested in the use of artificial neural networks (ANN) and other machine learning methods in the context of solving this problem, since, for example, neural networks are considered one of the best tools for approximating functions [3].

Currently, several types of ANN structures are known; however, in most studies on the subject of QT, a perceptron is used, characterized by one or more hidden layers and direct signal propagation. On each neuron, the products of the input data and the corresponding weight coefficients are summed up, which, after passing through the activation function as an argument, then become the input data for the neurons of the next layer.

According to theorems on approximation of functions given in [4], as well as in the review [3], any continuous function can be represented as a combination of linear operations and a single nonlinear element, which allows them to be applied to a multilayer perceptron, where as nonlinear element is an activation function.

There are many approaches to learning neural networks, which, in turn, affect the approximation process. Learning is understood as the process of finding the optimal weight coefficients that would fully reflect the relationship between the input values and the output of the neural network. One of the most famous and common learning methods is the backpropagation method.

Other widely used ANN learning methods include the Levenberg-Markart algorithm, the scaled conjugate gradient algorithm, and the Bayesian regularization algorithm. The question of choosing a specific algorithm from the entire set of existing ones can be solved experimentally when implementing a specific task. In this case, for example, one can be guided by the optimal ratio of such characteristics as the accuracy and labor intensity of the algorithms selected for comparison or by special metrics for machine algorithms.

The work [5] can be attributed to one of the first mentions of the application of machine learning methods, in particular, the algorithm for constructing a decision tree, to solving QT problems. It speaks in general about the potential of machine learning methods and discusses some new (thanks to these methods) possibilities of modeling the behavior of complex systems and the advantages of their use in the implementation of expert systems. The proposed approach is illustrated by the example of a simulation of one queuing system.

The work [6] presents the comparative analysis of the telecommunication network response time estimation, obtained using the classical methods of queuing theory (the analysis of the $M|M|2$ model) and the estimation obtained the neural network training based on the backpropagation method for the set of real data collected while monitoring the operation of the network under investigation. Naturally, the neural network gave a better estimate in the context of a lower value of the mean square error of the model. This was due to the simplicity of the proposed analytical model.

Existing works (articles) can be divided into several categories depending on the purpose of applying machine learning methods. On the one hand, machine learning can be directly used to simulate the operation of real queuing systems. On the other hand, their application is possible for the analysis of complex queuing models, for which the calculation of estimates of the characteristics is not always possible in a closed form, and numerical methods are not always productive. Let us analyze the available thematic publications in more detail in order to reflect one of the advantages of using machine learning, namely the variety of problems that can be solved using its methods.

3   An Approach to Solving Problems of Queuing Theory Using Machine Learning Methods

Despite the presence of articles in which machine learning methods are applied in solving various kinds of problems lying to one degree or another in the plane of the QT, none of them fully formulated the concept of a new methodology. All publications on this topic are rather scattered. It is rather difficult to identify some general idea from them and even more so to single it out as a separate new direction in solving complex problems of QT on a par with classical methods.

The main provisions of applying the new approach in the form that allows it to be used in the analysis of queuing models of any complexity were formulated in the works of the authors of this review. These publications, which outline the fundamental principles of the new approach, will be described below.

The main idea of the proposed approach is to combine traditional queuing theory methods with various data mining methods. In particular, we are talking about a combination of simulation and artificial neural networks. Simulation is one of the ways to obtain highly accurate estimates of the performance measures of queuing models. However, the time taken to obtain one value of the model characteristic can range from a few seconds to several minutes. It depends on the complexity of the modeled QS, software environment for simulation, hardware of the computing system.

The simulation time to obtain the required number of estimates to form a complete picture of their behavior can exceed all reasonable limits. Therefore, if using simulation to get a set of values of the characteristics of interest for individual values of the input parameters within a given numerical interval, then it is possible to train a neural network on the obtained data, which, with the required degree of accuracy, will provide an estimate already for any intermediate values of the input parameters from the same intervals without any restrictions on their number.

As a result, it will be necessary to spend time on simulation modeling not for all the required values of the input parameters but for their limited number, as well as on direct training of the neural network or some other intelligent model. The forecasting process itself does not actually require time.

As for the construction of a simulation model, there are several options. You can use specialized software applications developed for this purpose. Some of the more popular applications are GPSS World, AnyLogic, and Arena [7]. As an alternative to ready-made options, you can develop your own simulation model, for example, in the Python software environment with a fairly wide range of capabilities and many ready-made libraries, including for training artificial neural networks.

Summarizing, let us highlight the following main stages of the approach using machine learning methods:

1. obtaining, using simulation, the values of the characteristics of the analyzed system for a finite set of values from the given numerical intervals for the input parameters, on which the system performance depends;
2. training of an intelligent model on the data obtained by simulation using one of the machine learning methods to solve the forecasting problem;
3. almost instantaneous estimation of the required performance characteristics for any other intermediate values of the input parameters at the same numerical intervals using a trained, intelligent model.

The specifics of building simulation models of queuing systems can be found, for example, in [8]. However, the question of the length of the model run remains difficult, i.e., in this case, it means the number of requests that must be passed through the system to obtain one output value (or one set of output values). It should be noted that the data obtained during the implementation of one run to estimate the average value of almost any investigated random variable are correlated, which leads to the need for a significant increase in the number of realizations in comparison with the case of independent random variables. In addition, the run length is not a fixed value but is determined individually for each set of input parameters, which additionally increases the simulation time.

If a computational algorithm is developed for assessing the probabilistic-temporal characteristics of a certain network or queuing system, but it requires too much time and computational costs, then similarly using this algorithm, you can first obtain estimates of characteristics for a limited set of input parameters, and then train a neural network and solve the forecasting problem.

4   Review of Publications on the Application of Machine Learning Methods to Solving Problems of QT

Although machine learning methods and algorithms have found wide application in various fields of science and technology [9], including the study of modern broadband wireless networks of a new generation [10–13], their application in the queuing theory is weak reflected in world literature.

Physical queuing is a reality in many areas of human life, particularly in the service or sales industry. Therefore, research in this area aimed at reducing the waiting time and, accordingly, increasing service efficiency remains relevant today. the classical methods of the queuing theory remain one of the main tools for assessing the waiting time of a client in a queue (in a bank, a medical institution). Recently, however, there has been a lot of research on predicting queuing times using machine learning methods [14–17].

Works [18,19] are devoted to the construction of a model of the classical queuing system $M|M|1$ using an artificial neural network and the analysis of the adequacy of this simulation. In [18], the neural network was developed using a backpropagation algorithm with one hidden layer. Experiments have shown that the values simulated using the neural network coincide with the calculated values obtained by the classical mathematical approach.

In [19], an artificial neural network model was also developed to simulate the QS $M|M|1$. The input layer of neurons included such values as the intensity of the incoming flow and the intensity of service. The output layer consisted of eight neurons, which corresponded to the values of the probability of system downtime, the average number of requests in the queue and in the system, the average waiting time for the start of service and stay in the system, and others. The resulting model was tested, and the verification showed that the neural network is highly consistent with the analytical model and can predict the target parameters for the given input data with minimal insignificant error. In [20], similarly, a neural network simulating the classical QS is used to plan and optimize the queue at the airport runway.

The use of neural networks for the analysis of non-Markov QSs seems to be one of the most promising areas of research in the field of QT. At the moment, there is a small number of publications devoted to this topic. Nevertheless, this gap is beginning to be filled, including by the authors of this work, since it is through non-Markov systems that most real physical systems and processes in them are modeled, and their study using classical methods does not always give a satisfactory result.

One of the first works that touched on the application of the apparatus of neural networks to the analysis of non-Markov QS models is the article [21]. Here the non-Markov QS with a "warm-up" [22] is investigated, and this QS can be used to simulate the activation process of an empty system in the event that it receives the first request after a break in its operation. This system is successfully "markovized" by approximating it by means of the QS with the phase-type distribution of the incoming flow or the service time, which, as is known, can be used as an approximating function of general distribution. However, already in the case of QS with "warm-up" of the form $H_2|M|M|3$ and $M|H_2|M|3$, the numerical algorithms for calculating the stationary probabilities of states turn out to be very time-consuming and resource-consuming. The use of neural networks, in this case, made it possible to significantly reduce the complexity without losing the accuracy of the calculations.

A two-layer perceptron was chosen as the structure of the neural network. At the input of the neural network, the intensities of the incoming and serving flows, as well as the "warm-up" and the coefficient of variation, were fed. The output parameters in the case of the QS $H_2|M|M|3$ were the stationary distribution of the number of customers, and for $M|H_2|M|3$, the average waiting time for servicing and the average sojourn time in the system. Three algorithms were used to train the ANN: the Levenberg-Markart algorithm, the scaled conjugate gradient algorithm, and the Bayesian regularization method. The latter turned out to be the most accurate in the sense of the minimum mean square error of the approximation.

The article [23] examines the priority QS with a marked Markovian arrival process (MMAP). The system under consideration is multi-server with a finite capacity; the service time on the servers has a phase-type distribution, the parameters of which differ for customers belonging to different priority classes. The priorities are relative; in addition, probabilities, with which the customer can either stay in the system, taking its place in the queue, or leave it immediately, are specified for customers of different classes, depending on the total number of customers in the system.

MMAP is used to describe the processes occurring in modern information and computing systems since it allows modeling the correlation properties of traffic and for the case of an arbitrary number of priorities. However, there are not so many works devoted to the study of QS with MMAP due to the complexity of such an analysis.

In [23], an algorithm was presented for finding the stationary probabilities of states of the described QS $MMAP|PH|M|N$ with two priority classes ($R = 2$), including matrix calculations. At the same time, with an increase in the number of devices in the system, in particular at $M = 5$, the dimension of the matrices can be so large that even modern computers may not have enough power to calculate the main performance measures of the QS. Therefore, to analyze QS with a large number of priorities ($R > 2$), an approach based on a combination of simulation modeling with various machine learning methods (decision trees, random forest, gradient boosting, neural networks, etc.) was used. The results of numerical analysis for the response time and the loss probability indicate in favor of the new approach due to a significant reduction in the computation time (of the order of 10^5 in comparison with the analytical algorithm) without any loss in their accuracy.

In the paper [24] the end-to-end delay of a multiphase queuing network (QN) is investigated, in which the first node is the QS $G|G|1$, and the subsequent ($K-1$) nodes are systems of the form $\cdot|G|1$. There are no exact analysis methods for such networks, so approximate ones are used. One of the main approaches to studying such networks is the decomposition method together with diffusion approximation, which implies the estimation of the parameters of incoming flows (coefficients of variation). There are several variants of the formulas known to determine the coefficients of variation, as a result of which several estimates were obtained for the average end-to-end delay [8,25]. But all of them can give a relatively low accuracy of the approximation.

For a comparative analysis of the results of applying the decomposition method and a new approach using ANN, two types of service time distribution are considered—uniform and Pareto distributions. The value of the mean absolute percentage error (MAPE) for the mean end-to-end delay predicted by the neural network does not exceed 1%, while in the case of four variants of analytical formulas for determining the coefficient of variation, which, in turn, is involved in the analytical expression for the mean end-to-end delay latency, this value varies from 2.321% in the best case to 6.387% in the worst case, respectively.

In the articles [26,27] a new method is used to study the mean response time and its standard deviation (variance) for a fork-join QS. The main idea of the functioning of this system is as follows: at the moment of arrival, the request is split into several sub-requests, after which each of them enters the queue for service to the corresponding server. The time spent in the QS is the maximum of the times spent in the sub-system of each of its sub-requests. The complexity of analyzing the response time in a fork-join system lies in the correlation of the sojourn times due to their identical moments of appearance in the system.

In [26] the fork-join system with K subsystems of the type $M|M|1$ is considered. Five approximate formulas for the average response time $E[R_K]$ estimation of such a system are best known, and one of them gives the smallest error at $K \leq 32$. For the variance of the average response time $Var[R_K]$ not so many formulas are known; basically, one [26] is used. The values of the average relative error of MAPE approximations calculated on the test sample in the case of estimating the average response time and its standard deviation using a neural network are approximately 0.739% and 0.364% versus 1.592% and 6.896% calculated by the corresponding analytical formulas.

In [27] a fork-join system is also investigated, but with K subsystems like $M|G|1$. One of the main approaches for the average response time and its second moment estimations is based on the theory of order statistics, but the approximation is much poorer than the situation with an exponential distribution. The service time on the servers has a Pareto distribution with the α parameter. Thus, the number of indicators affecting the probabilistic-temporal characteristics of the system includes the load ρ, the number of subsystems $M|G|1$—K, and the value α. As a result, the MAPE values for the mathematical expectation and the standard deviation of the time spent in the fork-join of the QS are approximately 24.425% and 18.702% in the case of calculations using analytical formulas and 0.708% and 3.355% in the case of using a neural network.

On the example of the presented works, various measures of approximation errors when using machine learning methods take smaller values compared to the results of using analytical formulas, which testifies in favor of the new approach. However, it is also of interest what lies behind these average characteristics, i.e., how large the scatter of the specific values of the errors included in the averaged expressions can be, because they are more intuitive and more descriptive in the context of comparative analysis.

In work [28], the result of applying neural networks to the analysis of a closed QN is elucidated. Since simulation modeling can be time-consuming, for the detailed research of a significant number of elements that make up MAPE, the exponential closed QN was chosen, for the characteristics of which the exact analytical expressions are known. This allows, nevertheless, following all stages of the approach, to check a larger number of estimates. The well-known analytical expressions are the formulas for stationary probabilities of states, which include a normalizing constant. For its non-trivial calculation, Busen's algorithm, programmed in Python, is used.

To carry out the declared detailed analysis, by using the trained neural network and analytical formulas, the average number of customers N_i and the average time spent v_i in each of the five network nodes are calculated for sets of service intensities μ_i. A total of 59049 such sets are obtained. The structure of the relative errors of approximations for a set of almost sixty thousand input elements is such that in the worst case, the relative error does not exceed 5%, and for a small amount of data.

The new technique was applied to a wide class of stochastic polling systems [29]. Polling systems are QS with several queues and only one server. According to a certain rule, the server visits the queues and serves the requests located in them. Despite a significant number of works in this area, many unsolved problems remain, particularly the study of systems with correlated input flows or systems with limited queuing disciplines. The machine learning method using

ANN was first applied to calculate the characteristics of polling systems in the article [30].

For example, [30] describes machine learning results for polling systems like $M|M|1$ with cyclic polling, $MAP|M|1$ with correlated input flow, as well as systems like $M|M|1$ with adaptive cyclic polling. The results of analytical calculations of performance characteristics were used to train a machine model of a polling system of type $M|M|1$, and for other systems, for which it was not possible to develop an algorithm for calculating performance characteristics by known methods, the results of the simulation were used. Extensive computational experiments have shown that the ANN training results coincide with high accuracy with the results of analytical or simulation calculations, while the machine model can significantly reduce the time for calculating the characteristics of polling systems in comparison with simulation.

In article [31] the system k-out-of-$n : F$, $k < n$ from the point of view of reliability is studied. It is a repairable system with a single repair unit. The failure of this system occurs in the case of failure of k elements, each of which begins to be repaired immediately after the termination of functioning and after the completion of the reparation begins to work again. It is assumed that the lifetime of the system components has an exponential distribution with the parameter α, and the repair time has a general distribution with an average value of b. The described closed QS can be denoted in terms of Kendall's classification as $\langle M_{k<n}|G|1 \rangle$. The set of states of a given system is described by a two-dimensional Markov process. As a result, after compiling the Kolmogorov system of differential equations and its subsequent solution, expressions were obtained for the stationary probabilities of the states of the QS in terms of the Laplace transform. Since here an exact solution was obtained for the stationary probabilities of the system, which allows one to evaluate the most important characteristic of the reliability of its operation, called the availability factor, the neural network was trained not on simulation data but on data calculated using analytical expressions.

Further, to test the new method's performance within the framework of a numerical experiment, the comparative analysis of the learning outcomes of the neural network and analytics is carried out. As the structure of the neural network, the two-layer perceptron with two input neurons corresponding to the average lifetime and the average repair time was chosen; at the output of the neural network, there is one neuron that produces the system availability values, the hidden layer contains 16 neurons, and the hyperbolic tangent acts as activation functions and its derivative, the neural network learning algorithm is Adam's method. The results of the neural network operation on the test dataset indicate a good quality approximation of the availability factor, which opens up new perspectives in the study of system availability of a more general form $\langle G_{k<n}|G|1 \rangle$.

In [32], neural networks are used to find the optimal distribution policy for customers in a multi-server QS with an unlimited storage capacity, heterogeneous servers, and operating costs. The system receives a Poisson flow of customers with intensity λ, the service time of a customer on the jth server has an exponential distribution with the parameter μ_j, and the servers are ordered in increasing order of the average service time. In addition, the system assumes costs, namely: the holding cost of waiting in the queue is equal to $c_0 > 0$, and the operating cost of servicing a customer on the jth server are c_j per unit of time, that is, on average $c_j\mu_j^{-1}$ is consumed per one request when servicing on the jth server. The system has a controller, which, based on information about the state of the system, makes a decision on the distribution of requests between servers in accordance with some policy f. The controller at the time of a new request arrival in the system or at the time of the end of service of the request on the device can either send the first request in the queue for service or leave it in the queue. The optimal policy for distributing requests between servers in terms of minimizing long-term average costs for such a system has a threshold form.

In order to determine the optimal thresholds, the policy-iteration algorithm [33] is used, since the minimizing the average cost function directly may turn out to be too time-consuming. However, this algorithm also has some drawbacks. In particular, difficulties arise with its convergence under conditions of high system load, and there are also restrictions on the state space of the process under study. Therefore, the article proposes two ways to solve this difficult task. On the one hand, a heuristic solution for finding threshold levels has been formulated. On the other hand, a neural network-based solution is presented. The values λ, μ_j, c_0, c_j, $j = \overline{1, K}$ arrive at the input of the neural network, and the sought optimal thresholds are at the output of the neural network. The neural network (six-layer perceptron) was trained on 70% of the data obtained using the iterative algorithm, using the Adam method in the Mathematica software environment (Wolfram Research). A check on the remaining 30% of the data showed a low approximation error, which indicates a high potential for solving optimization problems in the field of QT.

5   Conclusion

The article describes the new, promising approach to the study of queuing systems and queuing networks based on a combination of machine learning and simulation methods. A review of works published in the world literature is given, where this method is effectively used to find the numerical characteristics of complex QS with a significant reduction in the complexity and computation time.

It is shown that the new method is promising for further research of unsolved QT problems, such as the analysis of priority QS or multiphase systems of large dimension with incoming correlated flows (BMAP), the study of queuing networks that do not satisfy the BCMP theorem [1]. Therefore, the new approach, due to its versatility, can generate considerable interest and provoke much new research in the field of queuing theory.

References

1. Dudin, A.N., Klimenok, V.I., Vishnevsky, V.M.: The Theory of Queuing Systems with Correlated Flows. Springer, Cham (2020). https://doi.org/10.1007/978-3-030-32072-0
2. Lakatos, L., Szeidl, L., Telek, M.: Introduction to Queueing Systems with Telecommunication Applications. 1st edn. Springer, Boston (2013). https://doi.org/10.1007/978-1-4614-5317-8
3. Shvedov, A.S.: Functions approximating by neural networks and fuzzy systems. Control Sci. **1**, 21–29 (2018)
4. Stone, M.N.: The generalized Weierstrass approximation theorem. Math. Mag. **21**(4), 167–184 (1948)
5. Khoshnevis, B., Parisay, S.: Machine learning and simulation: application in queuing systems. SIMULATION **61**(5), 294–302 (1993)
6. Merlo, G., Britos, P., Rossi, B., Garcia Martinez, R.: Neural networks applied to automatic estimation of networks performance. In: Proceedings of the International Conference on Intelligent Systems and Control, pp. 167–171. Santa Barbara, California (1999)
7. Dias, L.M.S., Vieira, A.A.C., Pereira, G.A.B., Oliveira, J.A.: Discrete simulation software ranking - a top list of the worldwide most popular and used tools. In: Proceedings of the 2016 Winter Simulation Conference (WSC), pp. 1060–1071. Washington, DC. IEEE (2016)
8. Bolch, G., Greiner, S., Meer, H., Trivedi, K.S.: Queueing networks and Markov chains: modeling and performance evaluation with computer science applications. 2nd edn. Wiley (2006)
9. Bonetto, R., Latzko, V.: Computing in communication networks. Chapter 8. Machine learning. Academic Press. In: Fitzek, F.H.P., Granelli, F., Seeling, P. (eds.) Computing in Communication Networks – From Theory to Practice, pp. 135–167. Academic Press (2020)
10. Morshedi, M., Noll, J.: Estimating PQoS of video streaming on Wi-Fi networks using machine learning. Sensors **21**(2), 621 (2021)
11. Laha, S., Chowdhury, N., Karmakar, R.: How can machine learning impact on wireless network and IoT? - a survey. In: 1th International Conference on Computing. Communication and Networking Technologies (ICCCNT), pp. 1–7. Kharagpur. IEEE (2020)
12. Luong, N.C., Hoang, D.T., Gong, S., et al.: Applications of deep reinforcement learning in communications and networking: a survey. IEEE Commun. Surv. Tutor. **21**(4), 3133–31741 (2019)
13. Mao, Q., Hu, F., Hao, Q.: Deep learning for intelligent wireless networks: a comprehensive survey. IEEE Commun. Surv. Tutor. **20**(4), 2595–2621 (2018)
14. Kyritsis, A.I., Deriaz, M.: A machine learning approach to waiting time prediction in queueing scenarios. In: Second IEEE International Conference on Artificial Intelligence for Industries, pp. 17–21. Laguna Hills, CA. IEEE (2019)
15. Hermanto, R.P.S., Suharjito, S., Nugroho, A.: Waiting-time estimation in bank customer queues using RPROP neural networks. Procedia Comput. Sci. **135**, 35–42 (2018)
16. Curtis, C., Liu, C., Bollerman, T.J., Pianykh, O.S.: Machine learning for predicting patient wait times and appointment delays. J. Am. Coll. Radiol. **15**(9), 1310–1316 (2018)

17. Mourõo, R.N., Carvalho, R.S., Carvalho, R.N., Ramos, G.N.: Predicting waiting time overflow on bank teller queues. In: Proceedings of the 16th IEEE International Conference on Machine Learning and Applications (ICMLA), pp. 842–847. Cancun, Mexico. IEEE (2018)

18. Sivakami Sundaria, M., Palaniammalb, S.: Simulation of $M|M|1$ queuing system using ANN. Malaya J. Matematik: Spec. Issue **1**, 279–294 (2015)

19. Sivakami Sundaria, M., Palaniammalb, S.: An ANN simulation of single server with infinite capacity queuing system. Int. J. Innovative Technol. Explor. Eng. **8**(12), 4067–4071 (2019)

20. Sivakami Sundari, M., Yamini, S., Kalicharan, R., Senthil Kumar, S., Palaniammalb, S.: Artificial neural network simulation for markovian queuing models in a busy airport. In: Proceedings of the International Conference on Computer Science, Engineering and Applications (ICCSEA), pp. 1–6. Gunupur. IEEE (2020)

21. Khomonenko, A.D., Yakovlev, E.L.: Nejrosetevaya approksimaciya harakteristik mnogokanal'nyh nemarkovskih sistem massovogo obsluzhivaniya [Neural network approximation of characteristics of multi-channel non-Markovian queuing systems]. SPIIRAS Proc. **4**(41), 81–93 (2015). [in Russian]

22. Gindin, S.I., Khomonenko, A.D., Adadurov, S.E.: CHislennyj raschet mnogokanal'noj sistemy massovogo obsluzhivaniya s rekurrentnym vhodyashchim potokom i "razogrevom" [Numerical calculations of multichannel queuing system with recurrent input and "warm up"]. Izvestiya Peterburgskogo universiteta putey soobscheniya [Proc. Petersburg Transp. Univ.] **37**(4), 92–101 (2013). [in Russian]

23. Vishnevsky, V., Klimenok, V., Sokolov, A., Larionov, A.: Performance evaluation of the priority multi-server system $MMAP|PH|M|N$ using machine learning methods. Mathematics **9**(24), 3236 (2021)

24. Gorbunova, A.V., Vishnevsky, V.M., Larionov, A.A.: Evaluation of the end-to-end delay of a multiphase queuing system using artificial neural networks. In: Vishnevskiy, V.M., Samouylov, K.E., Kozyrev, D.V. (eds.) DCCN 2020. LNCS, vol. 12563, pp. 631–642. Springer, Cham (2020). https://doi.org/10.1007/978-3-030-66471-8_48

25. Rabta, B.: A review of decomposition methods for open queueing networks. In: Reiner G. (eds.) Rapid Modelling for Increasing Competitiveness, pp. 25–42. Springer, London (2009). https://doi.org/10.1007/978-1-84882-748-6_3

26. Gorbunova, A.V., Vishnevsky, V.M.: Estimating the response time of a cloud computing system with the help of neural networks. Adv. Syst. Sci. Appl. **20**(3), 105–112 (2020)

27. Gorbunova, A.V., Lebedev, A.V.: Response time estimate for a fork-join system with pareto distributed service time as a model of a cloud computing system using neural networks. In: Vishnevskiy V.M., Samouylov K.E., Kozyrev D.V. (eds.) Distributed Computer and Communication Networks. DCCN 2021. Communications in Computer and Information Science. In print (2021)

28. Gorbunova, A.V., Vishnevsky, V.: Evaluation of the performance parameters of a closed queuing network using artificial neural networks. In: Vishnevskiy, V.M., Samouylov, K.E., Kozyrev, D.V. (eds.) DCCN 2021. LNCS, vol. 13144, pp. 265–278. Springer, Cham (2021). https://doi.org/10.1007/978-3-030-92507-9_22

29. Vishnevsky, V., Semenova, O.: Polling systems and their application to telecommunication networks. Mathematics **9**(2), 117 (2021)

30. Vishnevsky, V., Semenova, O., Bui, D.T.: Using a machine learning approach for analysis of polling systems with correlated arrivals. In: Vishnevskiy, V.M., Samouylov, K.E., Kozyrev, D.V. (eds.) DCCN 2021. LNCS, vol. 13144, pp. 336–345. Springer, Cham (2021). https://doi.org/10.1007/978-3-030-92507-9_27

31. Ivanova, N.M., Vishnevsky, V.M.: Ocenka nadezhnosti privyaznyh vysotnyh bespi-
    lotnyh platform s ispol'zovaniem modelej sistem k-iz-n i metodov mashinnogo
    obucheniya [On reliability of a tethered unmanned high-altitude platform using
    k-out-of-n system and machine learning methods]. Problemy informatiki [Probl.
    Inform.] **4**, 16–39 (2021). [in Russian]
32. Efrosinin, D., Stepanova, N.: Estimation of the optimal threshold policy in a queue
    with heterogeneous servers using a heuristic solution and artificial neural networks.
    Mathematics **9**(11), 1267 (2021)
33. Tijms, H.C.: Stochastic Models. An Algorithmic Approach. 1st edn. Wiley, New
    York (1994)

Resource Requirements Distribution Evaluation for Traffic Offloading Strategies in NR-U Networks

Anastasia Daraseliya[1]([⊠]) and Eduard Sopin[2]

[1] Peoples' Friendship University of Russia (RUDN University),
Moscow, Russian Federation
`daraselia-av@rudn.ru`
[2] Institute of Informatics Problems, Federal Research Center Computer Science
and Control of Russian Academy of Sciences, Moscow, Russian Federation
`sopin-es@rudn.ru`

Abstract. Today, the research community and standardization bodies seek for a systematic answer to address the effects of temporal variability in mobile traffic. One of a viable option for mitigating the impact of traffic fluctuations is offloading in unlicensed bands. In this paper, we have described model of offloading customers on unlicensed frequency range of wireless network. We proposed three offloading strategies and evaluated the effectiveness of their use. We obtained the resource requirement distribution of offloading customers onto unlicensed band.

Keywords: Wireless network · queuing theory

Introduction

5G New Radio (NR) technology, standardized as a part of 3GPP efforts, promises drastic boost in the access rate at the last mile [1]. This is specifically the case for NR operating in millimeter wave frequency band, where a large set of resources has been made available worldwide [2]. Similarly to the respective LTE specifications, NR-U documentation has been extended to include the possibility of operation over the unlicensed bands.

The unlicensed band can be used to boost bitrate of NR sessions, therefore, a number of ongoing studies have been devoted to its use. Most of the recent studies has considered the question of designing NR and WiGig coexistence in the unlicensed band including the duty cycle and pure random access.

The concept of integration between licensed and unlicensed mmWave bands was considered in [3,4]. The authors studied the coexistence of the two systems in terms of the downlink data rate, by comparing three different scenarios: WiGig only, coexistence of WiGig and NR-U, and NR-U only. The results indicated that the use of unlicensed bands by NR-U user equipment (UE) may dramatically degrade the performance of WiGig UEs in terms of their data rate.

The reported study was funded by RSF, project numbers 20-71-00124.

A. Dudin et al. (Eds.): ITMM 2021, CCIS 1605, pp. 317–329, 2022.
https://doi.org/10.1007/978-3-031-09331-9_25

Since NR-U operation may also increase the levels of interference, recent studies focused on the effective coexistence mechanisms between NR-U and WiGig systems [5,6], including the duty cycle and random access considerations.

In [7] the study further addressed the coexistence of NR-U and WiGig technologies in 60 GHz bands. In particular, the authors focused on determining whether NR-U fulfills its coexistence objective in terms of the fairness criterion. In [8], the authors studied the coexistence of cellular and WiGig users around 60 GHz bands. They proposed a sensing-based adaptive unlicensed channel sharing protocol. In [9] inspected the downlink performance of NR-U and WiGig technologies under inter-technology interference from each other in 60 GHz bands. Under a small-cell setting, that paper offered models for signal-to-interference-plus-noise ratio (SINR) and data rate.

In our previous study [10], we developed the model for collocated NR-U design explicitly capturing the random access behavior in unlicensed band and characterizing the NR-U customer loss probability. However, we utilized very simple M/M/K/0 queuing model to capture the specifics of resource allocation in the licensed band. In [11], we suggested a more accurate model of the service process in the licensed band. However, we described only a simple offloading strategy, where customer offloading onto unlicensed spectrum was determined by an only insufficient amount of resources in the licensed band. In this study, we propose a strategy in which, in addition, resource-intensive customers can be initially redirected to the unlicensed range based on their "weight".

1   Offloading Schemes

In this section, we describe three different strategies for offloading tasks to the unlicensed band. For the first considered strategy, called baseline, UE try to associate with the nearest base station (BS) and utilize licensed band. The task is redirected to the unlicensed band if there are no sufficient amount of resources in the licensed band to accept it for service, and it is dropped if the current data rate provided to the offloaded task in this band is less than the transmission rate threshold $R_{\min}$. The analysis of this strategy was presented in our previous study [11].

Then, we consider 2 strategies of offloading, based on the weight of incoming tasks, called "fat" and "slim" strategies. In the case of "fat" offloading, the task is directed to licensed or unlicensed band based on the amount of resources, needed to satisfy the minimum rate requirements $R_{\min}$. If the resources exceed the threshold R_1, the task is directed to the unlicensed band. In the other case, the task is initially sent to the licensed band and only if there is insufficient amount of resources it is offloaded to the unlicensed band. In contrast to the previous one, "slim" offloading assumes that if the amount of resources needed to achieve the minimum rate $R_{\min}$ is bellow the threshold R_2, then the task is directed to the unlicensed band.

Since the coverage radii of the licensed and unlicensed bands are different we consider two types of customers. The first type of customer can be served in the licensed band only. Therefore, if there is not enough resources to service this customer, the customer is lost. The second type of customers can be potentially offloaded to the unlicensed band.

The system receives customers arrival flow with rate λ, which can be represented as $\lambda = \lambda_1 + \lambda_2$, where λ_1 and λ_2 are the arrival rates of first and second types of customers, respectively.

For "weight" based offloading strategies (Fig. 1) the probability mass functions (pmfs) of resource requirements at BS depend on the threshold: R_1 for the fat strategy, and R_2 for the slim strategy, respectively. The arrival flow of the second type customers is divided according to the "weight" of the customer. For fat strategy, "heavier" customers are initially directed to the unlicensed band with a probability $\pi_{2,1}$, and with probability $(1 - \pi_{2,1})$ "lighter" customers are directed to licensed band. For the slim strategy, the principle of offloading is similar, but instead of heavy ones, light customers are initially directed to the licensed band. Thus, the overall rate of both types of customers to the licensed band is $\lambda_1 + \lambda_2 (1 - \pi_{2,1})$.

Observe, that the second type customers arrive to the unlicensed band in two cases: (i) when the "weight" of the customer is more than a threshold R_1 for "fat" strategy, or less than a threshold R_2 for "slim" strategy, respectively, (ii) when there are no sufficient amount of resources or servers available for a customer that has been initially routed to licensed band. The probability $\pi_{1,2}$ that the second type customer will be directed to the unlicensed spectrum is the sum of the probability $\pi_{2,1}$ that the customer is sent on a licensed spectrum according to its weight, i.e. it's "heavy" for fat strategy and is "light" for another one, and the probability $\pi_{2,2}$ that a customer cannot be handled at the licensed band and thus offloaded to unlicensed one, i.e.,

$$\pi_2 = \pi_{2,1} + (1 - \pi_{2,1})\pi_{2,2}. \tag{1}$$

In this way, the arrival rate to the unlicensed part of BS can be calculated as $\lambda_2\pi_2$.

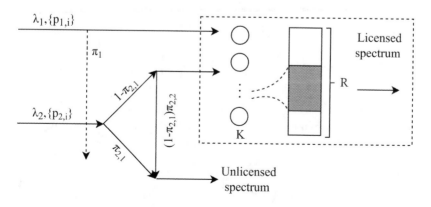

Fig. 1. Illustration of the queuing model.

2   Model Description

To model the customer service process in the licensed band we apply the framework of resources queuing systems [12–14]. For this purpose, consider a multiserver queuing system with $K < \infty$ servers and $R < \infty$ resource units, where K denotes the maximum number of UEs in the system, i.e., the maximum number of customers that can be simultaneously served in the licensed band. Customers of two types arrive to the system, both according to the Poisson processes with arrival rates λ_1 for the first type and λ_2 for the second one. Thus, the total arriving flow is Poisson with parameter $\lambda = \lambda_1 + \lambda_2$. The service time distribution is exponential with the rate μ.

Service process of each customer requires a server and a random amount of resources, $0 \leq r \leq R$. The distributions of resource requirements for considered customer types are given by $\{p_{l,j}\}_{j \geq 0}$, $l = 1, 2$, where $p_{l,j}$ is the probability that a customer of type l requires j resources. According to [12] resource-based queuing system with two flows can be analyzed as a system with one aggregated flow assuming the following

$$\tilde{p}_{j,1} = \frac{\rho_1^*}{\rho^*} p_{1,j} + \frac{\rho_2^*}{\rho^*} \tilde{p}_{j,1,2}, \tag{2}$$

where $\rho^* = \rho_1^* + \rho_2^*$, $\rho_1^* = \lambda_1/\mu$, $\rho_2^* = \lambda_2(1 - \pi_{2,1})/\mu$, and $\tilde{p}_{j,1,2}$ is the probability that the second type customer requires j resources in the licensed band. .

The system operates as follows. An arriving customer is accepted to the system if at the moment of arrival there are sufficient amount of resources available. Alternatively, an arriving customer is dropped. In this case, a first type of customer is lost while the customer of the second type is being redirected to the unlicensed band. When the service time of a customer is over, it leaves the system releasing all the occupied resources.

Denote by $P_k(r)$ the stationary probability that there are k customers in the system that totally occupy r resources. According to [15], the stationary distribution is given by

$$P_k(r) = P_0 \frac{\rho^i}{k!} \tilde{p}_{r,1}^{(k)}, \; k = 1, 2, \ldots, K,$$

$$P_0 = \left(1 + \sum_{k=1}^{K} \frac{\rho_s^k}{k!} \sum_{r=0}^{R} \tilde{p}_{r,1}^{(k)} \right)^{-1}, \tag{3}$$

where $\{\tilde{p}_{r,1}^{(k)}\}_{r \geq 0}$ is k-fold convolution of pmf $\{\tilde{p}_{r,1}\}_{r \geq 0}$.

The probability that the second type customer requires j resources in the licensed band for the "fat" strategy is given by

$$\tilde{p}_{j,1,2} = \left(\sum_{i=0}^{R_1} p_{2,i} \right)^{-1} p_{2,j}, \; 0 \leq j \leq R_1, \tag{4}$$

and for the "slim" strategy is given in a similar way by the following formula

$$\tilde{p}_{j,1,2} = \left(1 - \sum_{i=0}^{R_2} p_{2,i}\right)^{-1} p_{2,j}, \ 0 \le j \le R_1, \tag{5}$$

The probability that there is no sufficient amount of resources in the licensed band to serve a session of the first type is

$$\pi_1 = 1 - P_0 \sum_{k=0}^{K-1} \frac{(\rho^*)^k}{k!} \sum_{r=0}^{R} \tilde{p}_{r,1}^{(k+1)}. \tag{6}$$

In the case of large values of K and R according to (6) calculations are computationally demanding. For this reason, we can adopted a recurrent computational algorithm proposed in [13]. Let us introduce an auxiliary function $G(K, R)$ as

$$G(n, r) = \sum_{i=0}^{n} \frac{(\rho^*)^i}{i!} \sum_{j=0}^{r} \tilde{p}_{j,1}^{(i)}, \ P_0 = G^{-1}(K, R). \tag{7}$$

According to it, the the probability π_1 from (6) can be rewritten as

$$\pi_1 = 1 - G^{-1}(K, R) \sum_{i=0}^{R} p_{1,i} G(K-1, R-i). \tag{8}$$

3 Resource Requirement Distribution of Offloading Customers

In this section, we will specify the probability distribution of resource requirements and the intensities of second type customer offloads to the unlicensed band.

3.1 Fat Strategy

The customer is considered "heavy" if it requires more than R_1 resources, and is thus originally routed to the unlicensed spectrum. Then the probability $\pi_{2,1}$ that the customer is "heavy", can be calculated as follows

$$\pi_{2,1} = 1 - \sum_{i=0}^{R_1} p_{2,i}, \tag{9}$$

The probability $\pi_{2,2}$ that a "light" customer cannot be served in the licensed band and thus offloaded to the unlicensed band is calculated similarly to (6) as

$$\pi_{2,2} = 1 - P_0 \sum_{k=0}^{K-1} \frac{(\rho^*)^k}{k!} \sum_{r=0}^{R} \tilde{p}_{r,1}^{(k+1)}. \tag{10}$$

By analogy to (8), the probability $\pi_{2,2}$ (10) can be written using a recurrent algorithm as

$$\pi_{2,2} = 1 - G^{-1}(K, R) \sum_{i=0}^{R} \tilde{p}_{i,2} G(K - 1, R - i). \tag{11}$$

The probability that the customer requires j resources in the unlicensed band needs to be calculated separately for two cases: when a customer is "heavy" and thus initially routed to the unlicensed band, and when a customer is first routed to the licensed band but there are not enough of resources available for its service. Reflecting on these cases we arrive at

$$\frac{1 - \pi_{2,1}}{\pi_2} p_{2,j} \left(\sum_{r=0}^{R} P_K(r) + \sum_{k=0}^{K-1} \sum_{r=R-j+1}^{R} P_k(r) \right), j \leq R_1,$$

$$\frac{1}{\pi_2} p_{2,j}, j > R_1. \tag{12}$$

After substituting the function $G(n, r)$ into (12), the probability that the customer requires j resources in the unlicensed band is

$$\tilde{p}_{j,2} = \begin{cases} \frac{1 - \pi_{2,1}}{\pi_{1,2}} p_{2,j} \frac{G(K,R) - G(K-1,R-j)}{G(K,R)}, j \leq R_1, \\ \frac{1}{\pi_2} p_{2,j}, j > R_1. \end{cases} \tag{13}$$

3.2   Slim Strategy

For this strategy, the customer is considered "light" if it requires less than R_2 resources, and is thus originally routed to the unlicensed spectrum. Then, the probability $\pi_{2,1}$ that the customer is "light" can be calculated by the following formula

$$\pi_{2,1} = \sum_{i=0}^{R_2} p_{2,i}. \tag{14}$$

The probability $\pi_{2,2}$ that a "heavy" customer is offloaded to the unlicensed band due to lack of resources on the licensed one is found similarly to the formula is be calculated as (11) for the previous strategy.

Similarly to (12), the probability $\tilde{p}_{2,j}$ that the customer requires j resources in the unlicensed band is

$$\tilde{p}_{2,j} = \begin{cases} \frac{1 - \pi_{2,1}}{\pi_2} p_{2,j} \left(\sum_{r=0}^{R} P_K(r) + \sum_{k=0}^{K-1} \sum_{r=R-j+1}^{R} P_k(r) \right), j \geq R_2, \\ \frac{1}{\pi_2} p_{2,j}, j < R_2. \end{cases} \tag{15}$$

After applying the recurrent algorithm, the formula (15) for calculating the probability $\tilde{p}_{2,j}$ can be represented as

$$\tilde{p}_{2,j} = \begin{cases} \frac{1 - \pi_{2,1}}{\pi_2} p_{2,j} \frac{G(K,R) - G(K-1,R-j)}{G(K,R)}, j \geq R_2, \\ \frac{1}{\pi_{SU}} p_{2,j}, j < R_2. \end{cases} \tag{16}$$

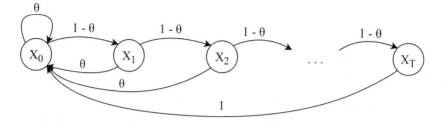

Fig. 2. State transition diagram of the Markov model.

4 Evaluation of the Efficiency of Offloading on Unlicensed Spectrum

NR-U UE sessions offloaded to the unlicensed band compete for transmission resources with WiGig UEs. Denote ζ as the collision probability and ψ as the probability that the LoS path is blocked. Then, the probability of successful transmission is the probability that there was no collision or LoS blocking and can be writen as

$$\theta = (1 - \zeta)(1 - \psi). \tag{17}$$

The behavior of system can be described by a Markov chain $\{X_n, n = 0, 1, \ldots, T\}$, where $X_n = n$ denotes transmission attempt when the back-off counter is in $[0, 2^n W - 1]$. According to the state transition diagram shown in Fig. 2, the formula for calculating stationary probabilities can be written as

$$q_i = \frac{\theta}{1 - (1 - \theta)^{T+1}} (1 - \theta)^i, \; i = 0, 1, ..T \tag{18}$$

Let now π_1 and π_2 be the probabilities that the NR-U and WiGig UEs transmit in arbitrarily chosen slots. Then, if there are n NR-U and m WiGig competing sessions, the collision probability is

$$\zeta = 1 - (1 - \pi_1)^n (1 - \pi_2)^m. \tag{19}$$

The unknowns π_1^* and π_2^* can be established by considering the LBT access procedure. Since UEs transmit only in states $X_j = j$, the transmission probability π_1^* can be calculated as a fraction of slot time divided by the mean number of time slots UE spends in any state. Thus, to find the probability that UE performs the transmission attempt, we need to sum up the mean number of time slots b_j that UE spends in state j, multiplied by probability q_j that UE is in the state j, i.e.,

$$\pi_1^* = \left[\sum_{i=0}^{T} q_j b_j \right]^{-1}, \tag{20}$$

where the mean number of slots b_j in state j is given by

$$b_j = \sum_{i=1}^{2^j W} \frac{1}{2^j W} i = \frac{2^j W + 1}{2}, j = 0, 1, .., T. \tag{21}$$

Finally, substituting (18), (21) into (20), and using simple algebraic manipulations, the transmission probability π_1 can be written in the following form

$$\pi_1^* = \left[\frac{\theta W \left(1 - 2^{T+1}(1-\theta)^{T+1}\right)}{2\left(1 - (1-\theta)^{T+1}\right)(2\theta - 1)} + \frac{1}{2} \right]^{-1}. \tag{22}$$

Having obtained the probability of transmission π_1 (22), we can determine the average successful transmission probability as a function of the number of NR and WiGig UEs competing for transmission, respectively, i.e.,

$$\Pi_1^* = \sum_{i=1}^{\infty} \frac{(\rho_1^*)^i}{i!} e^{-\rho_1^*} \sum_{j=0}^{\infty} \frac{(\rho_2^*)^j}{j!} e^{-\rho_2^*} \pi_1^*(i, j) \theta(i, j), \tag{23}$$

where $\rho_1^* = \lambda_2/\mu$ and ρ_2^* are the total offered load on the licensed and unlicensed bands.

Let η be a random variable of data rate on an unlicensed frequency range. The transmission and successful transmission probabilities for WiGig UE are calculated similarly. The random value v of the transmission rate in the unlicensed frequency range is a linear function of the random value of the spectral efficiency with the distribution $\tilde{p}_{j,1}$ can be represented as η_j is given by

$$v = \Pi_1^* B \eta. \tag{24}$$

Then, the expected value of the data rate achieved by UE in the unlicensed band is

$$E[v] = \sum_{j=0}^{R} \tilde{p}_{2,j} \Pi_1^* B \eta_j, \tag{25}$$

The rate achieved by WiGig UEs is obtained similarly.

To determine the eventual NR-U session loss probability we define $\tilde{Q}$ to be NR-U UE session loss probability, i.e., the probability that the minimum rate $R_{\min}$ is not satisfied in the unlicensed band.

By using v, (25) becomes limited by rate threshold $R_{\min}$ and the sought metric is given by

$$\tilde{Q} = \pi_2 P\{v < R_{min}\} = \pi_2 \sum_{j:\Pi_1^* B \eta_i < R_{min}} \tilde{p}_{2,j} \tag{26}$$

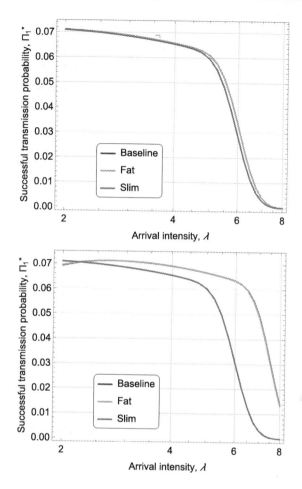

Fig. 3. Successful transmission probability as a function of arrival intensity.

5   Numerical Results

Now we are in position to proceed with comparison of the considered offloading strategies. To this aim, below we consider three variants of the strategies: (i) baseline, where a customer is offloaded when no resources are available at the licensed band, (ii) fat strategy in which "heavier" customers are initially directed to the unlicensed band , and (iii) slim strategy n which"lighter" customers are initially directed to the unlicensed band.

The parameters used for the calculation are provided in Table 1.

Figure 3 illustrates the successful transmission probability for all considered strategies and two different maximum numbers of sessions that can be simultaneously served in the licensed band, $K = 6$ and $K = 50$. Note that baseline and fat strategies give approximately the same probability of a successful data

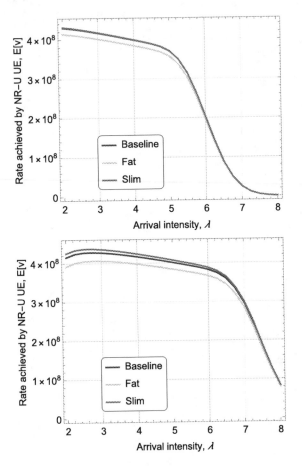

Fig. 4. Data rate achieved by UE in the unlicensed band as a function of arrival intensity.

transfer, approximately the same. For a small total offered load, all three strategies give approximately the same result, with an increase in the load, the gain remains for the base and fat strategies.

Figure 4 illustrates the data rate achieved by UE in the unlicensed band for all considered strategies. Here the situation is the opposite. The greatest gain in the rate is achieved with the slim strategy mainly for light loads. Note that this gain gradually disappears with increasing load along with a drop in the transmission rate.

Figure 5 illustrates the eventual session loss probability for all considered strategies. By analyzing the presented results one may observe that the baseline strategy, where a session is offloaded onto unlicensed band when no resources for its service are available in the licensed one, is associated with the minimal values of the eventual session drop probabilities. At the same point, offloading heavy sessions to the unlicensed band leads to the greatest gain stably for the

Table 1. Default system parameters.

Parameter	Value
Initial contention window, CW	16
Maximum number of customers, K	50
Number of retransmissions, T	10
Service rate, μ	0.02
Minimum requested session rate, R_{min}	50 Mbps
LoS blockage probability, ψ	0.166

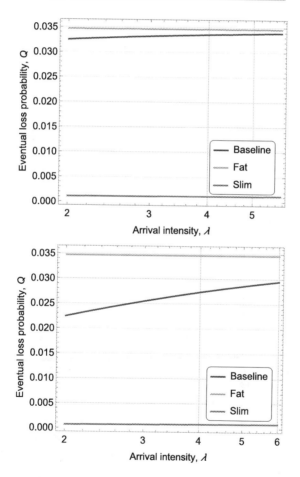

Fig. 5. The eventual session loss probability as a function of arrival intensity.

entire considered offered load interval, while "fat" strategy, where "heavy" sessions are offloaded onto unlicensed band is characterized by drastically the worse performance.

6   Conclusion

In this paper, by utilizing the tools of queuing theory we have described resource model with the "weight" based strategies of offloading customers onto unlicensed band and obtained a resource requirement distribution of offloaded customers. We have proposed and compared 3 offloading strategies: baseline, in which the session is offloaded when no resources are available at the licensed band, fat strategy in which "heavier" sessions are initially directed to the unlicensed band, and the slim strategy in which "lighter" sessions are initially directed to the unlicensed band. For the parameters presented in the numerical analysis, it was noted that the most gainful offloading strategy for this case is the slim strategy.

References

1. Dahlman, E., Parkvall, S.: NR - the new 5G radio-access technology. In: 2018 IEEE 87th Vehicular Technology Conference (VTC Spring), pp. 1–6 (2018). https://doi.org/10.1109/VTCSpring.2018.8417851
2. Wang, T., et al.: Spectrum analysis and regulations for 5G. In: Xiang, W., Zheng, K., Shen, X.S. (eds.) 5G Mobile Communications, pp. 27–50. Springer, Cham (2017). https://doi.org/10.1007/978-3-319-34208-5_2
3. Lu, X., Lema, M., Mahmoodi, T., Dohler, M.: Downlink data rate analysis of 5G-U (5G on unlicensed band): coexistence for 3GPP 5G and IEEE 802.11ad WiGig. In: European Wireless 2017; 23th European Wireless Conference. VDE, pp. 1–6 (2017)
4. Lu, X., et al.: Integrated use of licensed- and unlicensed-band mmWave radio technology in 5G and beyond. In: IEEE Access, vol. 7, pp. 24376–24391 (2019)
5. Semaan, E., Ansari, J., Li, G., Tejedor, E., Wiemann, H.: An outlook on the unlicensed operation aspects of NR. In: IEEE Wireless Communications and Networking Conference (WCNC). IEEE, vol. 2017, pp. 1–6 (2017)
6. Lagen, S., Giupponi, L.: Listen before receive for coexistence in unlicensed mmWave bands. In: IEEE Wireless Communications and Networking Conference (WCNC). IEEE vol. 2018, pp. 1–6 (2018)
7. Patriciello, N., et al.: NR-U and IEEE 80-211 technologies coexistence in unlicensed mmWave spectrum: models and evaluation. IEEE Access **8**, 71254–71271 (2020)
8. Wang, P., Di, B., Song, L.: Unlicensed spectrum sharing with WiGig in millimeter-wave cellular networks in 6G era. In: GLOBECOM 2020–2020 IEEE Global Communications Conference, pp. 1–6 (2020)
9. Verboom, J., Kim, S.: On the coexistence of WiGig and NR-U in 60 GHz band. In: 2021 IEEE 93rd Vehicular Technology Conference (VTC2021-Spring), pp. 1–5 (2020). https://doi.org/10.1109/VTC2021-Spring51267.2021.9448837
10. Daraseliya, A., Korshykov, M., Sopin, E., Moltchanov, D., Koucheryavy, Y., Samouylov, K.: Traffic, handling overflow, in millimeter wave 5G NR deployments using NR-U technology. In: IEEE 31st Annual International Symposium on Personal. Indoor and Mobile Radio Communications, vol. 2020, pp.1–7 (2020). https://doi.org/10.1109/PIMRC48278.2020.9217313
11. Daraseliya, A.V., Sopin, E.S.: Optimization of mobile device energy consumption in a fog-based mobile computing offloading mechanism. Discrete Continuous Models Appl. Comput. Sci. 29(1), 53–62 (2021). https://doi.org/10.22363/2658-4670-2021-29-1-53-62

12. Samouylov, K., Sopin, E., Vikhrova, O.: Analyzing blocking probability in LTE wireless network via queuing system with finite amount of resources. International Conference on Information Technologies and Mathematical Modelling, Springer, pp. 393–403 (2015). https://doi.org/10.22363/2312-9735-2017-25-3-209-216
13. Sopin, E.S., Ageev, K.A., Markova, E.V., Vikhrova, O.G., Gaidamaka, Y.V.: Performance analysis of M2M traffic in LTE network using queuing systems with random resource requirements. Autom. Control Comput. Sci. **52**(5), 345–353 (2018). https://doi.org/10.3103/S0146411618050127
14. Begishev, V., et al.: Quantifying the impact of guard capacity on session continuity in 3GPP new radio systems. IEEE Trans. Veh. Technol. **68**(12), 12345–12359 (2019). https://doi.org/10.1109/TVT.2019.2948702
15. Naumov, V.A., Samuilov, K.E., Samuilov, A.K.: On the total amount of resources occupied by serviced customers. Autom. Remote Control **77**, 1419–1427 (2016)

Stability Analysis for Retrial Queue with Collisions and r-Persistent Customers

Anatoly Nazarov and Olga Lizyura[✉]

Institute of Applied Mathematics and Computer Science,
National Research Tomsk State University, 36 Lenina Ave., Tomsk 634050, Russia
oliztsu@mail.ru

Abstract. We consider a single server retrial queue with general distribution of service times, collisions and r-persistent customers. The last phenomena describes the behaviour of customers that are leaving the system immediately if the server is busy upon arrival. We consider the system with customers, which leave the system without servicing with constant probability r. We provide the numerical stability analysis in such system using the following approach. First, we build the diffusion limit for the number of customers in the orbit and then analyze its drift coefficient. For different system parameters, we have different stability conditions.

Keywords: retrial queue · collisions · r-persistent customers · diffusion approximation

Introduction

Retrial queues arose as models of communication systems. The basic phenomenon of such systems is the retrial behavior of customers: if the server is busy upon arrival, the customer enters the orbit and repeats the attempt to access the server after a random amount of time.

There are several modifications of retrial queues that reflect the system features such as collisions and non-persistent customers, which appear in various switching communication systems and CSMA-based networks [1]. In recent years, queueing systems with collisions are of interest due to the reborn of IEEE 802.11 wireless LANs. In papers [8,9], authors describe the markovian retrial queue with collisions and shows applications of persistence to modeling CSMA-CD protocols. In paper [6], the author consider similar markovian model and takes into account the impatience of customers.

Nazarov and Sztrik with their research group have considered several models of finite-source retrial queues with collisions [7,13,14,17–19]. The phenomena of non-persistent customers in retrial queues was considered by [4,5]. Lakaour and his colleagues have considered markovian models with collisions, transmission errors and unreliable server [10,11].

© Springer Nature Switzerland AG 2022
A. Dudin et al. (Eds.): ITMM 2021, CCIS 1605, pp. 330–342, 2022.
https://doi.org/10.1007/978-3-031-09331-9_26

Retrial queues with collisions and impatient customers were considered in [2,3,16]. The phenomena of impatient customers is similar to the r-persistence due to the fact that a customer, which have not received the service upon arrival can leave the system. However, there is some difference, because non-persistent customers leave the system immediately with some probability and never join the orbit.

Another model of queueing system with collisions is considered by Phung-Duc and Fiems [15]. The model is markovian and has two phases of service. The authors study how the division into phases affects queueing performance.

We consider retrial queue with arbitrary distribution of service times, collisions and r-persistent customers. We build diffusion approximation for the number of customers in the orbit and construct the approximation of its probability distribution under the limit condition of growing delay in the orbit. Considering different sets of parameters, we show the numerical examples of system stability using the obtained approximation.

The rest of the paper is organized as follows. In Sect. 1, we describe the model structure and derive the equations for the probability distribution of system states. Section 2 is devoted to the asymptotic-diffusion analysis of the system under consideration. The approach is described in the paper [12]. After that, we show the results of numerical experiments in Sect. 3. Section 4 is dedicated to the conclusion.

1 Model Description and Problem Definition

We consider a retrial queue with an arbitrary distribution of service times defined by the distribution function $B(x)$. The input is stationary Poisson process with rate λ. If the server is idle upon arrival, the incoming customer occupies it for service. Otherwise, the collision occurs and one of the customers joins the orbit. The other customer can also join the orbit with probability r or leave the system with probability $(1 - r)$.

At the orbit, a customer waits for some random time and tries again to occupy the server. The duration of delay follows an exponential distribution with rate σ.

Let $k(t)$ denote the state of the server at instant t: 0, if the server is idle; 1, if the server is busy. Let $i(t)$ denote the number of customers in the orbit at instant t. We also introduce process $z(t)$, which represents the residual service time. Thus, process $\{k(t),\ i(t),\ z(t)\}$ has variable number of components and exhaustively describes the system state. We denote the probability distribution of process $\{k(t),\ i(t),\ z(t)\}$ as follows:

$$P_0(i,t) = P\{k(t) = 0, i(t) = i\}, \quad P_1(i,z,t) = P\{k(t) = 1, i(t) = i, z(t) < z\},$$

and introduce the partial characteristic functions

$$H_0(u,t) = \sum_{i=0}^{\infty} e^{jui} P_0(i,t), \quad H_1(u,z,t) = \sum_{i=0}^{\infty} e^{jui} P_1(i,z,t),$$

where j is the imaginary unit. The Kolmogorov system of differential equations for the partial characteristic functions has the following form:

$$\frac{\partial H_0(u,t)}{\partial t} = -\lambda H_0(u,t) + j\sigma\frac{\partial H_0(u,t)}{\partial u} + \frac{\partial H_1(u,0,t)}{\partial z}$$

$$+ \lambda e^{ju}(1 + r(e^{ju} - 1))H_1(u,t) - j\sigma(1 + r(e^{ju} - 1))\frac{\partial H_1(u,t)}{\partial u},$$

$$\frac{\partial H_1(u,z,t)}{\partial t} = \frac{\partial H_1(u,z,t)}{\partial z} - \frac{\partial H_1(u,0,t)}{\partial z} - \lambda H_1(u,z,t)$$

$$+ j\sigma\frac{\partial H_1(u,z,t)}{\partial u} + \lambda H_0(u,t)B(z) - j\sigma e^{-ju}\frac{\partial H_0(u,t)}{\partial u}B(z).$$

(1)

After that, we sum up the equations of system (1). Taking the limit by $z \to \infty$, we obtain

$$\frac{\partial H(u,t)}{\partial t} = (e^{ju} - 1)$$

$$\times \left\{ j\sigma e^{-ju}\frac{\partial H_0(u,t)}{\partial u} + \lambda(1 + re^{ju})H_1(u,t) - j\sigma r\frac{\partial H_1(u,t)}{\partial u} \right\}.$$

(2)

Solving system (1) and equation (2) in the limit by $\sigma \to 0$, we derive drift and diffusion coefficients of approximating diffusion process.

2   Asymptotic-Diffusion Analysis

In system (1) and equation (2), we introduce the following notations:

$$\sigma = \varepsilon, \ u = \varepsilon w, \ \tau = \varepsilon t,$$

$$H_0(u,t) = F_0(w,\tau,\varepsilon), \ H_1(u,z,t) = F_1(w,z,\tau,\varepsilon),$$

(3)

and obtain the system of equations

$$\varepsilon\frac{\partial F_0(w,\tau,\varepsilon)}{\partial \tau} = -\lambda F_0(w,\tau,\varepsilon) + j\frac{\partial F_0(w,\tau,\varepsilon)}{\partial w} + \frac{\partial F_1(w,0,\tau,\varepsilon)}{\partial z}$$

$$+ \lambda e^{jw\varepsilon}(1 + r(e^{jw\varepsilon} - 1))F_1(w,\tau,\varepsilon) - j(1 + r(e^{jw\varepsilon} - 1))\frac{\partial F_1(w,\tau,\varepsilon)}{\partial w},$$

$$\varepsilon\frac{\partial F_1(w,z,\tau,\varepsilon)}{\partial \tau} = \frac{\partial F_1(w,z,\tau,\varepsilon)}{\partial z} - \frac{\partial F_1(w,0,\tau,\varepsilon)}{\partial z} - \lambda F_1(w,z,\tau,\varepsilon)$$

$$+ j\frac{\partial F_1(w,z,\tau,\varepsilon)}{\partial w} + \lambda F_0(w,\tau,\varepsilon)B(z) - je^{-jw\varepsilon}\frac{\partial F_0(w,\tau,\varepsilon)}{\partial w}B(z),$$

$$\varepsilon\frac{\partial F(w,\tau,\varepsilon)}{\partial \tau} = (e^{jw\varepsilon} - 1)$$

$$\times \left\{ je^{-jw\varepsilon}\frac{\partial F_0(w,\tau,\varepsilon)}{\partial w} + \lambda(1 + re^{jw\varepsilon})F_1(w,\tau,\varepsilon) - jr\frac{\partial F_1(w,\tau,\varepsilon)}{\partial w} \right\}.$$

(4)

We solve system (4) in the limit by $\varepsilon \to 0$ and formulate the following theorem.

Theorem 1. *In considered retrial queue, under the limit condition $\sigma \to 0$, the following equality holds:*

$$\lim_{\sigma \to 0} \mathbb{E}e^{jw\sigma i(\frac{\tau}{\sigma})} = e^{jwx(\tau)},$$

where $x(\tau)$ is a solution of differential equation

$$x'(\tau) = -x(\tau)r_0 + [\lambda + (\lambda + x(\tau))r]r_1, \tag{5}$$

values r_0, r_1 have the following form:

$$r_0 = \frac{1}{2 - B^*(\lambda + x)}, \quad r_1 = \frac{1 - B^*(\lambda + x)}{2 - B^*(\lambda + x)}. \tag{6}$$

Here $B^*(s)$ is the Laplace-Stieltjes transform (LST) of the distribution function of the service times $B(x)$.

Proof. We assume that $\lim_{\varepsilon \to 0} F_k(w, z, \tau, \varepsilon) = F_k(w, z, \tau)$ and consider system (4) in the limit by $\varepsilon \to 0$. After that, we seek the solution in the form

$$F_0(w, \tau) = r_0 e^{jwx(\tau)}, \quad F_1(w, z, \tau) = r_1(z)e^{jwx(\tau)},$$

which give us the following system:

$$-(\lambda + x)r_0 + r_1'(0) + (\lambda + x)r_1 = 0,$$
$$r_1'(z) - r_1'(0) - (\lambda + x)r_1(z) + (\lambda + x)r_0 B(z) = 0, \tag{7}$$
$$x'(\tau) = -x(\tau)r_0 + [\lambda + (\lambda + x(\tau))r]r_1.$$

Here $r_1 = r_1(\infty)$. The last equation of system (7) coincides with (5). From the first equation of system (7), we have

$$r_1'(0) = (\lambda + x)(r_0 - r_1).$$

Substituting the equality into the second equation yields

$$r_1'(z) - (\lambda + x)(r_0 - r_1) - (\lambda + x)r_1(z) + (\lambda + x)r_0 B(z) = 0.$$

We apply the Laplace-Stieltjes transform to the obtained differential equation and obtain

$$r_1^*(s)(\lambda + x - s) = (\lambda + x)r_1 - (\lambda + x)r_0(1 - B^*(s)).$$

If we set $s = \lambda + x$ in the last equation, we can write

$$(\lambda + x)r_1 - (\lambda + x)r_0(1 - B^*(\lambda + x)) = 0,$$

which we finally consider as system together with the normalization condition $r_0 + r_1 = 1$. We have

$$r_0 = \frac{1}{2 - B^*(\lambda + x)}, \quad r_1 = \frac{1 - B^*(\lambda + x)}{2 - B^*(\lambda + x)},$$

which coincides with (6).

We note that r_0 and r_1 depend on τ since they depend on x. We omit the arguments to simplify the expressions.

From (5), we denote function

$$a(x) = -xr_0 + (\lambda + (\lambda + x)r)r_1. \tag{8}$$

For the second step of analysis, we make the following substitutions in equations (1)–(2):

$$H_0(u,t) = e^{j\frac{u}{\sigma}x(\sigma t)} H_0^{(2)}(u,t), \ \ H_1(u,z,t) = e^{j\frac{u}{\sigma}x(\sigma t)} H_1^{(2)}(u,z,t).$$

Thus, we obtain the equations for the partial characteristic functions of centered number of customers in the orbit. After that, we introduce the following substitutions:

$$\sigma = \varepsilon^2, \ u = w\varepsilon, \ \tau = t\varepsilon^2, \\ H_0^{(2)}(u,t) = F_0^{(2)}(w,\tau,\varepsilon), \ H_1^{(2)}(u,z,t) = F_1^{(2)}(w,z,\tau,\varepsilon), \tag{9}$$

and obtain the system of equations

$$\varepsilon^2 \frac{\partial F_0^{(2)}(w,\tau,\varepsilon)}{\partial \tau} + jw\varepsilon a(x)F_0^{(2)}(w,\tau,\varepsilon) = -(\lambda+x)F_0^{(2)}(w,\tau,\varepsilon)$$
$$+ j\varepsilon\frac{\partial F_0^{(2)}(w,\tau,\varepsilon)}{\partial w} + \frac{\partial F_1^{(2)}(w,0,\tau,\varepsilon)}{\partial z}$$
$$+ (\lambda e^{jw\varepsilon} + x)(1 + r(e^{jw\varepsilon} - 1))F_1^{(2)}(w,\tau,\varepsilon)$$
$$- j\varepsilon(1 + r(e^{jw\varepsilon} - 1))\frac{\partial F_1^{(2)}(w,\tau,\varepsilon)}{\partial w},$$

$$\varepsilon^2 \frac{\partial F_1^{(2)}(w,z,\tau,\varepsilon)}{\partial \tau} + jw\varepsilon a(x)F_1^{(2)}(w,z,\tau,\varepsilon) = \frac{\partial F_1^{(2)}(w,z,\tau,\varepsilon)}{\partial z}$$
$$- \frac{\partial F_1^{(2)}(w,0,\tau,\varepsilon)}{\partial z} - (\lambda+x)F_1^{(2)}(w,z,\tau,\varepsilon) + j\varepsilon\frac{\partial F_1^{(2)}(w,z,\tau,\varepsilon)}{\partial w} \tag{10}$$
$$+ (\lambda + xe^{-jw\varepsilon})F_0^{(2)}(w,\tau,\varepsilon)B(z) - j\varepsilon e^{-jw\varepsilon}\frac{\partial F_0^{(2)}(w,\tau,\varepsilon)}{\partial w}B(z),$$

$$\varepsilon^2 \frac{\partial F^{(2)}(w,\tau,\varepsilon)}{\partial \tau} + jw\varepsilon a(x)F^{(2)}(w,\tau,\varepsilon)$$
$$= (e^{jw\varepsilon} - 1)\left\{ j\varepsilon e^{-jw\varepsilon}\frac{\partial F_0^{(2)}(w,\tau,\varepsilon)}{\partial w} - xe^{-jw\varepsilon}F_0^{(2)}(w,\tau,\varepsilon) \right.$$
$$\left. + (\lambda + r(\lambda e^{jw\varepsilon} + x))F_1^{(2)}(w,\tau,\varepsilon) - j\varepsilon r\frac{\partial F_1^{(2)}(w,\tau,\varepsilon)}{\partial w} \right\}.$$

Solving system (10) in the limit by $\varepsilon \to 0$, we present Theorem 2.

Theorem 2. *Function* $\lim\limits_{\varepsilon \to 0} F_k^{(2)}(w, \tau, \varepsilon) = F_k^{(2)}(w, \tau)$ *has the following form:*

$$F_k^{(2)}(w, \tau) = \Phi(w, \tau) r_k,$$

where r_k *is given by* (6), *function* $\Phi(w, \tau)$ *is the solution of equation*

$$\frac{\partial \Phi(w, \tau)}{\partial \tau} = w \frac{\partial \Phi(w, \tau)}{\partial w} a'(x) + \frac{(jw)^2}{2} \Phi(w, \tau) b(x). \tag{11}$$

Function $a(x)$ *is defined by* (8), $b(x)$ *is determined as follows:*

$$b(x) = a(x) + 2[-(\lambda + x)(1 + r)g_0 + xr_0 + r\lambda r_1], \tag{12}$$

where

$$g_0 = \frac{(a(x) + x)(1 - B^*(\lambda + x)) + (\lambda + x)a(x)B^{*\prime}(\lambda + x)}{(\lambda + x)(2 - B^*(\lambda + x))^2}.$$

Proof. Making the following substitutions in the system (10):

$$F_0^{(2)}(w, \tau, \varepsilon) = \Phi(w, \tau)\{r_0 + jw\varepsilon f_0\} + O(\varepsilon^2),$$

$$F_1^{(2)}(w, z, \tau, \varepsilon) = \Phi(w, \tau)\{r_1(z) + jw\varepsilon f_1(z)\} + O(\varepsilon^2), \tag{13}$$

we obtain the system of equations for f_0 and $f_1(z)$.

$$-(\lambda + x)f_0 + f_1'(0) + (\lambda + x)f_1$$
$$= a(x)r_0 - \frac{\partial \Phi(w, \tau)/\partial w}{w\Phi(w, \tau)} r_0 - (\lambda + r(\lambda + x))r_1 + \frac{\partial \Phi(w, \tau)/\partial w}{w\Phi(w, \tau)} r_1,$$
$$f_1'(z) - f_1'(0) - (\lambda + x)f_1(z) + (\lambda + x)f_0 B(z) \tag{14}$$
$$= a(x)r_1(z) - \frac{\partial \Phi(w, \tau)/\partial w}{w\Phi(w, \tau)} r_1(z) + xr_0 B(z) + \frac{\partial \Phi(w, \tau)/\partial w}{w\Phi(w, \tau)} r_0 B(z),$$

We solve system (14) using the following substitutions:

$$f_0 = Cr_0 + g_0 - \frac{\partial \Phi(w, \tau)/\partial w}{w\Phi(w, \tau)} \varphi_0,$$

$$f_1(z) = Cr_1(z) + g_1(z) - \frac{\partial \Phi(w, \tau)/\partial w}{w\Phi(w, \tau)} \varphi_1(z),$$

which yield three systems of equations. The first system coincide with the system for r_0 and $r_1(z)$. It is easy to see that the second system for φ_0 and $\varphi_1(z)$ can be obtained by differentiating of system (7). Thus, we can conclude that $\varphi_k = r_k'(x)$. The last system is given by

$$-(\lambda + x)g_0 + g_1'(0) + (\lambda + x)g_1 = a(x)r_0 - (\lambda + r(\lambda + x))r_1,$$
$$g_1'(z) - g_1'(0) - (\lambda + x)g_1(z) + (\lambda + x)g_0 B(z) = a(x)r_1(z) + xr_0 B(z). \tag{15}$$

We add an additional condition $g_0 + g_1 = 0$ and obtain the solution of the system in the following form:

$$g_0 = \frac{(a(x)+x)(1-B^*(\lambda+x)) + (\lambda+x)a(x)B^{*'}(\lambda+x)}{(\lambda+x)(2-B^*(\lambda+x))^2}, g_1 = -g_0.$$

During the analysis, we also obtain equation for $\Phi(w,\tau)$:

$$\frac{\partial \Phi(w,\tau)}{\partial \tau} = w \frac{\partial \Phi(w,\tau)}{\partial w}a'(x) + \frac{(jw)^2}{2}\Phi(w,\tau)b(x),$$

which coincide with (11). Here $a(x)$ and $b(x)$ are given by (8) and (12), respectively.

Here equation (11) is the Fourier transform of the Fokker-Planck equation for the process approximating the number of customers in the orbit of considered retrial queue. If we make the inverse Fourier transform, we can see that the drift coefficient of the obtained diffusion limit is $a(x)$ and diffusion coefficient if $b(x)$.

Discrete function $PD(i)$ is the approximation of the probability distribution of the number of customers in the orbit and has the following form:

$$PD(i) = \frac{D(i\sigma)}{\sum\limits_{n=0}^{\infty} D(n\sigma)}, \tag{16}$$

where

$$D(z) = \frac{1}{b(z)} \int\limits_0^z \frac{2}{\sigma} \frac{a(x)}{b(x)} dx.$$

We have briefly prooven theorems 1 and 2. The approach is widely described in [12]. In this paper, we concentrate at analysis of drift coefficient of the diffusion limit $a(x)$, which is given by (8).

3 Numerical Examples

3.1 Bistability Case

For the numerical examples, we show the analysis of the drift coefficient $a(x)$. Based on the number of roots of the equation $a(x) = 0$, we can consider several modes of stability. The first case occurs (Fig. 1) when the parameters of the system are as follows:

$$\lambda = 0.258, \ \alpha = 2, \ \beta = \alpha, \sigma = 0.1, r = 0.98,$$

where α and β are the shape and scale parameters of Gamma distribution of the service times. We note that in all cases we show graphics of $a(\sigma x)$, because the number of calls in the orbit is normalized by σ. When $a(x) > 0$ the number of customers in the orbit grows. On the other hand, if $a(x) < 0$, the number of

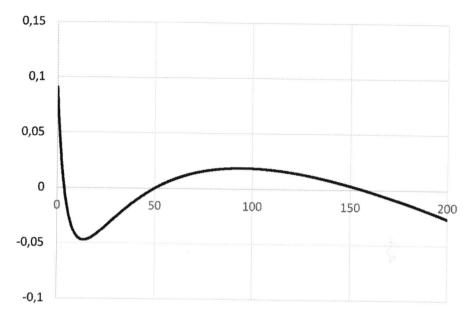

Fig. 1. Drift coefficient $a(x)$

customers in the orbit decreases. Here we have two stability areas around roots of equation $a(x) = 0$, when the sigh of $a(x)$ turns from plus to minus. In such case, the distribution of the number of customers in the orbit is bimodal (Fig. 2). We also note that if $a(x) < 0$ when $x \to \infty$, then the system is stable. If not, the steady state does not exist for the current set of parameters.

3.2 Standard Stability Case

The next case occurs when the parameters of the system are as follows:

$$\lambda = 0.258, \ \alpha = 1.8, \ \beta = \alpha, \sigma = 0.1, r = 0.98,$$

where α and β are the shape and scale parameters of Gamma distribution. In Fig. 3, we show that equation $a(x) = 0$ have only one root. Here we have the standard distribution with only one mode (Fig. 4) and $a(x) < 0$ when x grows to the infinity. Thus, the system is stable with such set of parameters.

3.3 Mixed Bistability Case

Another case occurs when the parameters of the system are as follows:

$$\lambda = 0.312, \ \alpha = 2, \ \beta = \alpha, \sigma = 0.1, r = 0.96,$$

where α and β are the shape and scale parameters of Gamma distribution. Here we also can observe the bistability phenomena (Fig. 5), but the modes are too close and affect on each other (Fig. 6).

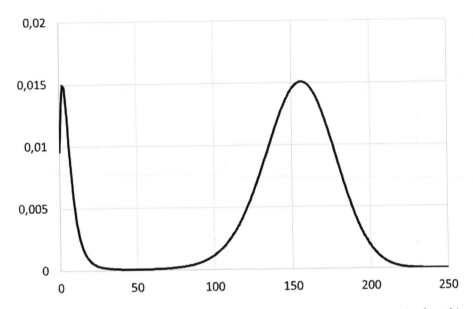

Fig. 2. Diffusion approximation of distribution of the number of customers in the orbit

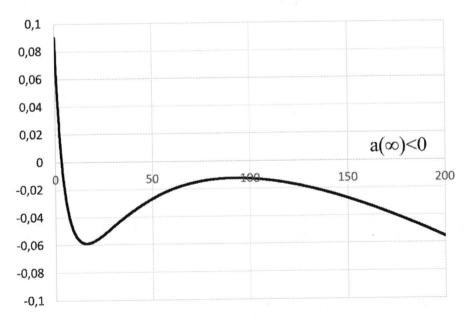

Fig. 3. Drift coefficient $a(x)$

3.4   Stabilization Area in Unstable System

The last case (Fig. 7) arise when the parameters of the system are given by

$$\lambda = 0.2, \ \alpha = 2, \ \beta = \alpha, \sigma = 0.1, r = 1,$$

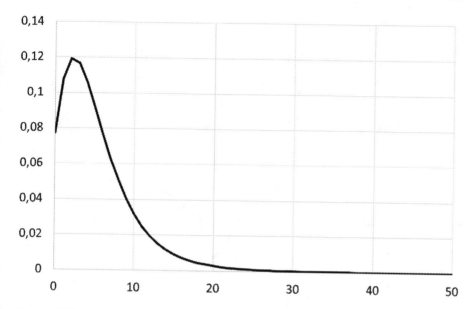

Fig. 4. Diffusion approximation of distribution of the number of customers in the orbit

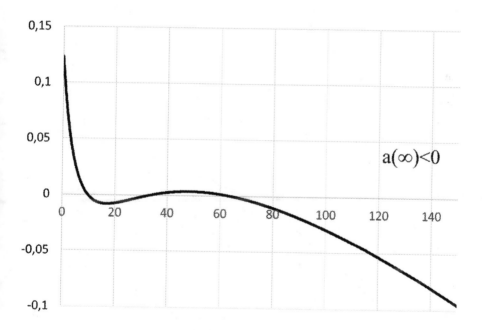

Fig. 5. Drift coefficient $a(x)$

where α and β are the shape and scale parameters of Gamma distribution. Even if $a(x) > 0$ when x grows to the infinity, the distribution has a stability area around the point where $a(x) = 0$. The process can spend a lot of time

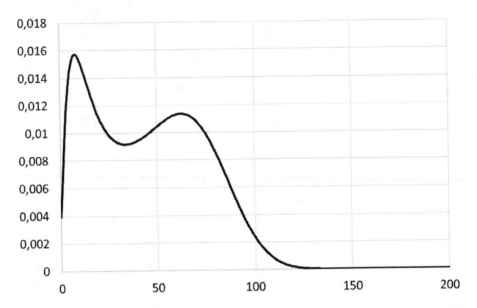

Fig. 6. Diffusion approximation of distribution of the number of customers in the orbit

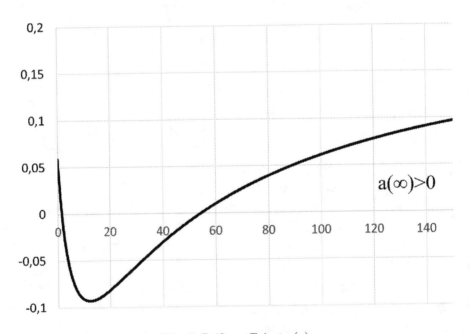

Fig. 7. Drift coefficient $a(x)$

before leaving the stability area. Thus, if we use zero of the function $a(x)$ as the truncation point, we can build an approximation (Fig. 8) for the distribution of the number of customers in the orbit using formula (16).

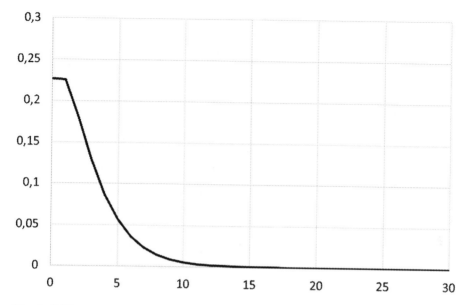

Fig. 8. Diffusion approximation of distribution of the number of customers in the orbit

4   Conclusion

We have considered the retrial queue with collisions and r-persistent customers. For the number of customers in the orbit, we have derived the approximation of the probability distribution (16). The analysis was prepared to show that there are several stability phenomena arise in such system. We show the numerical examples and the cases of stability for some sets of parameters based on the analysis of drift coefficient $a(x)$ of the obtained diffusion limit. For the future study, we plan to investigate the transition time between stability points in bistable retrial queue with collisions.

References

1. Choi, B.D., Shin, Y.W., Ahn, W.C.: Retrial queues with collision arising from unslotted CSMA/CD protocol. Queueing Syst. **11**(4), 335–356 (1992). https://doi.org/10.1007/BF01163860
2. Danilyuk, E.Y., Fedorova, E.A., Moiseeva, S.P.: Asymptotic analysis of an retrial queueing system M—M—1 with collisions and impatient calls. Autom. Remote Control **79**(12), 2136–2146 (2018). https://doi.org/10.1134/S0005117918120044
3. Danilyuk, E., Fedorova, E.: Negative binomial approximation in retrial queue M/M/1 with collisions and impatient calls. In: Vishnevskiy, V.M., Samouylov, K.E., Kozyrev, D.V. (eds.) DCCN 2019. CCIS, vol. 1141, pp. 461–471. Springer, Cham (2019). https://doi.org/10.1007/978-3-030-36625-4_37
4. Jailaxmi, V., Arumuganathan, R., Senthil Kumar, M.: Performance analysis of an M/G/1 retrial queue with general retrial time, modified M-vacations and collision. Oper. Res. **17**(2), 649–667 (2016). https://doi.org/10.1007/s12351-016-0248-7

5. Jailaxmi, V., Arumuganathan, R., Rathinasamy, A.: Performance analysis of an Mx/G/1 feedback retrial queue with non-persistent customers and multiple vacations with N-policy. Int. J. Oper. Res. **29**(2), 149–169 (2017)
6. Kim, J.S.: Retrial queueing system with collision and impatience. Commun. Korean Math. Soc. **25**(4), 647–653 (2010)
7. Kuki, A., Sztrik, J., Bérczes, T., Tóth, Á., Efrosinin, D.: Numerical analysis of non-reliable retrial queueing systems with collision and blocking of customers. J. Math. Sci. **248**(1), 1–13 (2020). https://doi.org/10.1007/s10958-020-04850-w
8. Kumar, B.K., Rukmani, R., Thangaraj, V., Krieger, U.R.: A single server retrial queue with bernoulli feedback and collisions. J. Stat. Theory Pract. **4**(2), 243–260 (2010). https://doi.org/10.1080/15598608.2010.10411984
9. Kumar, B.K., Vijayalakshmi, G., Krishnamoorthy, A., Basha, S.S.: A single server feedback retrial queue with collisions. Comput. Oper. Res. **37**(7), 1247–1255 (2010)
10. Lakaour, L., Aïssani, D., Adel-Aissanou, K., Barkaoui, K.: M/M/1 retrial queue with collisions and transmission errors. Methodol. Comput. Appl. Probab. **21**(4), 1395–1406 (2018). https://doi.org/10.1007/s11009-018-9680-x
11. Lakaour, L., Aissani, D., Adel-Aissanou, K., Barkaoui, K., Ziani, S.: An unreliable single server retrial queue with collisions and transmission errors. In: Communications in Statistics-Theory and Methods, pp. 1–25 (2020)
12. Nazarov, A., Phung-Duc, T., Paul, S., Lizura, O.: Asymptotic-diffusion analysis for retrial queue with batch poisson input and multiple types of outgoing calls. In: Vishnevskiy, V.M., Samouylov, K.E., Kozyrev, D.V. (eds.) DCCN 2019. LNCS, vol. 11965, pp. 207–222. Springer, Cham (2019). https://doi.org/10.1007/978-3-030-36614-8_16
13. Nazarov, A., Sztrik, J., Kvach, A.: Some features of a finite-source M/GI/1 retrial queuing system with collisions of customers. In: Vishnevskiy, V.M., Samouylov, K.E., Kozyrev, D.V. (eds.) DCCN 2017. CCIS, vol. 700, pp. 186–200. Springer, Cham (2017). https://doi.org/10.1007/978-3-319-66836-9_16
14. Nazarov, A., Sztrik, J., Kvach, A., Tóth, Á.: Asymptotic sojourn time analysis of finite-source M/M/1 retrial queueing system with collisions and server subject to breakdowns and repairs. Ann. Oper. Res. **288**(1), 417–434 (2020). https://doi.org/10.1007/s10479-019-03463-0
15. Phung-Duc, T., Fiems, D.: Exact performance analysis of retrial queues with collisions. In: Gribaudo, M., Sopin, E., Kochetkova, I. (eds.) ASMTA 2019. LNCS, vol. 12023, pp. 144–157. Springer, Cham (2020). https://doi.org/10.1007/978-3-030-62885-7_11
16. Sztrik, J., Tóth, Á., Danilyuk, E.Y., Moiseeva, S.P.: Analysis of retrial queueing system M/G/1 with impatient customers, collisions and unreliable server using simulation. In: Dudin, A., Nazarov, A., Moiseev, A. (eds.) ITMM 2020. CCIS, vol. 1391, pp. 291–303. Springer, Cham (2021). https://doi.org/10.1007/978-3-030-72247-0_22
17. Tóth, A., Bérczes, T., Sztrik, J., Kuki, A., Schreiner, W.: The simulation of finite-source retrial queueing systems with collisions and blocking. J. Math. Sci. **246**(4), 548–559 (2020). https://doi.org/10.1007/s10958-020-04759-4
18. Tóth, Á., Sztrik, J.: Simulation of finite-source retrial queuing systems with collisions, non-reliable server and impatient customers in the orbit. In: ICAI, pp. 408–419 (2020)
19. Tóth, Á., Sztrik, J., Pintér, Á., Bács, Z.: Reliability analysis of finite-source retrial queuing system with collisions and impatient customers in the orbit using simulation. In: 2021 International Conference on Information and Digital Technologies (IDT), pp. 230–234. IEEE (2021)

Asymptotic Analysis of Retrial Queueing System M/M/1 with Non-persistent Customers and Collisions

Anna Polkhovskaya, Svetlana Moiseeva, and Elena Danilyuk[✉]

National Research Tomsk State University, Tomsk, Russian Federation
daniluc.elena.yu@gmail.com

Abstract. A queuing system with repeated calls, one server, collisions (conflicts) of calls, H-persistence and rejections is considered. A call that found the device free occupies it, and service begins, which ends successfully if no other requests were received during it. If the server is busy, then a conflict (collision) arises between the call that have come for service and the ones being serviced, and in the general case, both calls instantly go to the orbit and repeat the attempt to successfully serve after a random time. In this article, in the event of a collision, one of the calls, for example, which was in service (on the device), goes into the orbit with probability H_1, the other goes into orbit with probability H_2, and with probability $(1 - H_1)$ and $(1 - H_2)$ respectively refuses service and leaves the system. The problem is to find asymptotic probabilities distribution of the calls number in the orbit.

Keywords: Retrial queueing system · Collisions · Rejections · Persistent calls · Asymptotic analysis

1 Introduction

Retrial queueing systems are characterized by the fact that incoming claims (calls, customers) that find a busy server join a given group of blocked clients' connections, called an orbit, to retry their requests in a random order and at random intervals. In modern networks such queueing systems, in which customers are allowed to make repeated attempts, are widely used to simulate many practical problems in telephone switching systems, telecommunication networks. Detailed overviews of the reconsideration queues can be found in [1–3].

This work is a natural continuation of the results obtained in [4], but it is a study of the probabilistic characteristics of a more complex system and is devoted to finding the probability distribution of the number of calls in the orbit of a queueing system with repeated calls $M/M/1$ with non-persistent calls, collisions and rejections.

The reported study was funded by RFBR and Tomsk region according to the research project № 19-41-703002.

Conflicts in the model, as a rule, arise in the problem of studying communication networks and presuppose the occurrence of situations when during the transmission of one message another arrives. Such messages collide, are considered garbled and go into orbit, from where they again turn to the device for service after a random delay [4–7, 12].

In real life, an impatience to wait (or persistence in trying to get a service) is the most noticeable feature of people, when they want to receive service, we always feel anxiety and impatience during a long wait for a service in real life. To characterize the behavior of impatient customers, the term "$(1 - H)$-non-persistence" is used, understood as a decision with a certain probability $(1 - H)$ not to join the line (device) after an unsuccessful attempt to get service followed by leaving the system (the problem can also be formulated in terms of "H - persistence". In [11–15], the authors use the terms "non-persistence" (non-persistence), "balking" (refusal to enter the queue), "reneging" (decision to leave the queue before the start of service). "Balking" and "reneging" are fundamental concepts in foreign literature on queuing theory, introduced by scientists Anker, Gafarian [16], Haight [17] and Bareer [18], as well as [19–23]. A fairly broad overview of systems with repeated calls with impatient calls is presented in [24]. Taking into account such features significantly complicates the mathematical model and limits the possibility of obtaining analytical expressions, therefore, to find the characteristics, numerical and approximation methods are used.

In the literature, the main methods for studying RQ-systems are matrix methods, numerical methods, and simulation, since exact analytical formulas can be obtained only for the simplest models. The Tomsk Scientific School develops asymptotic methods for studying queueing systems and networks [27] of various configurations, including for RQ-systems. Such methods make it possible to obtain asymptotic expressions acceptable for practice for the desired characteristics of the system in cases when their pre-limiting study is impossible. Various asymptotic methods and approaches in queuing theory are described in [4, 7–10] and others.

2   Mathematical Model

We consider an RQ-system with one server, at the input of which the simplest flow of calls with intensity arrives λ. A call that found the server free takes it up for service during a random time, exponentially distributed with a parameter μ. If the server is busy, then the arriving and being on the server claims enter into a conflict (collision), and the call that was in service (on the device) goes into orbit with probability H_1, and the call that arrived at the server and caused the conflict with probability H_2 goes into orbit, and with probability $(1 - H_1)$ and $(1 - H_2)$, respectively, claims refuse service and leave the system. In the orbit, each of the calls independently of each other carries out a random delay, the duration of which has an exponential distribution with a parameter σ, after which it again turns to the device with a repeated attempt to get service.

The problem is to find the stationary distribution of the number of calls in the orbit for the described system. To solve the problem, the method of asymptotic analysis is used, namely: its modification under the asymptotic condition of a large delay of calls in the orbit, which is consistent with the urgency of the problem. The research result is formulated for the case when $H_1 = 1$, and H_2 takes an arbitrary value from 0 to 1.

Let us $i(t)$, $i(t) = 0, 1, \ldots$, is the number of calls in the orbit at the time t. The random process $i(t)$ is not Markov, therefore, we introduce an additional process $k(t)$, which characterizes the state of the server at the moment of time t

$$k(t) = \begin{cases} 0, & \text{if the device is free,} \\ 1, & \text{if the server is busy serving a request.} \end{cases}$$

The two-dimensional process $\{i(t), k(t)\}$ forms a Markov chain with continuous time. We will assume that there is a stationary distribution of the probabilities of the states of this process.

We denote $P\{i(t) = i, k(t) = k\} = P_k(i, t)$ as the probability that at the time t the device is in state k, $k = \{0; 1\}$, and there is i, $i = 0, 1, \ldots$, calls in the orbit.

2.1   The System of Kolmogorov Equations

For the probability distribution $P(i, t) = \sum_{k=0}^{1} P_k(i, t)$ of states of the considered RQ-system (1), we compose a Kolmogorov system of differential equations (2). Using the formula for total probability, we obtain the system of equalities.

$$\begin{cases} P_0(i, t + \Delta t) = P_0(i, t)(1 - \lambda \Delta t)(1 - i\sigma \Delta t) + P_1(i, t)\mu \Delta t \\ + P_1(i - 2, t)\lambda H_2 \Delta t + P_1(i - 1, t)\lambda(1 - H_2)\Delta t \\ + P_1(i - 1, t)(i - 1)\sigma H_2 \Delta t + P_1(i, t)i\sigma(1 - H_2)\Delta t + o(\Delta t), \\ P_1(i, t + \Delta t) = P_1(i, t)(1 - \lambda \Delta t)(1 - \mu_1 \Delta t)(1 - i\sigma \Delta t) + \\ P_0(i, t)\lambda \Delta t + P_0(i + 1, t)(i + 1)\sigma \Delta t + o(\Delta t), \end{cases} \quad (1)$$

$$\begin{cases} \dfrac{\partial P_0(i, t)}{\partial t} = -(\lambda + i\sigma)P_0(i, t) + \mu P_1(i, t) + \lambda H_2 P_1(i - 2, t) \\ + \lambda(1 - H_2)P_1(i - 1, t) + (i - 1)\sigma H_2 P_1(i - 1, t) + i\sigma(1 - H_2)P_1(i, t), \\ \dfrac{\partial P_1(i, t)}{\partial t} = -(\lambda + \mu + i\sigma)P_1(i, t) + \lambda P_0(i, t) + \lambda P_0(i, t) \\ + (i + 1)\sigma P_0(i + 1, t). \end{cases} \quad (2)$$

We denote $\pi_k(i) = \lim\limits_{t \to \inf} P_k(i, t)$, $k = \{0; 1\}$, as stationary probabilities of the process $\{i(t), k(t)\}$ then (2) can be written as

$$\begin{cases} -(\lambda + i\sigma)\pi_0(i) + \mu\pi_1(i) + \lambda H_2\pi_1(i-2) + \lambda(1-H_2)\pi_1(i-1) \\ +(i-1)\sigma H_2\pi_1(i-1) + i\sigma(1-H_2)\pi_1(i) = 0, \\ -(\lambda + \mu + i\sigma)\pi_1(i) + \lambda\pi_0(i) + \lambda\pi_0(i) + (i+1)\sigma\pi_0(i+1) = 0. \end{cases} \quad (3)$$

2.2   The Characteristic Functions

We introduce the partial characteristic functions as follows

$$h_k(u) = \sum_{i=0}^{\infty} e^{jui}\pi_k(i,t), \quad k = \overline{0,1}, \quad (4)$$

where $j = \sqrt{-1}$.

Using (4) and $h'_k(u) = \dfrac{dh_k(u)}{du} = j\sum_{i=0}^{\infty} i e^{jui}\pi_k(i)$, $k = \{0,1\}$, we can write the system (3) as

$$\begin{cases} -\lambda h_0(u) + \left(\mu + \lambda(1-H_2)e^{ju} + H_2 e^{2ju}\right) h_1(u) \\ +j\sigma h'_0(u) - j\sigma\left((1-H_2) + H_2 e^{ju}\right) h'_1(u) = 0, \\ \lambda h_0(u) - (\mu + \lambda) h_1(u) - j\sigma e^{-ju} h'_0(u) + j\sigma h'_1(u) = 0. \end{cases} \quad (5)$$

Let us add to the system (5) one more equation obtained by summing the first equation and the second one, multiplied by e^{ju}. After simple transformations we have

$$\lambda h_0(u) - \left(\mu - \lambda H_2 e^{ju}\right) h_1(u) + j\sigma(1-H_2)h' h_1(u) = 0. \quad (6)$$

3   Asymptotic Analysis Method

The method of asymptotic analysis in queuing theory is the method of research of the equations determining some characteristics of an queuing system under some limit (asymptotic) condition, which is specific for any model and solving problem. To find the solution of system of equations we propose another approach by using the method of asymptotic analysis under the assumption that there is a long delay between customers from the orbit, i.e. when $\varepsilon \to 0$. We summarize the results of our study in the next Theorem 1

Theorem 1. *The stationary distribution of the number of calls in orbit in the RQ-system M/M/1 with non-persistent customers and collisions with the Poisson arrival process of intensity λ, exponential servicing distribution with parameter μ, exponential distribution law of the random delay with parameter σ is an asymptotically normal distribution with mean κ_1/σ and variance κ_2/σ, where*

$$\kappa_1 = \frac{\mu r_1}{(1-2r_1)} - \lambda, \quad (7)$$

$$\kappa_2 = -\frac{(\kappa_1 + \lambda)g_0 - \kappa_1 r_0 - (\lambda + \mu + \kappa_1)g_1}{r_1 - r_0}, \tag{8}$$

r_1 and r_0 are the probabilities that the device is occupied or free respectively in the stationary mode of system operation, which are determined by equations

$$\begin{cases} (1 + \rho)r_1^2 - (1 + 2\rho)r_1 + \rho = 0, \quad \rho = \lambda/\mu, \\ r_0 + r_1 = 1, \end{cases} \tag{9}$$

g_1 and g_0 are defined as follows

$$\begin{cases} \dfrac{(\kappa_1 + \lambda)g_0 - \kappa_1 r_0 - \dfrac{(\kappa_1 + \lambda)r_0}{r_1}g_1}{r_1 - r_0} = \dfrac{\lambda g_0 - \dfrac{\lambda r_0}{r_1}g_1 + \lambda H_2 r_1}{(1 - H_2)r_1}, \\ g_0 + g_1 = 0. \end{cases} \tag{10}$$

The Theorem 1 proving will carried out in two stages.

3.1   Stage 1. Finding First-Order Asymptotic

In the basic system (3) and (4), we make the substitutions

$$\sigma = \varepsilon, \; u = w\varepsilon, \; h_k(u) = f_k(w, \varepsilon), \; k = \overline{0, 1}, \tag{11}$$

where ε is infinitesimal value ($\varepsilon \to 0$).

Since according (11) $h_k'(u) = \dfrac{1}{\varepsilon}\dfrac{\partial f_k(w, \varepsilon)}{\partial w}$, $k = \{0, 1\}$, the equations system (5) and (6) can be written as

$$\begin{cases} -\lambda f_0(w, \varepsilon) + \left(\mu + \lambda(1 - H_2)e^{j\varepsilon w} + H_2 e^{2j\varepsilon w}\right) f_1(w, \varepsilon) + j\dfrac{\partial f_0(w, \varepsilon)}{\partial w} \\ \quad -j\left(1 - H_2(1 - e^{j\varepsilon w})\right)\dfrac{\partial f_1(w, \varepsilon)}{\partial w} = 0, \\ \lambda f_0(w, \varepsilon) - (\mu + \lambda) f_1(w, \varepsilon) - je^{-j\varepsilon w}\dfrac{\partial f_0(w, \varepsilon)}{\partial w} + j\dfrac{\partial f_1(w, \varepsilon)}{\partial w} = 0, \\ \lambda f_0(w, \varepsilon) - \left(\mu - \lambda H_2 e^{j\varepsilon w}\right) f_1(w, \varepsilon) + j(1 - H_2)\dfrac{\partial f_1(w, \varepsilon)}{\partial w} = 0. \end{cases} \tag{12}$$

We will find the solution $f_k(w)$, $k = \{0, 1\}$, of the (12) in the form

$$f_k(w) = r_k \Phi(w), \quad k = \{0, 1\}, \tag{13}$$

where $r_0 + r_1 = 1$, $r_k = h_k(0) = f_k(0)$, $k = \{0, 1\}$, and $\Phi(w)$ is unknown function.

Substituting (13) in (12) we have a system of differential equations with respect to the function $\Phi(w)$

$$\begin{cases} \lambda r_0 \Phi(w) - (\mu + \lambda) r_1 \Phi(w) - jr_0\dfrac{\partial \Phi(w)}{\partial w} + jr_1\dfrac{\partial \Phi(w)}{\partial w} = 0, \\ \lambda r_0 \Phi(w) - (\mu - H_2\lambda) r_1 \Phi(w) + j(1 - H_2)r_1\dfrac{\partial \Phi(w)}{\partial w} = 0, \\ r_0 + r_1 = 1. \end{cases} \tag{14}$$

The solution to the system (14) is the function $\Phi(w) = \exp\{jw\kappa_1\}$, where κ_1 is defined by (7) and r_1 is the positive root of the (9).

Let us return to the original characteristic function by means of inverse changes and put $\varepsilon = \sigma$. Then

$$h_k(u) = f_k(w, \varepsilon) = f_k(w) + o(\varepsilon) \approx f_k(w) = f_k\left(\frac{u}{\varepsilon}\right) = r_k \exp\left\{\frac{\kappa_1}{\sigma}ju\right\}, \quad (15)$$

Taking into account the normalization condition $r_0 + r_1 = 1$, we have that the asymptotic characteristic function of the first order has the form

$$h^{(1)}(u) = h_0(u) + h_1(u) = \exp\left\{\frac{\kappa_1}{\sigma}ju\right\}. \quad (16)$$

The resulting value $\dfrac{\kappa}{\sigma}$ (5) determines the asymptotic average value of the number of calls in an orbit in a shared access system with collisions and rejections.

3.2   Stage 2. Finding the Second-Order Asymptotic

In the basic system of Eqs. (3) and (4) with (16) we let

$$h_k^{(2)}(u) = \exp\left\{\frac{\kappa_1}{\sigma}ju\right\} h_k^{(2)}(u), \quad k = \{0, 1\}, \quad (17)$$

and the equations system (3) and (4) can be formulated as

$$
\begin{cases}
-(\lambda+\kappa_1)h_0^{(2)}(u)+\left(\mu+\lambda(1-H_2)e^{ju} + H_2e^{2ju}+\right.\\
\left.+\left((1-H_2)+H_2e^{ju}\right)\kappa_1\right)h_1^{(2)}(u) + j\sigma h_0'^{\,2}(u) \\
-j\sigma\left((1-H_2) + H_2e^{ju}\right)h_1'^{\,(2)}(u) = 0, \\
(\lambda + \kappa_1 e^{-ju})h_0^{(2)}(u) - (\mu+\lambda+\kappa_1)h_1^{(2)}(u) - j\sigma e^{-ju}h_0'^{\,2}(u) \\
+j\sigma h_1'^{\,(2)}(u) = 0, \\
\lambda h_0^{(2)}(u) + \left(\lambda H_2 e^{ju} - \mu - (1-H_2)\kappa_1\right)h_1^{(2)}(u) \\
+j\sigma(1-H_2)h_1'^{\,(2)}(u) = 0.
\end{cases}
\quad (18)
$$

In the system (18) we make the substitutions (19)

$$\sigma = \varepsilon^2, \quad u = w\varepsilon, \quad h_k^{(2)}(u) = f_k^{(2)}(w, \varepsilon), \quad k = \overline{0, 1}, \quad (19)$$

where ε is infinitesimal value ($\varepsilon \to 0$), and obtain

$$
\begin{cases}
-(\lambda + \kappa_1)f_0^{(2)}(\varepsilon, w) \\
+\left(\mu + \lambda(1-H_2)e^{jw\varepsilon} + H_2e^{2jw\varepsilon} + \left((1-H_2) + j\varepsilon f_0'^{\,2}((\varepsilon,w))+\right.\right.\\
\left.\left.+H_2e^{jw\varepsilon}\right)\kappa_1\right)f_1^{(2)}(\varepsilon, w) - j\varepsilon\left((1-H_2) + H_2e^{jw\varepsilon}\right)f_1'^{\,(2)}(\varepsilon, w) = 0, \\
(\lambda+\kappa_1 e^{-jw\varepsilon})f_0^{(2)}(\varepsilon, w) - (\mu+\lambda+\kappa_1)f_1^{(2)}(\varepsilon, w) - j\sigma e^{-jw\varepsilon}f_0'^{\,2}(\varepsilon, w) \\
+j\varepsilon f_1'^{\,(2)}(\varepsilon, w) = 0, \\
\lambda f_0^{(2)}(\varepsilon, w) + \left(\lambda H_2 e^{jw}\varepsilon - \mu - (1-H_2)\kappa_1\right)f_1^{(2)}(\varepsilon, w) \\
+j\varepsilon(1-H_2)f_1'^{\,(2)}(\varepsilon, w) = 0.
\end{cases}
\quad (20)
$$

We substitute into the system (20) the decomposition (21)

$$f_k^{(2)}(w) = (r_k + j\varepsilon g_k)\Phi_2(w) + o\left(\varepsilon^2\right), \quad k = \{0,1\},$$
(21)

where g_k, $k = \{0,1\}$, are some constants.

Using (21) in (20) we have

$$
\begin{cases}
-(\lambda + \kappa_1)(r_0 + j\varepsilon g_0)\,\Phi_2(w) + \left(\mu + \lambda(1 - H_2)e^{jw\varepsilon} + H_2 e^{2jw\varepsilon}\right. \\
+ \left((1 - H_2) + H_2 e^{jw\varepsilon}\right)\kappa_1)(r_1 + j\varepsilon g_1)\,\Phi_2(w) + j\varepsilon\left((r_0 + j\varepsilon g_0)\,\Phi_2'(w)\right. \\
+ j\varepsilon g_0 \Phi_2(w)) \\
- j\varepsilon\left((1 - H_2) + H_2 e^{jw\varepsilon}\right)(r_1 + j\varepsilon g_1)\,\Phi_2'(w)) + j\varepsilon g_1 \Phi_2(w)) = o\left(\varepsilon^2\right), \\
\left(\lambda + \kappa_1 e^{-jw\varepsilon}\right)(r_0 + j\varepsilon g_0)\,\Phi_2(w) - (\mu + \lambda + \kappa_1)(r_1 + j\varepsilon g_1)\,\Phi_2(w) \\
- j\varepsilon e^{-jw\varepsilon}\left((r_0 + j\varepsilon g_0)\,\Phi_2'(w)) + j\varepsilon g_0 \Phi_2(w)) + j\varepsilon\left((r_1 + j\varepsilon g_1)\,\Phi_2'(w)\right) \\
+ j\varepsilon g_1 \Phi_2(w)) = o\left(\varepsilon^2\right), \\
\lambda(r_0 + j\varepsilon g_0)\,\Phi_2(w) + \left(\lambda H_2 e^{jw\varepsilon} - \mu - (1 - H_2)\kappa_1\right)(r_1 + j\varepsilon g_1)\,\Phi_2(w) \\
+ j\varepsilon(1 - H_2)\left((r_1 + j\varepsilon g_1)\,\Phi_2'(w)) + j\varepsilon g_1 \Phi_2(w)) = o\left(\varepsilon^2\right).
\end{cases}
$$
(22)

The solution of system (22) has the form

$$\Phi_2(w) = \exp\left\{\kappa_2 \frac{(jw)^2}{2}\right\},$$
(23)

where κ_2 is the same in (8).

Using the same transformation as for the first-order asymptotic and additional conditions $g_1 + g_0 = 0$ we finally obtain expressions of system solution (20) existence.

$$
\begin{cases}
-(\lambda + \kappa_1)g_0 + (\lambda(1 + H_2) + H_2\kappa_1)r_1) + (\mu + \lambda + \kappa_1)g_1 \\
= (r_1 - r_0)\,\kappa_2, \\
(\lambda + \kappa_1)\,g_0 - (\mu + \lambda + \kappa_1)\,g_1 = -(r_1 - r_0)\,\kappa_2, \\
\lambda g_0 + (\lambda H_2 - \mu - (1 - H_2)\kappa_1)\,g_1 + \lambda H_2 r_1 = -(1 - H_2)\,r_1\kappa_2, \\
g_1 + g_0 = 0.
\end{cases}
$$
(24)

Making the reverse substitutions in (21) with (24) we get

$$h_k^{(2)}(u) = f_k^{(2)}(w, \varepsilon) = (r_k + jw\varepsilon g_k)\exp\left\{\kappa_2 \frac{(jw)^2}{2}\right\} + o\left(\varepsilon^2\right) \approx R_k \exp\left\{\frac{\kappa_2}{\sigma} \frac{(ju)^2}{2}\right\},$$
(25)

then using (25) expressions (17) can be written as

$$h_k^{(2)}(u) = \exp\left\{\frac{\kappa_1}{\sigma}ju\right\} h_k^{(2)}(u) \approx R_k \exp\left\{\frac{\kappa_1}{\sigma}ju + \frac{\kappa_2}{\sigma} \frac{(ju)^2}{2}\right\}, \quad k = \{0,1\}.$$
(26)

Taking into account (26), the characteristic function $h^{(2)}(u) = h_0^{(2)}(u) + h_1^{(2)}(u)$, provided that the customers in orbit have long delays and the patience is high, is a Gaussian

$$h_2^{(2)}(u) = \exp\left\{\frac{\kappa_1}{\sigma}ju + \frac{\kappa_2}{\sigma}\frac{(ju)^2}{2}\right\}. \tag{27}$$

The Theorem 1 is proved.

4   Numerical Results

To accompany the theoretical conclusions, numerical results are obtained showing the convergence of asymptotic results to pre-limit ones (obtained using the recurrent algorithm), and the boundaries of the field of application of the presented approximation are determined depending on the values of the system parameters.

Asymptotic distributions of the probabilities of the number of calls in the orbit for the given service parameters $\mu = 1$ and the persistence probability H_1 for different values of the intensity λ of the incoming flow of calls, the persistence probability H_2 and the delay parameter of calls σ are constructed; they are compared with the pre-limit probability distributions obtained by the recurrent method.

Figures 1, 2 show the implementations for the cases $\lambda = 0.4$, $H_2 = 0.9$ and $\sigma = 0.1$, $\sigma = 0.01$ respectively.

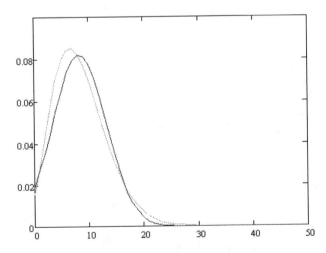

Fig. 1. Asymptotic (blue line) and pre-limit (red line) probability distributions of the number of claims in the orbit for $\sigma = 0.1$ (Color Figure Online)

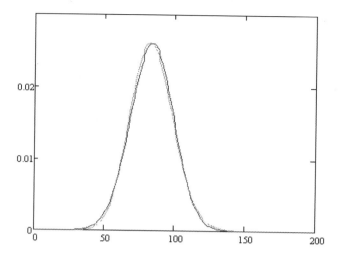

Fig. 2. Asymptotic (blue line) and pre-limit (red line) probability distributions of the number of claims in the orbit for $\sigma = 0.01$ (Color Figure Online)

Figures 3, 4 show the implementations for the cases $\lambda = 0.9$, $\sigma = 0.1$ and $H_2 = 0.4$, $H_2 = 0.95$ respectively.

Figures 5, 6 show the implementations for the cases $\lambda = 0.4$, $H_2 = 0.4$ and $\sigma = 0.1$, $\sigma = 0.01$ respectively.

As can be seen from the figures, the asymptotic distribution already at $\sigma < 0.1$ rather well approximates the pre-limit distribution at a high load of the system $\rho = \lambda/\mu$ and high persistence of calls H_2. For a small load $\rho < 0.5$, the Gaussian approximation gives a good result at $\sigma = 0.01$.

As a criterion for the proximity of distributions (asymptotic and pre-limit), the Kolmogorov distance was measured (28)

$$\Delta = \max_{n \geq 0} \left| \sum_{i=0}^{n} P_{requrrent}(i) - \sum_{i=0}^{n} P_{asympt}(i) \right|, \tag{28}$$

where $P_{requrrent}(i)$ is the probability distribution of the number of calls in the orbit obtained using the recurrent algorithm, and $P_{asympt}(i)$ is the probability distribution of the number of calls in the orbit obtained by the method of asymptotic analysis.

Table 1 shows that the value of the Kolmogorov distance decreases with an increase in the delay time of claims in orbit ($\sigma \to 0$) at a fixed value of the system load λ/μ. For a fixed value of the delay time of claims in the orbit, the Kolmogorov distance grows with increasing system load, but remains in the admissible (< 0.05) range for using the approximation

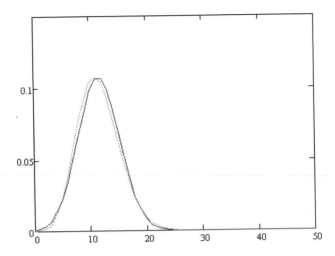

Fig. 3. Asymptotic (blue line) and pre-limit (red line) probability distributions of the number of claims in the orbit for $H_2 = 0.4$ (Color Figure Online)

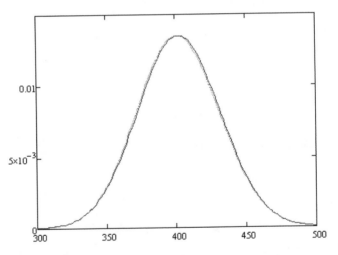

Fig. 4. Asymptotic (blue line) and pre-limit (red line) probability distributions of the number of claims in the orbit for $H_2 = 0.95$ (Color Figure Online)

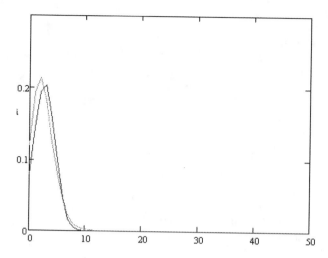

Fig. 5. Asymptotic (blue line) and pre-limit (red line) probability distributions of the number of claims in the orbit for $\sigma = 0.1$ (Color Figure Online)

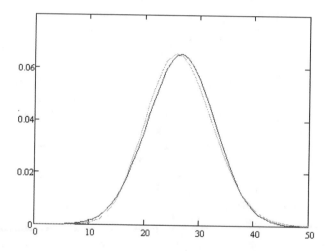

Fig. 6. Asymptotic (blue line) and pre-limit (red line) probability distributions of the number of claims in the orbit for $\sigma = 0.1$ (Color Figure Online)

Table 1. Distance values of Kolmogorov

$\lambda/\mu, H_2$	$\sigma = 0.5$	$\sigma = 0.1$	$\sigma = 0.05$	$\sigma = 0.01$
$\lambda/\mu = 0.4, H_2 = 0.9$	0.218	0.015	0.007	0.0013
$\lambda/\mu = 0.4, H_2 = 0.5$	0.287	0.042	0.019	0.0035
$\lambda/\mu = 0.9, H_2 = 0.9$	0.0062	0.0013	0.0006	0.0001
$\lambda/\mu = 0.9, H_2 = 0.4$	0.052	0.0094	0.0048	0.0009
$\lambda/\mu = 2.0, H_2 = 0.4$	0.016	0.0030	0.0015	0.0003

5   Conclusion

In the present paper, retrial queueing system of M/M/1 type with collisions and H_1, H_2-persistence, collisions and rejections is considered. It is proved that the probability distribution of the customers number in the orbit can be approximated by the Gaussian distribution under a long delay of customers in orbit condition.

References

1. Artalejo, J., Gomez-Corral, A.: Retrial Queueing Systems: A Computational Approach. Springer-Verlag, Berlin (2008)
2. Artalejo, J., Falin, G.: Standard and retrial queueing systems: a comparative analysis. Rev. Matematica Complutense **15**, 101–129 (2002)
3. Falin, G.L., Templeton, J.G.C.: Retrial Queues. Chapman & Hall, London (1997)
4. Danilyuk, E.Y., Fedorova, E.A., Moiseeva, S.P.: Asymptotic analysis of an retrial queueing system M—M—1 with collisions and impatient calls. Autom. Remote Control **79**(12), 2136–2146 (2018). https://doi.org/10.1134/S0005117918120044
5. Nazarov, A., Sztrik, J., Kvach, A., Bérczes, T.: Asymptotic analysis of finite-source M/M/1 retrial queueing system with collisions and server subject to breakdowns and repairs. Ann. Oper. Res. **277**(2), 213–229 (2018). https://doi.org/10.1007/s10479-018-2894-z
6. Lakaour, L., Aïssani, D., Adel-Aissanou, K., Barkaoui, K.: M/M/1 retrial queue with collisions and transmission errors. Methodol. Comput. Appl. Probab. **21**(4), 1395–1406 (2018). https://doi.org/10.1007/s11009-018-9680-x
7. Danilyuk, E., Moiseeva, S., Nazarov, A.: Asymptotic analysis of retrial queueing system M/GI/1 with collisions and impatient calls. In: Dudin, A., Nazarov, A., Moiseev, A. (eds.) ITMM 2019. CCIS, vol. 1109, pp. 230–242. Springer, Cham (2019). https://doi.org/10.1007/978-3-030-33388-1_19
8. Vygovskaya, O., Danilyuk, E., Moiseeva, S.: Retrial queueing system of MMPP/M/2 type with impatient calls in the orbit. In: Dudin, A., Nazarov, A., Moiseev, A. (eds.) ITMM/WRQ -2018. CCIS, vol. 912, pp. 387–399. Springer, Cham (2018). https://doi.org/10.1007/978-3-319-97595-5_30
9. Danilyuk, E., Vygoskaya, O., Moiseeva, S.: Retrial queue M/M/N with impatient customer in the orbit. In: Vishnevskiy, V.M., Kozyrev, D.V. (eds.) DCCN 2018. CCIS, vol. 919, pp. 493–504. Springer, Cham (2018). https://doi.org/10.1007/978-3-319-99447-5_42
10. Fedorova, E., Danilyuk, E., Nazarov, A., Melikov, A.: Retrial queueing system MMPP/M/1 with impatient calls under heavy load condition. In: Phung-Duc, T., Kasahara, S., Wittevrongel, S. (eds.) QTNA 2019. LNCS, vol. 11688, pp. 3–15. Springer, Cham (2019). https://doi.org/10.1007/978-3-030-27181-7_1
11. D'Arienzo, M.P., Dudin, A.N., Dudin, S.A., Manzo, R.: Analysis of a retrial queue with group service of impatient customers. J. Ambient Intell. Humanized Comput. **11**(6), 2591–2599 (2019). https://doi.org/10.1007/s12652-019-01318-x
12. Kim, J.: Retrial queueing system with collision and impatience. Commun. Korean Math. Soc. **4**, 647–653 (2010)
13. Kim, B., Kim, J.: Extension of the loss probability formula to an overloaded queue with impatient customers. Stat. Probab. Lett. **134**, 54–62 (2018)

14. Aissani, A., Lounis, F., Hamadouche, D., Taleb, S.: Analysis of customers impatience in a repairable retrial queue under postponed preventive actions. Am. J. Math. Manag. Sci. **38**(2), 125–150 (2019)

15. Dudin, A.N.: Operations research perspectives. Oper. Res. **5**, 245–255 (2018)

16. Ancker, C.J., Gafarian, C.J.: Some queuing problems with balking and reneging. I. Oper. Res. **11**(1), 88–100 (1963)

17. Haight, F.A.: Queueing with balking. Biometrika **44**(3/4), 360–369 (1957)

18. Barrer, D.Y.: Queuing with impatient customers and indifferent clerks. Oper. Res. **5**(5), 644–649 (1957)

19. Kumar, R., Sharma, S.K.: Economic analysis of M/M/c/N queue with retention of impatient customers. Int. J. Math. Oper. Res. **5**(6), 709–720 (2013)

20. Kumar, R., Som, B.K.: An M/M/1/N queuing system with reverse balking and reverse reneging. Adv. Model. Optim. **16**(2), 339–353 (2014)

21. Som, B.K.: A stochastic feedback queuing model with encouraged arrivals and retention of impatient customers. In: Laha, A.K. (ed.) Advances in Analytics and Applications. SPBE, pp. 261–272. Springer, Singapore (2019). https://doi.org/10.1007/978-981-13-1208-3_20

22. Santhakumaran, A., Thangaraj, B.: A single server queue with impatient and feedback customers. Inf. Manag. Sci. **11**(3), 71–79 (2010)

23. Som, B.K.: Cost-profit analysis of stochastic heterogeneous queue with reverse balking, feedback and retention of impatient customers. Reliab. Theory Appl. **14**(52), 87–101 (2019)

24. Wang, K., Li, N., Jiang, Z.: Queueing system with impatient customers: a review. In: Proceedings of 2010 IEEE International Conference on Service Operations and Logistics, and Informatics, pp. 82–87 (2010)

25. Nazarov, A., Sztrik, J., Kvach, A.: Comparative analysis of methods of residual and elapsed service time in the study of the closed retrial queuing system M/GI/1//N with collision of the customers and unreliable server. In: Dudin, A., Nazarov, A., Kirpichnikov, A. (eds.) ITMM 2017. CCIS, vol. 800, pp. 97–110. Springer, Cham (2017). https://doi.org/10.1007/978-3-319-68069-9_8

26. Bérczes, T., Sztrik, J., Tóth, Á., Nazarov, A.: Performance modeling of finite-source retrial queueing systems with collisions and non-reliable server using MOSEL. In: Vishnevskiy, V.M., Samouylov, K.E., Kozyrev, D.V. (eds.) DCCN 2017. CCIS, vol. 700, pp. 248–258. Springer, Cham (2017). https://doi.org/10.1007/978-3-319-66836-9_21

27. Nazarov, A.A., Moiseeva, S.P.: Method of Asymptotic Analysis In Queueing Theory. Publishing house NTL, Tomsk (2006). (in Russian)

Asymptotic Analysis of MMPP/M/1 Retrial Queueing System with Unreliable Server

N. M. Voronina[1]($^\boxtimes$) (ID), S. V. Rozhkova[1,2] (ID), and Ekaterina Fedorova[2] (ID)

[1] National Research Tomsk Polytechnic University, Lenin Avenue 30, Tomsk, Russia
{vnm,rozhkova}@tpu.ru
[2] National Research Tomsk State University, Lenin Avenue 36, Tomsk, Russia
moiskate@mail.ru

Abstract. In this paper, we study a single-server retrial queueing system with arrival Markov Modulated Poisson Process and an exponential law of the service time on an unreliable server. If the server is idle, an arrival customer occupies it for the servicing. When the server is busy, a customer goes into the orbit and waits a random time distributed exponentially. It is assumed that the server is unreliable, so it may fail. The server's repairing and working times are exponentially distributed. The method of asymptotic analysis is proposed to find the stationary distribution of the number of customers in the orbit. It is shown that the asymptotic probability distribution under the condition of a long delay has the Gaussian form with obtained parameters.

Keywords: Retrial queue · Markov Modulated Poisson Process · Asymptotic analysis · Unreliable server · Long delay

1 Introduction

Retrial queueing systems are widely used as models of call centers, cellular networks and random access protocols in local networks [1,2]. A characteristic feature of such models is the presence of repeated attempts to get service after unsuccessful one. Such situations can be caused not only by the lack of free servers at the arrival time, but also by technical reasons [3,4].

The most complete description of retrial queueing systems and their detailed comparison with classical queuing systems was published in [3–5]. The authors in [6–14] describe the main results for various retrial queue, as well as various methods for studying such systems.

Sometimes networks are overloaded, so servers may fail. Numerous breakdowns and limited repair options have a significant impact on networks performance. The identification and analysis of such aspects allow us to construct and optimize networks for minimizing the delay time and reducing the loss of calls.

Nowadays, a large number of works is devoted to the study of queuing systems with unreliable servers, the review is given in [15]. At the same time, the

© Springer Nature Switzerland AG 2022
A. Dudin et al. (Eds.): ITMM 2021, CCIS 1605, pp. 356–370, 2022.
https://doi.org/10.1007/978-3-031-09331-9_28

disadvantage of most works is very strict assumptions about the exponential distribution of all describing the system behavior characteristics (the time between the arrival moments of customers and breakdowns, the service time, the duration of the server repair). In [16–19], various types of retrial queues with impatient calls, collisions and unreliable servers are studied.

In this paper, we study a retrial queueing system with an unreliable server and arrival Markov Modulated Poisson process. Section 2 is devoted to the description of the considered model and the process under study. In Sect. 3, the system of Kolmogorov equations in the steady-state regime is written. In Sect. 4, the first order asymptotic method is proposed for the equation system solving. The theorem about the asymptotic mean of the number of customers in the orbit in the considered retrial queueing system under the limiting condition of the long delay is proved. In Sect. 5, the second order asymptotics is derived. The Gaussian form of the asymptotic distribution of the number of customers in the orbit under the long delay condition is proved. In Sect. 6, there are numerical analysis of the results in a particular case and the comparison of asymptotic and exact probability distributions. In Sect. 7, there is a conclusion.

2 Description of the Mathematical Model

Let's consider a single-server retrial queueing system with an unreliable server (see Fig. 1) and arrival Markov Modulated Poisson Process of customers (MMPP). A customer is serviced during random time distributed exponentially with parameter μ_1. We assume that the server is unreliable. An unreliable server may be in the following states: idle, busy or under repair. If the server is idle, and customer arrives, then the servicing immediately begins. If the server is busy at an arrival moment, then the customer goes into the orbit and waits a random time distributed exponentially with parameter σ, and then the customer tries to occupy the server again. The working time is distributed exponentially with parameter γ_1, if server is idle and with parameter γ_2, if the server is busy. As soon as a breakdown occurs, the server is sent to repair and the servicing customer goes into the orbit. During repairing, all incoming customers go into the orbit. The recovery time is distributed exponentially with parameter μ_2. The goal of the research is to find a stationary probability distribution of the number of customers in the orbit.

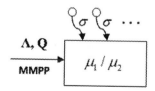

Fig. 1. Model of MMPP/M/1 retrial queueing system with unreliable server

Let $i(t)$ be the number of customers in the orbit at time t, $n(t)$ be the underlying process of the arrival MMPP and $k(t)$ determine the state of the server as follows

$$k(t) = \begin{cases} 0, \text{if the server is idle}, \\ 1, \text{if the server is busy}, \\ 2, \text{if the server is under repair}. \end{cases}$$

The MMPP is given by generator $\mathbf{Q} = [q_{\nu n}]$ and diagonal matrix $\mathbf{\Lambda} = \text{diag}[\lambda_n]$ of conditional rates λ_n, where $n = \overline{1, N}$. Three-dimensional random process $\{i(t), k(t), n(t)\}$ is the Markov chain with continuous time.

Denote the probability that at time t the server is in state k, there are i customers in the orbit and the underlying process $n(t)$ takes value n by $P_k(i, n, t) = P\{i(t) = i, k(t) = k, n(t) = n\}$, where $k = \{0, 1, 2\}$, $i = \overline{0, \infty}$, $n = \overline{1, N}$.

3   Kolmogorov Equations

Let us compose the system of Kolmogorov equations for probabilities $P_k(i, n, t)$

$$\begin{cases} \dfrac{\partial P_0(i, n, t)}{\partial t} = -(\lambda_n + i\sigma + \gamma_1) P_0(i, n, t) + \mu_1 P_1(i, n, t) \\ \quad + \mu_2 P_2(i, n, t) + \sum\limits_{\nu} P_0(i, \nu, t) \cdot q_{\nu n}, \\ \dfrac{\partial P_1(i, n, t)}{\partial t} = -(\lambda_n + \mu_1 + \gamma_2) P_1(i, n, t) + \lambda_n P_0(i, n, t) \\ \quad + (i + 1)\sigma P_0(i + 1, n, t) + \lambda_n P_1(i - 1, t) + \sum\limits_{\nu} P_1(i, \nu, t) \cdot q_{\nu n}, \\ \dfrac{\partial P_2(i, n, t)}{\partial t} = -(\lambda_n + \mu_2) P_2(i, n, t) + \gamma_1 P_0(i, n, t) \\ \quad + \gamma_2 P_1(i - 1, n, t) + \lambda_n P_2(i - 1, n, t) + \sum\limits_{\nu} P_2(i, \nu, t) \cdot q_{\nu n}. \end{cases} \quad (1)$$

Denoting $P_k(i, n) = \lim\limits_{t \to \infty} P_k(i, n, t)$ in the steady-state regime, we can rewrite System (1) in the following form:

$$\begin{cases} -(\lambda_n + i\sigma + \gamma_1) P_0(i, n) + \mu_1 P_1(i, n) + \mu_2 P_2(i, n) \\ \qquad\qquad\qquad\qquad + \sum\limits_{\nu} P_0(i, \nu) \cdot q_{\nu n} = 0, \\ -(\lambda_n + \mu_1 + \gamma_2) P_1(i, n) + \lambda_n P_0(i, n) + (i + 1)\sigma P_0(i + 1, n) \\ \qquad\qquad\qquad + \lambda_n P_1(i - 1, n) + \sum\limits_{\nu} P_1(i, \nu) \cdot q_{\nu n} = 0, \\ -(\lambda_n + \mu_2) P_2(i) + \gamma_1 P_0(i, n) + \gamma_2 P_1(i - 1, n) + \lambda_n P_2(i - 1, n) \\ \qquad\qquad\qquad\qquad + \sum\limits_{\nu} P_2(i, \nu) \cdot q_{\nu n} = 0. \end{cases} \quad (2)$$

Introducing the partial characteristic function

$$H_k(u, n) = \sum_{i=0}^{\infty} e^{jui} P_k(i, n),$$

where $j = \sqrt{-1}$, System (2) can be rewritten as

$$
\begin{cases}
-\lambda_n H_0(u, n) + \sigma j \dfrac{\partial H_0(u, n)}{\partial u} - \gamma_1 H_0(u, n) + \mu_1 H_1(u, n) \\
\qquad\qquad + \mu_2 H_2(u, n) + \sum_\nu H_0(u, \nu) \cdot q_{\nu n} = 0, \\
-\lambda_n H_1(u, n) - \mu_1 H_1(u, n) - \gamma_2 H_1(u, n) + \lambda_n H_0(u, n) \\
- j\sigma \cdot e^{-ju} \dfrac{\partial H_0(u, n)}{\partial u} + \lambda_n \cdot e^{ju} H_1(u, n) + \sum_\nu H_1(u, \nu) \cdot q_{\nu n} = 0, \\
-\lambda_n H_2(u, n) - \mu_2 H_2(u, n) + \gamma_1 H_0(u, n) + \gamma_2 \cdot e^{ju} H_1(u, n) \\
\qquad\qquad + \lambda_n \cdot e^{ju} H_2(u, n) + \sum_\nu H_2(u, \nu) \cdot q_{\nu n} = 0.
\end{cases}
\tag{3}
$$

We denote row-vector $\mathbf{H}_n(u) = \{H_n(u, 1), H_n(u, 2), \ldots, H_n(u, N)\}$.
Then we rewrite System (3) in the matrix form as

$$
\begin{cases}
\mathbf{H}_0(u)(\mathbf{Q} - \boldsymbol{\Lambda} - \gamma_1 \mathbf{I}) + j\sigma \dfrac{\partial \mathbf{H}_0(u)}{\partial u} + \mu_1 \mathbf{H}_1(u) + \mu_2 \mathbf{H}_2(u) = \mathbf{0}, \\
\mathbf{H}_0(u)\boldsymbol{\Lambda} + \mathbf{H}_1(u)\left[(\mathbf{Q} - (1 - e^{ju})\boldsymbol{\Lambda}) - (\mu_1 + \gamma_2)\mathbf{I}\right] \\
\qquad\qquad\qquad\qquad - j\sigma e^{-ju} \dfrac{\partial \mathbf{H}_0(u)}{\partial u} = \mathbf{0}, \\
\gamma_1 \mathbf{H}_0(u) \mathbf{I} + \gamma_2 e^{ju} \mathbf{H}_1(u) + \mathbf{H}_2(u)\left[\mathbf{Q} - (1 - e^{ju})\boldsymbol{\Lambda} - \mu_2 \mathbf{I}\right] = \mathbf{0},
\end{cases}
\tag{4}
$$

where $\mathbf{I}$ is the identity matrix.

We sum up all equations of System (4) and multiply the result by unit column-vector $\mathbf{e}$.

After some transformation and taking into account that

$$\mathbf{Q}\mathbf{e} = \mathbf{0}, \mathbf{I}\mathbf{e} = \mathbf{e},$$

we obtain the following additional equation

$$-\mathbf{H}_1(u)(\boldsymbol{\Lambda} + \gamma_2 \mathbf{I})\mathbf{e} - \mathbf{H}_2(u)\boldsymbol{\Lambda}\mathbf{e} - j\sigma e^{-ju} \dfrac{\partial \mathbf{H}_0(u)}{\partial u}\mathbf{e} = 0. \tag{5}$$

We will find a solution of System (4) and Eq. (5) under the condition of long delay of customers in the orbit $(\sigma \to 0.)$.

4   First Order Asymptotics

Theorem 1. *Let $i(t)$ be the number of customers in the orbit in MMPP/M/1 retrial queueing system with unreliable server. Then, for a sequence of characteristic functions, the following limiting equality holds*

$$\lim_{\sigma \to 0} M\{\exp\{jw\sigma i(t)\}\} = \exp\{jwG_1\},$$

where

$$G_1 = \lambda(\mu_1\gamma_1 + \gamma_1\gamma_2 + \gamma_2\mu_2 + \lambda\mu_2 + \lambda\gamma_2)/(\mu_1\mu_2 - \lambda\mu_2 - \lambda\gamma_2) \qquad (6)$$

and

$$\lambda = \boldsymbol{\Lambda} \cdot \boldsymbol{R} \cdot e,$$

vector $\boldsymbol{R}$ is the row-vector of the stationary distribution of the underlying process $n(t)$ determined by the following equations

$$\boldsymbol{R}\boldsymbol{Q} = \boldsymbol{0}, \quad \boldsymbol{R}e = 1.$$

Proof. In System (4) and Eq. (5), we introduce the following substitutions

$$\sigma = \varepsilon, \quad u = \varepsilon w, \quad \mathbf{H}_k(u) = \mathbf{F}_k(w, \varepsilon)$$

in order to obtain the following system of equations

$$
\begin{cases}
\mathbf{F}_0(w, \varepsilon)(\mathbf{Q} - \boldsymbol{\Lambda} - \gamma_1\mathbf{I}) + j\dfrac{\partial \mathbf{F}_0(w, \varepsilon)}{\partial w} + \mu_1\mathbf{F}_1(w, \varepsilon) + \mu_2\mathbf{F}_2(w, \varepsilon) = \mathbf{0}, \\[2mm]
\mathbf{F}_0(w, \varepsilon)\boldsymbol{\Lambda} + \mathbf{F}_1(w, \varepsilon)\left[(\mathbf{Q} - (1 - e^{jw\varepsilon})\boldsymbol{\Lambda}) - (\mu_1 + \gamma_2)\mathbf{I}\right] \\[2mm]
\hspace{4cm} - je^{-jw\varepsilon}\dfrac{\partial \mathbf{F}_0(w, \varepsilon)}{\partial w} = \mathbf{0}, \qquad (7) \\[2mm]
\gamma_1\mathbf{F}_0(w, \varepsilon) + \gamma_2 e^{jw\varepsilon}\mathbf{F}_1(w, \varepsilon) + \mathbf{F}_2(w, \varepsilon)\left[\mathbf{Q} - (1 - e^{jw\varepsilon})\boldsymbol{\Lambda} - \mu_2\mathbf{I}\right] = \mathbf{0}, \\[2mm]
-\mathbf{F}_1(w, \varepsilon)(\boldsymbol{\Lambda} + \gamma_2\mathbf{I})e - \mathbf{F}_2(w, \varepsilon)\boldsymbol{\Lambda}e - j\dfrac{\partial \mathbf{F}_0(w, \varepsilon)}{\partial w}e = 0.
\end{cases}
$$

Let us consider System (7) in a limit form for $\varepsilon \to 0$.

$$
\begin{cases}
\mathbf{F}_0(w)(\mathbf{Q} - \boldsymbol{\Lambda} - \gamma_1\mathbf{I}) + j\dfrac{\partial \mathbf{F}_0(w)}{\partial w} + \mu_1\mathbf{F}_1(w) \\[2mm]
\hspace{5cm} + \mu_2\mathbf{F}_2(w) = \mathbf{0}, \\[2mm]
\mathbf{F}_0(w)\boldsymbol{\Lambda} + \mathbf{F}_1(w)\left[\mathbf{Q} - (\mu_1 + \gamma_2)\mathbf{I}\right] - j\dfrac{\partial \mathbf{F}_0(w)}{\partial w} = \mathbf{0}, \qquad (8) \\[2mm]
\gamma_1\mathbf{F}_0(w) + \gamma_2\mathbf{F}_1(w) + \mathbf{F}_2(w)\left[\mathbf{Q} - \mu_2\mathbf{I}\right] = \mathbf{0}, \\[2mm]
-\mathbf{F}_1(w)(\boldsymbol{\Lambda} + \gamma_2\mathbf{I})e - \mathbf{F}_2(w)\boldsymbol{\Lambda}e - j\dfrac{\partial \mathbf{F}_0(w)}{\partial w}e = 0.
\end{cases}
$$

We assume the solution of System (8) be in the following form

$$\mathbf{F}_k(w) = \Phi(w)\mathbf{R}_k, \quad k = \overline{0, N},$$

where $\mathbf{R}_k$ is the stationary probability distribution two-dimensional process $\{k(t), n(t)\}$.

Substituting $\mathbf{F}_k(w)$ in System (8), we obtain

$$
\begin{cases}
\Phi(w)\mathbf{R}_0(\mathbf{Q} - \Lambda - \gamma_1\mathbf{I}) + j\Phi'(w)\mathbf{R}_0 + \Phi(w)\mathbf{R}_1\mu_1\mathbf{I} + \Phi(w)\mu_2\mathbf{R}_2 = \mathbf{0}, \\
\Phi(w)\mathbf{R}_0\Lambda + \Phi(w)\mathbf{R}_1[\mathbf{Q} - (\mu_1 + \gamma_2)\mathbf{I}] - j\Phi'(w)\mathbf{R}_0 = \mathbf{0}, \\
\Phi(w)\gamma_1\mathbf{R}_0 + \Phi(w)\gamma_2\mathbf{R}_1 + \Phi(w)\mathbf{R}_2[\mathbf{Q} - \mu_2\mathbf{I}] = \mathbf{0}, \\
-\Phi(w)\mathbf{R}_1(\Lambda + \gamma_2\mathbf{I})\mathbf{e} - \Phi(w)\mathbf{R}_2\Lambda\mathbf{e} - j\Phi'(w)\mathbf{R}_0\mathbf{e} = 0.
\end{cases}
$$

(9)

Dividing equations of System (9) by $\Phi(w)$, we have the following system

$$
\begin{cases}
\mathbf{R}_0(\mathbf{Q} - \Lambda - \gamma_1\mathbf{I}) + j\dfrac{\Phi'(w)}{\Phi(w)}\mathbf{R}_0 + \mu_1\mathbf{R}_1 + \mu_2\mathbf{R}_2 = \mathbf{0}, \\[2mm]
\mathbf{R}_0\Lambda + \mathbf{R}_1[\mathbf{Q} - (\mu_1 + \gamma_2)\mathbf{I}] - j\dfrac{\Phi'(w)}{\Phi(w)}\mathbf{R}_0 = \mathbf{0}, \\[2mm]
\gamma_1\mathbf{R}_0 + \gamma_2\mathbf{R}_1 + \mathbf{R}_2[\mathbf{Q} - \mu_2\mathbf{I}] = \mathbf{0}, \\[2mm]
-\mathbf{R}_1(\Lambda + \gamma_2\mathbf{I})\mathbf{e} - \mathbf{R}_2\Lambda\mathbf{e} - j\dfrac{\Phi'(w)}{\Phi(w)}\mathbf{R}_0\mathbf{e} = 0.
\end{cases}
$$

(10)

Since expression $\dfrac{\Phi'(w)}{\Phi(w)}$ does not depend on w, function $\Phi(w)$ can be expressed as

$$\Phi(w) = \exp\{jwG_1\},$$

where G_1 is an unknown variable.

Taking into account that $j\dfrac{\Phi'(w)}{\Phi(w)} = -G_1$, we rewrite System (10) in the following form

$$
\begin{cases}
\mathbf{R}_0(\mathbf{Q} - \Lambda - \gamma_1\mathbf{I}) - G_1\mathbf{R}_0 + \mu_1\mathbf{R}_1 + \mu_2\mathbf{R}_2 = \mathbf{0}, \\
\mathbf{R}_0\Lambda + \mathbf{R}_1[\mathbf{Q} - (\mu_1 + \gamma_2)\mathbf{I}] + G_1\mathbf{R}_0 = \mathbf{0}, \\
\gamma_1\mathbf{R}_0 + \gamma_2\mathbf{R}_1 + \mathbf{R}_2[\mathbf{Q} - \mu_2\mathbf{I}] = \mathbf{0}, \\
-\mathbf{R}_1(\Lambda + \gamma_2\mathbf{I})\mathbf{e} - \mathbf{R}_2\Lambda\mathbf{e} + G_1\mathbf{R}_0\mathbf{e} = 0.
\end{cases}
$$

(11)

Let us multiply all equations of System (11) by unit column-vector $\mathbf{e}$. Then we have the following system of equation

$$
\begin{cases}
-\mathbf{R}_0\Lambda\mathbf{e} - \gamma_1\mathbf{R}_0\mathbf{e} - G_1\mathbf{R}_0\mathbf{e} + \mu_1\mathbf{R}_1\mathbf{e} + \mu_2\mathbf{R}_2\mathbf{e} = 0, \\
\mathbf{R}_0\Lambda\mathbf{e} - \mu_1\mathbf{R}_1\mathbf{e} - \gamma_2\mathbf{R}_1\mathbf{e} + G_1\mathbf{R}_0\mathbf{e} = 0, \\
\gamma_1\mathbf{R}_0\mathbf{e} + \gamma_2\mathbf{R}_1\mathbf{e} - \mu_2\mathbf{R}_2\mathbf{e} = 0, \\
-\mathbf{R}_1\Lambda\mathbf{e} - \gamma_2\mathbf{R}_1\mathbf{e} - \mathbf{R}_2\Lambda\mathbf{e} + G_1\mathbf{R}_0\mathbf{e} = 0.
\end{cases}
$$

(12)

Let us write down the condition for the consistency of multivariate distributions for the stationary distribution of the states of the server

$$\sum_{k=0}^{2} \mathbf{R}_k = \mathbf{R},$$

where $\mathbf{R}$ is the row-vector of the stationary distribution of the underlying process $n(t)$ determined by the following equations

$$\mathbf{RQ} = \mathbf{0}, \quad \mathbf{Re} = 1.$$

Denoting $\mathbf{R}_0\mathbf{e} = r_0, \mathbf{R}_1\mathbf{e} = r_1, \mathbf{R}_2\mathbf{e} = r_2$ in System (12), we can find expressions for r_0, r_1, r_2, G_1:

$$r_0 = \frac{\mu_1\mu_2 - \lambda\mu_2 - \lambda\gamma_2}{\mu_1(\gamma_1 + \mu_2)}, \quad r_1 = \frac{\lambda}{\mu_1}, \quad r_2 = \frac{\mu_1\gamma_1 - \lambda\gamma_1 + \lambda\gamma_2}{\mu_1(\gamma_1 + \mu_2)},$$
$$G_1 = \lambda(\mu_1\gamma_1 + \gamma_1\gamma_2 + \gamma_2\mu_2 + \lambda\mu_2 + \lambda\gamma_2)/(\mu_1\mu_2 - \lambda\mu_2 - \lambda\gamma_2).$$

So Theorem 1 is proved.

In case of the probabilities r_0, r_1, r_2 must be positive, inequalities $\mu_1 > \lambda$ and $\mu_2 > \dfrac{\lambda\gamma_2}{\mu_1 - \lambda}$ must be true. This is condition of the asymptotic analysis application.

Theorem 1 determines the asymptotic average of the number of customers in the orbit in the studied retrial queueing system under the limiting condition of a long delay of customers in the orbit.

Let us consider the second-order asymptotics for more complete study.

5   Second Order Asymptotics

Theorem 2. *Let $i(t)$ be the number of customers in the orbit in MMPP/M/1 retrial queueing system with unreliable server, then for a sequence of characteristic functions, the following limiting equality holds*

$$\lim_{\sigma \to 0} M\{\exp\{jw\sqrt{\sigma}\,(i(t) - G1/\sigma\,)\}\} = \exp\left\{\frac{(jw)^2}{2}G2\right\},$$

where

$$G_2 = (r_0 G_1 \mathbf{I} - G_1 \mathbf{u}_0 + \mathbf{\Lambda}\mathbf{u}_1 + \gamma_2 \mathbf{u}_1 + \mathbf{\Lambda}\mathbf{u}_2)(G_1 \mathbf{z}_0 - \mathbf{\Lambda}\mathbf{z}_2 + r_0 \mathbf{I})^{-1} \tag{13}$$

and

$$\mathbf{z}_0 = -r_0(\mathbf{\Lambda} + G_1\mathbf{I})^{-1}, \quad \mathbf{z}_2 = \frac{-r_0\gamma_1(\mathbf{\Lambda} + G_1\mathbf{I})^{-1}}{\mu_2},$$

$$\mathbf{u}_0 = \left[r_0 G_1 \mathbf{I} - r_1\lambda_1\mathbf{I} - \frac{(\mu_1 + \gamma_2)(r_1\gamma_2\mathbf{I} + r_2\lambda_2\mathbf{I} - r_0 G_1\mathbf{I})}{\mu_1}\right](\mathbf{\Lambda} + G_1\mathbf{I})^{-1},$$

$$\mathbf{u}_1 = \frac{(-r_1\gamma_2\mathbf{I} - r_2\lambda_2\mathbf{I} + r_0 G_1\mathbf{I})}{\mu_1}, \quad \mathbf{u}_2 = \frac{(r_1\gamma_2\mathbf{I} + r_2\lambda_2\mathbf{I} + \gamma_1\mathbf{u}_0 + \gamma_2\mathbf{u}_1)}{\mu_2}.$$

Proof. In System of Eq. (12), we substitute the characteristic function in the following form

$$\mathbf{H}_k(u) = e^{ju\frac{G_1}{\sigma}}\mathbf{H}_k^{(2)}(u).$$

Then we reduce all equations by a common multiplier $e^{ju\frac{G_1}{\sigma}}$ and after some transformation, we have the following system

$$
\begin{cases}
\mathbf{H}_0^{(2)}(u)(\mathbf{Q} - \boldsymbol{\Lambda} - \gamma_1\mathbf{I}) - G_1\mathbf{H}_0^{(2)}(u) + j\sigma\dfrac{\partial \mathbf{H}_0^{(2)}(u)}{\partial u} + \mu_1\mathbf{H}_1^{(2)}(u) \\
\qquad\qquad\qquad\qquad\qquad\qquad\qquad\qquad + \mu_2\mathbf{H}_2^{(2)}(u) = \mathbf{0}, \\
\mathbf{H}_0^{(2)}(u)\boldsymbol{\Lambda} + \mathbf{H}_1^{(2)}(u)\left[(\mathbf{Q} - (1 - e^{ju})\boldsymbol{\Lambda}) - (\mu_1 + \gamma_2)\mathbf{I}\right] + e^{-ju}G_1\mathbf{H}_0^{(2)}(u) \\
\qquad\qquad\qquad\qquad\qquad\qquad\qquad - j\sigma e^{-ju}\dfrac{\partial \mathbf{H}_0^{(2)}(u)}{\partial u} = \mathbf{0}, \\
\gamma_1\mathbf{H}_0^{(2)}(u) + \gamma_2 e^{ju}\mathbf{H}_1^{(2)}(u) + \mathbf{H}_2^{(2)}(u)\left[\mathbf{Q} - (1 - e^{ju})\boldsymbol{\Lambda} - \mu_2\mathbf{I}\right] = \mathbf{0}, \\
-\mathbf{H}_1^{(2)}(u)(\boldsymbol{\Lambda} + \gamma_2\mathbf{I})\mathbf{e} - \mathbf{H}_2^{(2)}(u)\boldsymbol{\Lambda}\mathbf{e} + G_1 e^{-ju}\mathbf{H}_0^{(2)}(u)\mathbf{e} \\
\qquad\qquad\qquad\qquad\qquad\qquad - j\sigma e^{-ju}\dfrac{\partial \mathbf{H}_0^{(2)}(u)}{\partial u}\mathbf{e} = 0.
\end{cases}
$$

Let us introduce the following substitutions

$$\sigma = \varepsilon^2, \quad u = \varepsilon w, \quad \mathbf{H}_k^{(2)}(u) = \mathbf{F}_k^{(2)}(w, \varepsilon).$$

We have the following equations for functions $\mathbf{F}_k^{(2)}(w, \varepsilon)$:

$$
\begin{cases}
\mathbf{F}_0^{(2)}(w, \varepsilon)(\mathbf{Q} - \boldsymbol{\Lambda} - \gamma_1\mathbf{I} - G_1\mathbf{I}) + j\varepsilon\dfrac{\partial \mathbf{F}_0^{(2)}(w, \varepsilon)}{\partial w} + \mu_1\mathbf{F}_1^{(2)}(w, \varepsilon) \\
\qquad\qquad\qquad\qquad\qquad\qquad\qquad\qquad + \mu_2\mathbf{F}_2^{(2)}(w, \varepsilon) = \mathbf{0}, \\
\mathbf{F}_0^{(2)}(w, \varepsilon)(\boldsymbol{\Lambda} + e^{-jw\varepsilon}G_1\mathbf{I}) + \mathbf{F}_1^{(2)}(w, \varepsilon)((\mathbf{Q} - (1 - e^{jw\varepsilon})\boldsymbol{\Lambda}) \\
\qquad\qquad\qquad\qquad - (\mu_1 + \gamma_2)\mathbf{I}) - j\varepsilon e^{-jw\varepsilon}\dfrac{\partial \mathbf{F}_0^{(2)}(w, \varepsilon)}{\partial w} = \mathbf{0}, \quad (14) \\
\gamma_1\mathbf{F}_0^{(2)}(w, \varepsilon) + \gamma_2 e^{jw\varepsilon}\mathbf{F}_1^{(2)}(w, \varepsilon) + \mathbf{F}_2^{(2)}(w, \varepsilon)(\mathbf{Q} - (1 - e^{jw\varepsilon})\boldsymbol{\Lambda} \\
\qquad\qquad\qquad\qquad\qquad\qquad\qquad\qquad - \mu_2\mathbf{I}) = \mathbf{0}, \\
-\mathbf{F}_1^{(2)}(w, \varepsilon)(\boldsymbol{\Lambda} + \gamma_2\mathbf{I})\mathbf{e} - \mathbf{F}_2^{(2)}(w, \varepsilon)\boldsymbol{\Lambda}\mathbf{e} + G_1 e^{-jw\varepsilon}\mathbf{F}_0^{(2)}(w, \varepsilon)\mathbf{e} \\
\qquad\qquad\qquad\qquad\qquad - j\varepsilon e^{-jw\varepsilon}\dfrac{\partial \mathbf{F}_0^{(2)}(w, \varepsilon)}{\partial w}\mathbf{e} = 0.
\end{cases}
$$

We will find a solution of System (14) in the form

$$\mathbf{F}_k(w, \varepsilon) = \Phi_2(w)\{\mathbf{R}_k + jw\varepsilon\mathbf{f}_k\} + \mathbf{O}(\varepsilon^2).$$

Substituting this expression into System (14), we obtain

$$
\begin{cases}
\Phi_2(w)\mathbf{R}_0(\mathbf{Q} - \mathbf{\Lambda} - \gamma_1\mathbf{I} - G_1\mathbf{I}) + jw\varepsilon\Phi_2(w)\mathbf{f}_0(\mathbf{Q} - \mathbf{\Lambda} - \gamma_1\mathbf{I} - G_1\mathbf{I}) \\
\quad + j\varepsilon j\varepsilon\Phi_2(w)\mathbf{f}_0 + \Phi'_2(w)\left(j\varepsilon\mathbf{R}_0 + j\varepsilon jw\varepsilon\mathbf{f}_0\right) + \mu_1\Phi_2(w)\mathbf{R}_1 \\
\quad + jw\varepsilon\mu_1\Phi_2(w)\mathbf{f}_1 + \mu_2\Phi_2(w)\mathbf{R}_2 + jw\varepsilon\mu_2\Phi_2(w)\mathbf{f}_2 = \mathbf{0}, \\[4pt]
\Phi_2(w)\mathbf{R}_0\left(\mathbf{\Lambda} + e^{-jw\varepsilon}G_1\mathbf{I}\right) + jw\varepsilon\Phi_2(w)\mathbf{f}_0\left(\mathbf{\Lambda} + e^{-jw\varepsilon}G_1\mathbf{I}\right) \\
\quad - j\varepsilon e^{-jw\varepsilon}j\varepsilon\Phi_2(w)\mathbf{f}_0 - \Phi'_2(w)\left(j\varepsilon e^{-jw\varepsilon}\mathbf{R}_0 + j\varepsilon e^{-jw\varepsilon}jw\varepsilon\mathbf{f}_0\right) \\
\quad + \Phi_2(w)\mathbf{R}_1\left[(\mathbf{Q} - (1 - e^{jw\varepsilon})\mathbf{\Lambda}) - (\mu_1 + \gamma_2)\mathbf{I}\right] \\
\quad + jw\varepsilon\Phi_2(w)\mathbf{f}_1\left[(\mathbf{Q} - (1 - e^{jw\varepsilon})\mathbf{\Lambda}) - (\mu_1 + \gamma_2)\mathbf{I}\right] = \mathbf{0}, \quad (15)\\[4pt]
\gamma_1\Phi_2(w)\mathbf{R}_0 + jw\varepsilon\gamma_1\Phi_2(w)\mathbf{f}_0 + \gamma_2 e^{jw\varepsilon}\Phi_2(w)\mathbf{R}_1 + jw\varepsilon\gamma_2 e^{jw\varepsilon}\Phi_2(w)\mathbf{f}_1 \\
\quad + \Phi_2(w)\mathbf{R}_2\left[\mathbf{Q} - (1 - e^{jw\varepsilon})\mathbf{\Lambda} - \mu_2\mathbf{I}\right] \\
\quad + jw\varepsilon\Phi_2(w)\mathbf{f}_2\left[\mathbf{Q} - (1 - e^{jw\varepsilon})\mathbf{\Lambda} - \mu_2\mathbf{I}\right] = \mathbf{0}, \\[4pt]
- \Phi_2(w)\mathbf{R}_1(\mathbf{\Lambda} + \gamma_2\mathbf{I})\mathbf{e} - jw\varepsilon\Phi_2(w)\mathbf{f}_1(\mathbf{\Lambda} + \gamma_2\mathbf{I})\mathbf{e} - \Phi_2(w)\mathbf{R}_2\mathbf{\Lambda}\mathbf{e} \\
- jw\varepsilon\Phi_2(w)\mathbf{f}_2\mathbf{\Lambda}\mathbf{e} + e^{-jw\varepsilon}G_1\Phi_2(w)\mathbf{R}_0\mathbf{e} + jw\varepsilon e^{-jw\varepsilon}G_1\mathbf{f}_0\Phi_2(w)\mathbf{e} \\
- j\varepsilon e^{-jw\varepsilon}j\varepsilon\Phi_2(w)\mathbf{f}_0\mathbf{e} - \Phi'_2(w)\left(j\varepsilon e^{-jw\varepsilon}\mathbf{R}_0 + j\varepsilon e^{-jw\varepsilon}jw\varepsilon\mathbf{f}_0\right)\mathbf{e} = 0.
\end{cases}
$$

Using Taylor's series in System (15), we have

$$
\begin{cases}
\Phi_2(w)\mathbf{R}_0(\mathbf{Q} - \mathbf{\Lambda} - \gamma_1\mathbf{I} - G_1\mathbf{I}) + jw\varepsilon\Phi_2(w)\mathbf{f}_0(\mathbf{Q} - \mathbf{\Lambda} - \gamma_1\mathbf{I} - G_1\mathbf{I}) \\
- \varepsilon^2\Phi_2(w)\mathbf{f}_0 + \Phi'_2(w)\left(j\varepsilon\mathbf{R}_0 - w\varepsilon^2\mathbf{f}_0\right) + \mu_1\Phi_2(w)\mathbf{R}_1 + jw\varepsilon\mu_1\Phi_2(w)\mathbf{f}_1 \\
\hspace{5cm} + \mu_2\Phi_2(w)\mathbf{R}_2 + jw\varepsilon\mu_2\Phi_2(w)\mathbf{f}_2 = \mathbf{0}, \\[4pt]
\Phi_2(w)\mathbf{R}_0\left(\mathbf{\Lambda} + (1 - jw\varepsilon)G_1\mathbf{I}\right) + jw\varepsilon\Phi_2(w)\mathbf{f}_0\left(\mathbf{\Lambda} + (1 - jw\varepsilon)G_1\mathbf{I}\right) \\
+ \varepsilon^2(1 - jw\varepsilon)\Phi_2(w)\mathbf{f}_0 - \Phi'_2(w)\left(j\varepsilon(1 - jw\varepsilon)\mathbf{R}_0 + j\varepsilon(1 - jw\varepsilon)jw\varepsilon\mathbf{f}_0\right) \\
+ \Phi_2(w)\mathbf{R}_1\left[(\mathbf{Q} + jw\varepsilon\mathbf{\Lambda}) - (\mu_1 + \gamma_2)\mathbf{I}\right] \\
\hspace{4cm} + jw\varepsilon\Phi_2(w)\mathbf{f}_1\left[(\mathbf{Q} + jw\varepsilon\mathbf{\Lambda}) - (\mu_1 + \gamma_2)\mathbf{I}\right] = \mathbf{0}, \\[4pt]
\gamma_1\Phi_2(w)\mathbf{R}_0 + jw\varepsilon\gamma_1\Phi_2(w)\mathbf{f}_0 + \gamma_2(1 + jw\varepsilon)\Phi_2(w)\mathbf{R}_1 \\
+ jw\varepsilon\gamma_2(1 + jw\varepsilon)\Phi_2(w)\mathbf{f}_1 + \Phi_2(w)\mathbf{R}_2(\mathbf{Q} + jw\varepsilon\mathbf{\Lambda} - \mu_2\mathbf{I}) \\
\hspace{5cm} + jw\varepsilon\Phi_2(w)\mathbf{f}_2(\mathbf{Q} + jw\varepsilon\mathbf{\Lambda} - \mu_2\mathbf{I}) = \mathbf{0}, \\[4pt]
- \Phi_2(w)\mathbf{R}_1(\mathbf{\Lambda} + \gamma_2\mathbf{I})\mathbf{e} - jw\varepsilon\Phi_2(w)\mathbf{f}_1(\mathbf{\Lambda} + \gamma_2\mathbf{I})\mathbf{e} - \Phi_2(w)\mathbf{R}_2\mathbf{\Lambda}\mathbf{e} \\
- jw\varepsilon\Phi_2(w)\mathbf{f}_2\mathbf{\Lambda}\mathbf{e} + G_1(1 - jw\varepsilon)\Phi_2(w)\mathbf{R}_0\mathbf{e} + jw\varepsilon G_1(1 - jw\varepsilon)\Phi_2(w)\mathbf{f}_0\mathbf{e} \\
+ \varepsilon^2(1 - jw\varepsilon)\Phi_2(w)\mathbf{f}_0\mathbf{e} - \Phi'_2(w)(j\varepsilon(1 - jw\varepsilon)\mathbf{R}_0 - w\varepsilon^2(1 - jw\varepsilon)\mathbf{f}_0)\mathbf{e} = 0.
\end{cases}
$$

Let us write equations for terms with ε^1.

$$\begin{cases} jw\varepsilon\Phi_2(w)\mathbf{f}_0(\mathbf{Q} - \mathbf{\Lambda} - \gamma_1\mathbf{I} - G_1\mathbf{I}) + j\varepsilon\Phi'_2(w)\mathbf{R}_0 + jw\varepsilon\mu_1\Phi_2(w)\mathbf{f}_1 \\ \qquad\qquad\qquad\qquad\qquad\qquad + jw\varepsilon\mu_2\Phi_2(w)\mathbf{f}_2 = \mathbf{0}, \\ -jw\varepsilon G_1\Phi_2(w)\mathbf{R}_0 + jw\varepsilon\Phi_2(w)\mathbf{f}_0\,(\mathbf{\Lambda} + G_1\mathbf{I}) - j\varepsilon\Phi'_2(w)\mathbf{R}_0 \\ \qquad\qquad + jw\varepsilon\Phi_2(w)\mathbf{R}_1\mathbf{\Lambda} + jw\varepsilon\Phi_2(w)\mathbf{f}_1\,[\mathbf{Q} - (\mu_1 + \gamma_2)\mathbf{I}] = \mathbf{0}, \\ jw\varepsilon\gamma_1\Phi_2(w)\mathbf{f}_0 + jw\varepsilon\gamma_2\Phi_2(w)\mathbf{R}_1 + jw\varepsilon\gamma_2\Phi_2(w)\mathbf{f}_1 + jw\varepsilon\Phi_2(w)\mathbf{R}_2\mathbf{\Lambda} \\ \qquad\qquad\qquad\qquad\qquad\qquad + jw\varepsilon\Phi_2(w)\mathbf{f}_2(\mathbf{Q} - \mu_2\mathbf{I}) = \mathbf{0}, \\ -jw\varepsilon\Phi_2(w)\mathbf{f}_1(\mathbf{\Lambda} + \gamma_2\mathbf{I})\mathbf{e} - jw\varepsilon\Phi_2(w)\mathbf{f}_2\mathbf{\Lambda}\mathbf{e} - jw\varepsilon G_1\Phi_2(w)\mathbf{R}_0\mathbf{e} \\ \qquad\qquad\qquad\qquad + jw\varepsilon G_1\Phi_2(w)\mathbf{f}_0\mathbf{e} - j\varepsilon\Phi'_2(w)\mathbf{R}_0\mathbf{e} = 0. \end{cases}$$

Dividing by expression $jw\varepsilon\Phi_2(w)$, we obtain the following system of equations in case of $\varepsilon \to 0$

$$\begin{cases} \mathbf{f}_0(\mathbf{Q} - \mathbf{\Lambda} - \gamma_1\mathbf{I} - G_1\mathbf{I}) + \dfrac{\Phi'_2(w)}{w\Phi_2(w)}\mathbf{R}_0 + \mu_1\mathbf{f}_1 + \mu_2\mathbf{f}_2 = \mathbf{0}, \\ -\mathbf{R}_0 G_1\mathbf{I} + \mathbf{f}_0\,(\mathbf{\Lambda} + G_1\mathbf{I}) - \dfrac{\Phi'_2(w)}{w\Phi_2(w)}\mathbf{R}_0 + \mathbf{R}_1\mathbf{\Lambda} \\ \qquad\qquad\qquad\qquad\qquad + \mathbf{f}_1\,[\mathbf{Q} - (\mu_1 + \gamma_2)\mathbf{I}] = \mathbf{0}, \qquad\qquad (16) \\ \gamma_1\mathbf{f}_0 + \gamma_2\mathbf{R}_1 + \gamma_2\mathbf{f}_1 + \mathbf{R}_2\mathbf{\Lambda} + \mathbf{f}_2(\mathbf{Q} - \mu_2\mathbf{I}) = \mathbf{0}, \\ -\mathbf{f}_1(\mathbf{\Lambda} + \gamma_2\mathbf{I})\mathbf{e} - \mathbf{f}_2\mathbf{\Lambda}\mathbf{e} - G_1\mathbf{R}_0\mathbf{e} + G_1\mathbf{f}_0\mathbf{e} - \dfrac{\Phi'_2(w)}{w\Phi_2(w)}\mathbf{R}_0\mathbf{e} = 0. \end{cases}$$

Note that expression $\dfrac{\Phi'_2(w)}{w\Phi(w)}$ in System (16) does not depend on w, so we can write function $\Phi_2(w)$ as

$$\Phi_2(w) = \exp\left\{ \frac{(jw)^2}{2}G_2 \right\},$$

where G_2 is an unknown variable .

Taking into account that $\dfrac{\Phi'_2(w)}{w\Phi_2(w)} = -G_2$, we have

$$\begin{cases} \mathbf{f}_0(\mathbf{Q} - \mathbf{\Lambda} - \gamma_1\mathbf{I} - G_1\mathbf{I}) - G_2\mathbf{R}_0 + \mu_1\mathbf{f}_1 + \mu_2\mathbf{f}_2 = \mathbf{0}, \\ -G_1\mathbf{R}_0 + \mathbf{f}_0\,(\mathbf{\Lambda} + G_1\mathbf{I}) + G_2\mathbf{R}_0 + \mathbf{R}_1\mathbf{\Lambda} + \mathbf{f}_1\,[\mathbf{Q} - (\mu_1 + \gamma_2)\mathbf{I}] = \mathbf{0}, \\ \gamma_1\mathbf{f}_0 + \gamma_2\mathbf{R}_1 + \gamma_2\mathbf{f}_1 + \mathbf{R}_2\mathbf{\Lambda} + \mathbf{f}_2(\mathbf{Q} - \mu_2\mathbf{I}) = \mathbf{0}, \qquad (17) \\ -\mathbf{f}_1(\mathbf{\Lambda} + \gamma_2\mathbf{I})\mathbf{e} - \mathbf{f}_2\mathbf{\Lambda}\mathbf{e} - G_1\mathbf{R}_0\mathbf{e} + G_1\mathbf{f}_0\mathbf{e} + G_2\mathbf{R}_0\mathbf{e} = 0. \end{cases}$$

Let us multiply the first, the second and the third equations of System (17) by $\mathbf{e}$:

$$\begin{cases} -\mathbf{f}_0\mathbf{\Lambda}\mathbf{e} - \gamma_1\mathbf{f}_0\mathbf{e} - G_1\mathbf{f}_0\mathbf{e} - G_2\mathbf{R}_0\mathbf{e} + \mu_1\mathbf{f}_1\mathbf{e} + \mu_2\mathbf{f}_2\mathbf{e} = 0, \\ -G_1\mathbf{R}_0\mathbf{e} + \mathbf{f}_0\mathbf{\Lambda}\mathbf{e} + G_1\mathbf{f}_0\mathbf{e} + G_2\mathbf{R}_0\mathbf{e} + \mathbf{R}_1\mathbf{\Lambda}\mathbf{e} - \mu_1\mathbf{f}_1\mathbf{e} - \gamma_2\mathbf{f}_1\mathbf{e} = 0, \\ \gamma_1\mathbf{f}_0\mathbf{e} + \gamma_2\mathbf{R}_1\mathbf{e} + \gamma_2\mathbf{f}_1\mathbf{e} + \mathbf{R}_2\mathbf{\Lambda}\mathbf{e} - \mu_2\mathbf{f}_2\mathbf{e} = 0, \\ -\mathbf{f}_1\mathbf{\Lambda}\mathbf{e} - \gamma_2\mathbf{f}_1\mathbf{e} - \mathbf{f}_2\mathbf{\Lambda}\mathbf{e} - G_1\mathbf{R}_0\mathbf{e} + G_1\mathbf{f}_0\mathbf{e} + G_2\mathbf{R}_0\mathbf{e} = 0. \end{cases}$$

System (17) is heterogeneous, but it is similar to homogeneous System (11). Therefore, the solution of this system can be written in the form: $\mathbf{f}_k = C\mathbf{R}_k + G_2\mathbf{g}_k + \mathbf{y}_k$, so we obtain the following equations

$$
\begin{cases}
(-\varLambda\mathbf{e} - \gamma_1\mathbf{e} - G_1\mathbf{e})\left[C\mathbf{R}_0 + G_2\mathbf{g}_0 + \mathbf{y}_0\right] + \mu_1\mathbf{e}\left[C\mathbf{R}_1 + G_2\mathbf{g}_1 + \mathbf{y}_1\right] \\
\qquad\qquad + \mu_2\mathbf{e}\left[C\mathbf{R}_2 + G_2\mathbf{g}_2 + \mathbf{y}_2\right] = G_2\mathbf{R}_0\mathbf{e}, \\
(\varLambda\mathbf{e} + G_1\mathbf{e})\left[C\mathbf{R}_0 + G_2\mathbf{g}_0 + \mathbf{y}_0\right] \\
\qquad - (\mu_1\mathbf{e} + \gamma_2\mathbf{e})\left[C\mathbf{R}_1 + G_2\mathbf{g}_1 + \mathbf{y}_1\right] = \mathbf{R}_0 G_1\mathbf{e} - \mathbf{R}_0 G_2\mathbf{e} - \mathbf{R}_1\varLambda\mathbf{e}, \\
\gamma_1\mathbf{e}\left[C\mathbf{R}_0 + G_2\mathbf{g}_0 + \mathbf{y}_0\right] + \gamma_2\mathbf{e}\left[C\mathbf{R}_1 + G_2\mathbf{g}_1 + \mathbf{y}_1\right] \\
\qquad\qquad - \mu_2\mathbf{e}\left[C\mathbf{R}_2 + G_2\mathbf{g}_2 + \mathbf{y}_2\right] = -\mathbf{R}_1\gamma_2\mathbf{e} - \mathbf{R}_2\varLambda\mathbf{e}, \\
G_1\mathbf{e}\left[C\mathbf{R}_0 + G_2\mathbf{g}_0 + \mathbf{y}_0\right] - (\varLambda\mathbf{e} + \gamma_2\mathbf{e})\left[C\mathbf{R}_1 + G_2\mathbf{g}_1 + \mathbf{y}_1\right] \\
\qquad\qquad - \varLambda\mathbf{e}\left[C\mathbf{R}_2 + G_2\mathbf{g}_2 + \mathbf{y}_2\right] = G_1\mathbf{R}_0\mathbf{e} - G_2\mathbf{R}_0\mathbf{e}.
\end{cases}
\tag{18}
$$

In System (18), the terms with C are reduced.

Let us combine the coefficients with different degrees of G_2 into two systems: for $\mathbf{g}_k$:

$$
\begin{cases}
\mathbf{g}_0(-\varLambda\mathbf{e} - \gamma_1\mathbf{e} - G_1\mathbf{e}) + \mu_1\mathbf{g}_1\mathbf{e} + \mu_2\mathbf{g}_2\mathbf{e} = \mathbf{R}_0\mathbf{e}, \\
\mathbf{g}_0(\varLambda\mathbf{e} + G_1\mathbf{e}) - \mathbf{g}_1(\mu_1\mathbf{e} + \gamma_2\mathbf{e}) = -\mathbf{R}_0\mathbf{e}, \\
\gamma_1\mathbf{g}_0\mathbf{e} + \gamma_2\mathbf{g}_1\mathbf{e} - \mu_2\mathbf{g}_2\mathbf{e} = 0, \\
G_1\mathbf{g}_0\mathbf{e} - \mathbf{g}_1(\varLambda\mathbf{e} + \gamma_2\mathbf{e}) - \mathbf{g}_2\varLambda\mathbf{e} = -\mathbf{R}_0\mathbf{e}.
\end{cases}
\tag{19}
$$

and for $\mathbf{y}_k$:

$$
\begin{cases}
\mathbf{y}_0(-\varLambda\mathbf{e} - \gamma_1\mathbf{e} - G_1\mathbf{e}) + \mu_1\mathbf{y}_1\mathbf{e} + \mu_2\mathbf{y}_2\mathbf{e} = 0, \\
\mathbf{y}_0(\varLambda\mathbf{e} + G_1\mathbf{e}) - \mathbf{y}_1(\mu_1\mathbf{e} + \gamma_2\mathbf{e}) = G_1\mathbf{R}_0\mathbf{e} - \mathbf{R}_1\varLambda\mathbf{e}, \\
\gamma_1\mathbf{y}_0\mathbf{e} + \gamma_2\mathbf{y}_1\mathbf{e} - \mu_2\mathbf{y}_2\mathbf{e} = -\gamma_2\mathbf{R}_1\mathbf{e} - \mathbf{R}_2\varLambda\mathbf{e}, \\
G_1\mathbf{y}_0\mathbf{e} - \mathbf{y}_1(\varLambda\mathbf{e} + \gamma_2\mathbf{e}) - \mathbf{y}_2\varLambda\mathbf{e} = G_1\mathbf{R}_0\mathbf{e}.
\end{cases}
\tag{20}
$$

Also we write additional conditions for both systems as follows

$$
\sum_{k=0}^{N}\mathbf{g}_k\mathbf{e} = 0, \quad \sum_{k=0}^{N}\mathbf{y}_k\mathbf{e} = 0.
$$

In System (19), we denote $\mathbf{g}_0\mathbf{e} = \mathbf{z}_0$, $\mathbf{g}_1\mathbf{e} = \mathbf{z}_1$, $\mathbf{g}_2\mathbf{e} = \mathbf{z}_2$, then we obtain

$$
\mathbf{z}_0 = -r_0(\varLambda + G_1\mathbf{I})^{-1}, \quad \mathbf{z}_1 = 0, \quad \mathbf{z}_2 = \frac{-r_0\gamma_1(\varLambda + G_1\mathbf{I})^{-1}}{\mu_2}.
$$

In System (20), we denote $\mathbf{y}_0\mathbf{e} = \mathbf{u}_0$, $\mathbf{y}_1\mathbf{e} = \mathbf{u}_1$, $\mathbf{y}_2\mathbf{e} = \mathbf{u}_2$. So we obtain

$$
\mathbf{u}_1 = \frac{(-r_1\gamma_2\mathbf{I} - r_2\lambda_2\mathbf{I} + r_0 G_1\mathbf{I})}{\mu_1}.
$$

From the second equation of the system (20), we express $\mathbf{u}_0$

$$\mathbf{u}_0 = \left[r_0 G_1 \mathbf{I} - r_1 \lambda_1 \mathbf{I} - \frac{(\mu_1 + \gamma_2)(r_1 \gamma_2 \mathbf{I} + r_2 \lambda_2 \mathbf{I} - r_0 G_1 \mathbf{I})}{\mu_1} \right] (\mathbf{\Lambda} + G_1 \mathbf{I})^{-1}.$$

By substituting $\mathbf{u}_0$ and $\mathbf{u}_1$ into the third equation of System (20), we find that

$$\mathbf{u}_2 = \frac{(r_1 \gamma_2 \mathbf{I} + r_2 \lambda_2 \mathbf{I} + \gamma_1 \mathbf{u}_0 + \gamma_2 \mathbf{u}_1)}{\mu_2}.$$

We express G_2 from the fourth equation of System (20) as follows

$$G_2 = (r_0 G_1 \mathbf{I} - G_1 \mathbf{u}_0 + \mathbf{\Lambda} \mathbf{u}_1 + \gamma_2 \mathbf{u}_1 + \mathbf{\Lambda} \mathbf{u}_2)(G_1 \mathbf{z}_0 - \mathbf{\Lambda} \mathbf{z}_2 + r_0 \mathbf{I})^{-1}.$$

Theorem 2 is proved.

Theorem 2 shows that the distribution of the number of customers in the orbit has variance G_2/σ.

Combining the results of the first and the second order asymptotics, we have obtained that the asymptotic probability distribution of the number of customers in the orbit in studied retrial queue under the a long delay condition has the Gaussian form

$$h(u) = \exp \left\{ ju \frac{G_1}{\sigma} + \frac{(ju)^2}{2} \frac{G_2}{\sigma} \right\}.$$

6 Numerical Example

For the numerical analysis, let us consider a particular case of the considered model—a retrial queue with Poisson arrivals. Then $\mathbf{Q} = 0, \mathbf{\Lambda} = \lambda$ in obtained expressions.

So we have the following formulas for the parameters:

$$R_0 = \frac{\mu_1 \mu_2 - \mu_2 \lambda - \gamma_2 \lambda}{\mu_1 \mu_2 + \mu_1 \gamma_1}, \quad R_1 = \frac{\lambda}{\mu_1}, \quad R_2 = \frac{\mu_1 \gamma_1 + \gamma_2 \lambda - \lambda \gamma_1}{\mu_1 \mu_2 + \mu_1 \gamma_1},$$

$$G_1 = \frac{\lambda(\gamma_2 \mu_2 + \mu_1 \gamma_1 + \gamma_1 \gamma_2 + \mu_2 \lambda + \gamma_2 \lambda)}{\mu_1 \mu_2 - \mu_2 \lambda - \gamma_2 \lambda},$$

$$G_2 = \frac{G_1 R_0 + (G_1 + \gamma_1 + \lambda) f_0 - \mu_1 f_1 - \mu_2 f_2}{R_0},$$

where

$$f_0 = (-G_1 R_0 \gamma_2 - G_1 R_0 \mu_2 - R_1 \gamma_2 \lambda - R_1 \lambda \mu_2 - R_2 \gamma_2 \lambda - 2R_2 \lambda \mu_1$$
$$+ R_2 \lambda \mu_2)/[2(G_1 \gamma_2 + G_1 \mu_2 + \gamma_1 \gamma_2 + \gamma_1 \mu_1 + \gamma_2 \lambda + \gamma_2 \mu_2 + \lambda \mu_2 + \mu_1 \mu_2)],$$

$$f_1 = (G_1 R_0 \gamma_1 + G_1 R_0 \mu_2 - 2G_1 R_2 \lambda + R_1 \gamma_1 \lambda + R_1 \lambda \mu_2 - R_2 \gamma_1 \lambda - 2R_2 \lambda^2$$
$$- R_2 \lambda \mu_2)/[2(G_1 \gamma_2 + G_1 \mu_2 + \gamma_1 \gamma_2 + \gamma_1 \mu_1 + \gamma_2 \lambda + \gamma_2 \mu_2 + \lambda \mu_2 + \mu_1 \mu_2)],$$

$$f_2 = (-G_1 R_0 \gamma_1 + G_1 R_0 \gamma_2 + 2G_1 R_2 \lambda - R_1 \gamma_1 \lambda + R_1 \gamma_2 \lambda + R_2 \gamma_1 \lambda$$
$$+ R_2 \gamma_2 \lambda + 2R_2 \lambda^2 + 2R_2 \lambda \mu_1)/[2(G_1 \gamma_2 + G_1 \mu_2 + \gamma_1 \gamma_2 + \gamma_1 \mu_1 + \gamma_2 \lambda$$
$$+ \gamma_2 \mu_2 + \lambda \mu_2 + \mu_1 \mu_2)].$$

All these expressions coincide with [20].

Let us compare the probability distribution of the number of customers in the orbit obtained by the matrix method [20] with the results of asymptotic analysis.

In a numerical example, we take $\mu_1 = 5$, $\mu_2 = 2$, $\gamma_1 = 0.01$, $\gamma_2 = 0.01$, $\lambda = 3$.

Figure 2 shows a comparison of the asymptotic and the exact (calculated by matrix method) probability distributions of the number of customers for $\sigma = 0.1$ and $\sigma = 0.01$.

Let us study the range of applicability of the asymptotic method based on numerical analysis.

To determine the accuracy of the method, we use the Kolmogorov distance:

$$\Delta = \max_{0 \le k \le N} \left| \sum_{i=0}^{k} [P_{matrix}(i) - P_{asimp}(i)] \right|,$$

where $P_{matrix}(i)$ is a distribution obtained by the matrix method and $P_{asimp}(i)$ is a distribution obtained by the asymptotic method.

Table 1. Kolmogorov distance.

σ	0.5	0.1	0.05	0.01
Δ	0.289	0.039	0.033	0.024

In Table 1, there are values of Kolmogorov distance for different values of parameter σ. As you can see Kolmogorov distance decreases with decreasing of σ. We conclude that the proposed asymptotic method has good accuracy for $\sigma < 0.1$.

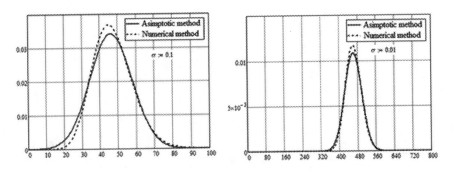

Fig. 2. Asymptotic and exact probability distributions comparison

7   Conclusion

In this paper, we consider the retrial queueing system with an unreliable server. During the study, the asymptotic characteristic function of the distribution of the number of customers in the orbit was obtained in a long delay condition. It is shown that the stationary probability distribution of the number of customers in the orbit can be approximated by the Gaussian distribution with obtained parameters. The numerical analysis is presented.

References

1. Gosztony, G.: Repeated call attempts and their effect on traffic engineering. Budavox Telecommun. Rev. **2**, 16–26 (1976)
2. Cohen, J.W.: Basic problems of telephone traffic and the influence of repeated calls. Philips Telecommun. Rev. **18**(2), 49–100 (1976)
3. Artalejo, J.R., Gomez-Corral, A.: Retrial Queueing Systems: A Computational Approach. Springer, Heidelberg (2008). https://doi.org/10.1007/978-3-540-78725-9
4. Falin, G.I., Templeton, J.G.C.: Retrial Queues. Chapman and Hall, London (1997)
5. Shortle, J.F., Thompson, J.M., Gross, D., Harris, C.M.: Fundamentals of Queueing Theory. Wiley, Hoboken (2018)
6. Dimitriou, I.: A queueing model with two classes of retrial customers and paired services. Ann. Oper. Res. **238**(1), 123–143 (2015). https://doi.org/10.1007/s10479-015-2059-2
7. Fiems, D., Phung-Duc, T. : Light-traffic analysis of queues with limited heterogenous retrials. In: Proceedings of 11th International Conference on Queueing Theory and Network Application (QTNA2016) (2016)
8. Gao, S., Niu, X., Li, T.: Analysis of a constant retrial queue with joining strategy and impatient retrial customers. Math. Probl. Eng. **2017**, 1–8 (2017)
9. Chakravarthy, S.R., Ozkar, S., Shruti, S.: Analysis of M/M/C retrial queue with thresholds, PH distribution of retrial times and unreliable servers. J. Appl. Math. Inf. **39**(1–2), 173–196 (2021)
10. Kim, C., Klimenok, V.I., Orlovsky, D.S.: The BMAP/PH/N retrial queue with Markovian flow of breakdowns. Eur. J. Oper. Res. **189**(3), 1057–1072 (2008)
11. Dudin, S., Dudina, O.: Retrial multi-server queuing system with PHF service time distribution as a model of a channel with unreliable transmission of information. Appl. Math. Model. **65**, 676–695 (2019)
12. Dragieva, V.: System state distributions in one finite source unreliable retrial queue. http://elib.bsu.by/handle/123456789/35903
13. Dudin, A.N., Sun', B.: A multiserver MAP/PH/N system with controlled broadcasting by unreliable servers. Autom. Control Comput. Sci. **43**(247), 32–44 (2009). https://doi.org/10.3103/S0146411609050046
14. Fedorova, E., Danilyuk, E., Nazarov, A., Melikov, A.: Retrial queueing system MMPP/M/1 with impatient calls under heavy load condition. In: Phung-Duc, T., Kasahara, S., Wittevrongel, S. (eds.) QTNA 2019. LNCS, vol. 11688, pp. 3–15. Springer, Cham (2019). https://doi.org/10.1007/978-3-030-27181-7_1

15. Dudin, A.N., Dudin, S.A.: A brief overview of works in the field of research of queuing systems with unreliable service devices. In: The Collection: Queueing Theory and Network Applications, International Congress on Informatics: Information Systems and Technologies, Materials of the International Scientific Congress, pp. 612–616 (2016)
16. Aissani, A., Lounis, F., Hamadouche, D., Taleb, S.: Analysis of customers impatience in a repairable retrial queue under postponed preventive actions. Am. J. Math. Manag. Sci. **38**(2), 125–150 (2009)
17. Nazarov, A.A., Paul, S.V., Lizyura, O.D.: Two-way communication retrial queue with unreliable server and multiple types of outgoing calls. Discrete Continuous Models Appl. Comput. Sci. **28**, 49–61 (2020)
18. Lakaour, L., Aissani, D., Adel-Aissanou, K., Barkaoui, K., Ziani, S.: An unreliable single server retrial queue with collisions and transmission errors. Commun. Stat. - Theor. Methods Taylor Francis **51**, 1–25 (2020)
19. Tóth, Á., Sztrik, J.: Simulation of finite-source retrial queuing systems with collisions, non-reliable server and impatient customers in the orbit. In: Proceedings of 11th International Conference on Applied Informatics, pp. 408–419. CEUR-WS (2020)
20. Rozhkova, S.V., Voronina, N.M., Semashko, A.V.: Investigation of the M/M/1 RQ system with unreliable server by asymptotic and matrix methods. In: Discrete and Continuous Models and Applied Computational Science. Information Technologies and mathematical modeling (ITMM-2020), Materials of the XIX International Scientific and Practical Conference Named After A.F. Terpugov, Tomsk, NTL, pp. 244–250 (2021)

Author Index

Printed in the United States
by Baker & Taylor Publisher Services